Lecture Notes on Multidisciplinary Industrial Engineering

Series editor

J. Paulo Davim, Department of Mechanical Engineering, University of Aveiro, Aveiro, Portugal

More information about this series at http://www.springer.com/series/15734

Marek Ochowiak · Szymon Woziwodzki
Michał Doligalski · Piotr Tomasz Mitkowski
Editors

Practical Aspects
of Chemical Engineering

Selected Contributions from PAIC 2017

 Springer

Editors
Marek Ochowiak
Institute of Chemical Technology
  and Engineering
Poznan University of Technology
Poznań
Poland

Szymon Woziwodzki
Institute of Chemical Technology
  and Engineering
Poznan University of Technology
Poznań
Poland

Michał Doligalski
Institute of Metrology, Electronics
  and Computer Science
University of Zielona Góra
Zielona Góra
Poland

Piotr Tomasz Mitkowski
Institute of Chemical Technology
  and Engineering
Poznan University of Technology
Poznań
Poland

ISSN 2522-5022   ISSN 2522-5030   (electronic)
Lecture Notes on Multidisciplinary Industrial Engineering
ISBN 978-3-319-73977-9   ISBN 978-3-319-73978-6   (eBook)
https://doi.org/10.1007/978-3-319-73978-6

Library of Congress Control Number: 2017963523

Printed on acid-free paper

This Springer imprint is published by the registered company Springer International Publishing AG part of Springer Nature
The registered company address is: Gewerbestrasse 11, 6330 Cham, Switzerland

Preface

Chemical and process engineering is a field of knowledge that deals with processes in which the composition and properties of substances change. Despite the fact that it emerged a long time ago, it has continued to develop and embrace new areas of interest such as biotechnology and nanotechnology. The aim of chemical and process engineering is to manufacture products of the chemical industry or the wider range of products of the whole processing industries. Its most important task is to describe these processes, which is the basis for design, proper exploitation, optimization and control of production installations, and to solve technical problems of process execution. Therefore, it is extremely important to transfer the results of scientific experiments to industrial practice and production of finished goods.

The book aims at presenting the reader a combination of scientific research with process practice and product development, as well as the practical aspects of chemical and process engineering. It contains selected issues with a detailed description of the use of research results, i.e.:

- Rheological properties of liquids and complex systems,
- Mixing,
- Distribution of mixtures,
- Single- and multiphase flows,
- Reactors,
- Momentum transfer processes,
- Heat and mass and their intensification,
- Design and modeling of chemical processes and devices,
- Measurement and process control,
- Environmental protection engineering, and
- Other issues of chemical and process engineering.

The editors of this book hope that it will be a valuable piece of reading for both researchers and students of chemical and process engineering, and engineers working in the area of design of chemical processes and equipment. The reader is presumed to have a basic knowledge of chemical equipment design and the theory of chemical and process engineering.

Poznań, Poland   Marek Ochowiak
September 2017

Contents

Silage of Switchgrass (*Panicum virgatum*) as a Bioenergy Feedstock in Poland

Mariusz Adamski, Waldemar Szaferski, Piotr Gulewicz and Włodzimierz Majtkowski

1 Introduction

Poland lies within the moderate climatic zone with influence of the continental climate from the east and the ocean climate from the west. The large variability and the variety of the weather types remain the characteristic features of this climate. Nowadays one can observe the increase in the insolation level, which influences the changes of agronomic conditions, particularly with regards to the temperature and the deficit of water. This circumstances influence significantly the vegetations of plants. Many species of cultivated plants are not adapted to such conditions within a period of the vegetation. Grasses belonging to C_4 plants, in comparison to native grasses of C-3 carbon fixation pathway, are better adapted to such conditions and can be the nutritious fodder and the substratum for the biogas production (Majtkowski et al. 2004).

By implementing Farming Common Policy, European Union funds the in-creasing participation of the energy gained from renewable sources as biomass (Pisarek et al. 2000; Majtkowski and Majtkowska 2000). The utilization of the biomass for the purpose of energy production causes independence from traditional

M. Adamski (✉)
Faculty of Agriculture and Bioengineering, Institute of Biosystems
Engineering, Poznan University of Life Sciences, Poznań, Poland
e-mail: mariusz.adamski@up.poznan.pl

W. Szaferski
Institute of Chemical Technology and Engineering, Faculty of Chemical
Technology, Poznan University of Technology, Poznań, Poland

P. Gulewicz
Department of Animal Nutrition and Feed Management,
Bydgoszcz University of Technology and Life Sciences, Bydgoszcz, Poland

W. Majtkowski
Plant Breeding and Acclimatization Institute, Botanical Garden, Bydgoszcz, Poland

© Springer International Publishing AG, part of Springer Nature 2018
M. Ochowiak et al. (eds.), *Practical Aspects of Chemical Engineering*,
Lecture Notes on Multidisciplinary Industrial Engineering,
https://doi.org/10.1007/978-3-319-73978-6_1

1

sources of energy, positive influence on carbon balance in the environment and the state of it, creates additional workplaces (Pisarek et al. 2000). It was estimated that until the year 2013 the production of the biogas in Poland would have reached the level of 1,000,000,000 m^3 and until 2020 it will have duplicated. Thus, it is necessary to cultivate plants with the purpose of energy production in mind.

Most frequently waste material after the animal production, i.e. excrements, by-products of farming and alimentary industry as well as energy plants cultivated are used for this purpose. It is recommended to use the plants that achieve high biomass crop between April and October and remain possessory of high content of easily fermenting components. The gathered green forage ought to be in silage and kept being utilized till the production of the biogas. In Polish agronomic conditions recommended plants cultivated for energy production are: corn, cereals in the pure sowing, cereal mixtures, mixtures cereal-leguminous, sunflower, Jerusalem artichoke, grasses, lucerne and clover. Energy-plants utilised for biogas production should fulfil the same requirements that the plants in-tended for ruminant nourishment. This results from the fact that methane fermentation process is similar to the processes in the gastrointestinal tract of ruminants. The cultivation of these plants for energy purposes demands the usage of the same agrotechnical endeavours as in the production of the fodder.

The millet belongs to C$_4$ plants and is one of oldest cultivated plants. It appears on considerable areas of Northern America mainly in the flora of the prairie. It can be grown on sandy, light soils. Within a period of summer American cultivars of switchgrass deliver valuable fodder for ruminants, while other fodder-grasses are rather into the standstill.

The significance of millet has decreased along with the development of the agriculture. Nowadays the greatest millet producers in the world are: India, China, countries of Africa and Russia. In Europe the switchgrass tillage has unique significance because of low repeatability of yield stability and the reduction of mil-let consumption.

Specific climatic and soil requirements cause that the millet tilled in Poland give abundant yields in eastern, south eastern, southwest and in central regions of Poland. The growing season of millet lengthens out in cooler regions of Poland having weaker insolation. Switchgrass has also poor conditions for growing and its growing season is longer in the mountain-foot regions with high precipitations and low temperatures at night. In the last decade one observes a growing interest with this plant due to its large potential for energy production (Parish and Fike 2005; Sanderson et al. 1996). Mainstreams of the investigations with regards to the utilization as energy source are: the combustion, the thermochemical conversion, the production of the ethanol or the production of the biogas (McLaughlin and Kszos 2005; Ahn et al. 2001).

2 Materials and Methods

2.1 Plant Material

Switchgrass *Panicum virgatum* was cultivated Botanic Garden of the Plant Breeding and Acclimatization Institute in Bydgoszcz.

2.2 Silage Production

The harvested switchgrass was cut up into pieces of 10 cm in length and exactly pressed with the pneumatic press in polythene microsiloes with 8.65 dm^3 of volume (the diameter 15 cm, the height 49 cm). The difference of the press level be-tween containers was under 4%. The microsiloes were closed and gasketed with gum covers with installed fermentative tubes filled with glycerine to eliminate the excess of gases. The silaging process proceeded at a room temperature. After 8 weeks microsiloes were opened. The silages were subjected to the Weende analysis, i.e. the content of short-chain of fatty acids (AOAC 1995; Van Soest et al. 1991).

2.3 The Methane Fermentation

Research into the production of biogas was performed on the test stand, using eudiometric tanks (Fig. 1). The biogas yield tests were carried out in accordance to DIN 38 414-S8 in a multi-chamber fermentation station (Fig. 2), based on an eu-genic system that stores the biogas generated on a 1 dm^3 fermentation tank (KTBL-Heft-84 2009; Eder and Schulz 2007). A measurement station for methane, carbon dioxide, hydrogen sulphide, oxygen, ammonia, nitric oxide and nitrogen dioxide was used for the biogas gas concentration tests. For the preparation of inoculum was used methanogenic thermostatic biostat with a capacity of 1650 ml. The fermentation station was equipped with a thermostatic tank, keeping the set thermal parameters of the process, fermentation tanks of 1 dm^3 and tanks for the storage of biogas with a capacity of 1200 ml. Biogas tanks are equipped with valves and connectors, which allow the removal of stored gas and the injection into the gas route equipped with biogas gas concentration analysers (DIN 38414 S.8).

Measurements of concentration and volume of secreted gas were carried out at 24-hour intervals. A mixture of identical composition was in two biofermentors to improve the correctness of the results. MG-72 and MG-73 series measuring heads have been used for the measurement of the composition of the biogas produced with measuring ranges 0–100% of volume and measuring resolution in the order of 0.1 ppm to 1% volume.

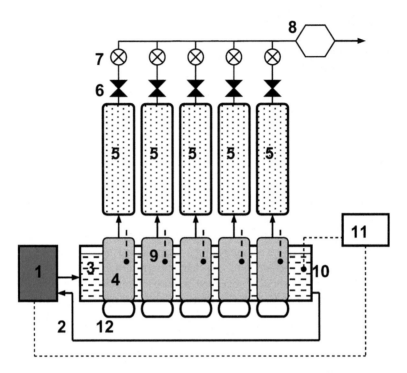

Fig. 1 Diagram of a eudiometric system for the research of biogas productivity of substrates: *1* water heater with temperature regulator, *2* insulated tubes for heating fluid, *3* water jacket with temperature control, *4* biofermentor with a capacity of 1 or 2 dm³, *5* biogas tank, *6* shut-off valves, *7* gas flowmeters, *8* gas analysers (CH_4, CO_2, NH_3, H_2S, O_2, NO_x), *9* pH sensors, *10* temperature sensor, *11* registration control unit, *12* magnetic stirrers of contents

Based on our own research and literature analysis (Jędrczak 2007; Myczko et al. 2011; Steppa 1988), these factors have been identified that characterize the fer-mentation pulp. Factors that may have a significant impact on the biogas pro-duction process include, but are not limited to, the dry substance content, organic matter content, batch weight, reaction rate, percentage of ingredients in the fermenting mix, and time from the start of the experiment (Görisch and Helm 2006). The following standards were used: PN-74/C-04540/00, PN-75/C-04616/ 01–04 and PN-90 C-04540/01. The parameters that are also evaluated are the volume of generated biogas and the cumulative value (Dach et al. 2009). During the study the process temperature was set to 6°C.

The object of the study was a mixture of solid and liquid substrates, subjected to anaerobic degradation. Cattle slurry and inoculum were also used for the study. The content of dry matter was set between 6 and 8% m/m for introduction into the process of increased dose of substrate representing the lignocellulose complex (Fugol and Szlachta 2010).

Fig. 2 Research stand for the study of biogas productivity of substrates according to DIN 38414 s.8 (left), inoculum station for quasi continuous fermentation work (right)

Mixture (approximately 10% of the dry matter) of loose bovine slurry and switchgrass were the substrates for methane fermentation. Before preparing the mixture, silage of the switchgrass was chopped into 3 cm pieces. Fresh switchgrass silage and switchgrass silage exposed to air were used. The mixture was inoculated with postfermentative pulp from biogas plant in Liszkowo (Amon 2007). Content of mixtures used for fermentation is presented in Tables 1 and 2.

The methane fermentation was carried out in water jacked biofermentor with thermostat in temperature 36 °C with 1 min mixing every 2 h. Biogas was collected in polymethyl methacrylate (PMMA) containers filled with neutral liquid. The level of the liquid decreased with increasing volume of collected biogas. Each container was connected gas analysers set (methane, ammonia, carbon dioxide and hydrogen sulphide detectors). Experiments were made in three replications.

2.4 Statistical Analysis

Data were expressed as the mean ± standard deviation of three independent replicates. Data were subjected to multifactor analysis of variance (ANOVA) using last-squares differences (LSD) test with the Statistica 8.0.

Table 1 Parameters of mixture with switchgrass fresh silage (N = 3)

Constituent	Dry mass of constituent	Contribution in mixture [%]	Contribution in mixture fresh mass [g]		Contribution in mixture dry mass [g]		pH		Cond [mS]	
Mixture	7.40	100.00	1969.95	2.22	145.72	0.34	7.63	0.10	14.85	0.73
Slurry	2.9	69	1366.4	0.9341	39.62	0.02				
Inoculate	5.8	18	351.5	1.1623	20.38	0.06				
Switchgrass silage	34	13	252.1	0.7848	85.70	0.26				

Table 2 Parameters of mixture with switchgrass silage exposed on air operation (N = 3)

Constituent	Dry mass of constituent	Contribution in mixture [%]	Contribution in mixture fresh mass [g]		Contribution in mixture dry mass [g]		pH		Cond [mS]	
Mixture	8.00	100.00	1961.50	0.71	156.88	0.02	7.65	0.24	12.84	0.38
Slurry	2.90	69.35	1360.88	0.66	39.48	0.02				
Inoculate	5.80	17.85	350.38	0.18	20.33	0.01				
Switchgrass silage	38.80	12.80	250.18	0.10	97.08	0.03				

3 Results

Nutritional content of switchgrass fresh forage and silage is presented in Table 3. The fatty acid content is shown in Table 4. This parameter did not differ significantly for fresh and exposed samples for atmospheric air. Cumulative biogas production is presented on Fig. 3. In both cases the production of biogas was increasing fast to 16–17th day and subsequently dropped.

Average daily production of biogas is depicted in Fig. 4. The production of biogas from test mixture with switchgrass fresh silage strongly increase since the beginning of the experiments. On the third and fourth day production was on the same level and in the fifth day it achieved the peak of 2.83 dm^3. Afterwards, the biogas production dramatically decreased to 0.82 dm^3 on sixth day later increased to 2.39 dm^3. Next, the gas production day by day went out irregularly achieving the value of 0.05 dm^3 on 99th day. The curve of average daily gas production form mixture with switchgrass silage exposed to air, shows that in this case the production of biogas was increased even higher. The production achieved its peak at the level of 3.5 dm^3 in fifth and sixth day and decreased irregularly but more steadily achieving the value of 0.06 dm^3 in the 99th day.

Methane concentration in biogases produced in both cases is depicted in Fig. 5. Concentration of methane in biogas from mixture with silage exposed to air rapidly exceeded 60% on the 3rd day of experiment and maintained above this value to the end of the experiment. 60% content of methane in SS biogas was achieved on the

Table 3 Chemical composition of switchgrass roughage, silage and silage exposed on air operation

Material		Dry matter	Crude ash	Organic matter	Crude protein	Ether extract	Crude fibber	Nitrogen free extract
Switchgrass roughage	Mean	29.00	4.98	95.02	6.09	2.08	38.76	48.63
	SD*	–	0.05	0.06	0.29	0.15	0.58	1.32
Switchgrass silage	Mean	34.00	6.81	92.99	5.93	2.57	46.95	37.54
	SD	–	0.21	0.19	0.47	0.08	1.37	0.23
Switchgrass silage exposed to air	Mean	38.80	8.31	91.69	–	–	–	–
	SD	–	1.50	1.50	–	–	–	–

*The same superscript in the same column row means no significant difference $p < 0.05$

Table 4 Short chain fatty acid content in fresh silage and silage exposed on air exposed

Material	Lactic acid [mg/g f.m.]*	Acetic acid [mg/g f.m.]	Butyric acid [mg/g f.m.]
Switchgrass silage	0.39 ± 0.02*	1.19 ± 0.14	0.91 ± 0.11

*The same superscript in the same column row means no significant difference $p < 0.05$

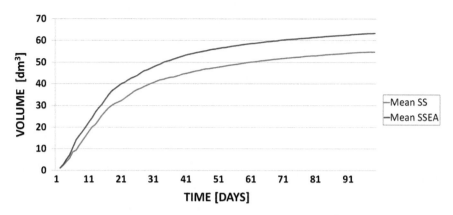

Fig. 3 Cumulative biogas production on the sample mass of mixture [dm³] (*SS* silage switchgrass, *SSEA* silage switchgrass exposed air)

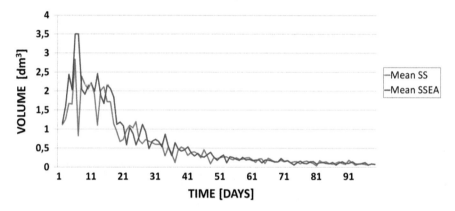

Fig. 4 Average daily production of biogas [dm³] (*SS* silage switchgrass, *SSEA* silage switchgrass exposed air)

6th day and maintained above this value till the end of experiment. The concentrations of methane reached their peaks in the days 16, 17, 18 and 14, 15, 16, 17, 18, 19 respectively for SS and SSEA biogas. The average concentration of methane in biogas was 69.40 and 70.15%.

Cumulative methane production is presented in Fig. 6. There can be observed a correlation between cumulative biogas production and cumulative methane production—correlation coefficients 0.98 and 0.99 for SS and SSEA. Total methane volume produced during fermentation was 38.6 and 46.6 dm³ respectively for SS and SSEA biogas.

Concentration of carbon dioxide in biogases is depicted in Fig. 7. In SS biogas the concentration of carbon dioxide ranged from 18.67 to 47% and from 17.67 to 31.61% in SS biogas and SSEA, respectively. The highest concentration in SS

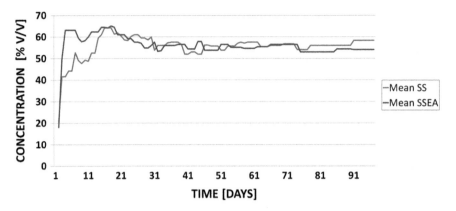

Fig. 5 Methane concentration in biogases [%] (*SS* silage switchgrass, *SSEA* silage switchgrass exposed air)

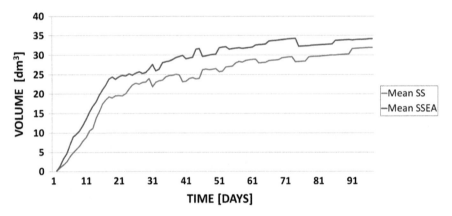

Fig. 6 Cumulative methane production [dm³] (*SS* silage switchgrass, *SSEA* silage switchgrass exposed air)

biogas carbon dioxide (47%) was reached on 2nd and 3rd day of fermentation. In case of SSEA a peak (31.67%) of carbon dioxide was reached in 34th day.

Figure 8 presents the concentrations of hydrogen sulphide. The highest concentrations of H_2S were observed in the first two weeks of fermentation. In SS biogas, the concentration achieved the highest value on the 7th day (475.67 ppm) and in SSEA on the 4th day (642.33 ppm). Total cumulative production of hydrogen sulphide in case of SSEA (0.79 cm³) was over twice as high as in SS (0.37 cm³).

Oxygen concentration in biogases (Fig. 9) were the highest on the first day— 1.17% and 1.87% respectively for SS and SSEA biogases. Afterwards, the concentration dropped reaching the minimal value between 80th and 88th day for SS (0.27%) and two minima for SSEA(%): between 67th to 75th and 80th to 86th day.

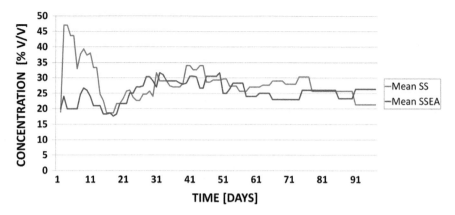

Fig. 7 Concentration of carbon dioxide in biogases [%]

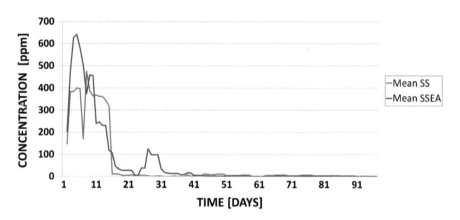

Fig. 8 Concentration of hydrogen sulphide [ppm] (*SS* silage switchgrass, *SSEA* silage switchgrass exposed air)

At the end of the fermentation, the concentrations of oxygen maintained at the level of 0.3 and 0.4% for SS and SSEA biogases.

Figure 10 shows the ammonia concentration in biogases. At the beginning of the fermentation ammonia reached the highest concentration—19.67 and 29.67 ppm. Afterwards, the concentration dropped to 0 on the last days of the process. Cumulatively ammonia amount in SS biogas was lower almost two times than in SSEA biogas.

NO and NO_2 concentrations in biogases produced in both cases is shown in Fig. 11. Content of NO in both biogases increased to 19 ppm and 33.33 ppm and steadily decreased to reach the level 0–1 at the end of the concentration. Concentration of NO_2 increased from the beginning of the fermentation till the values 2.8 and 4.67 ppm for SS and SSEA biogas and dropped to 0.

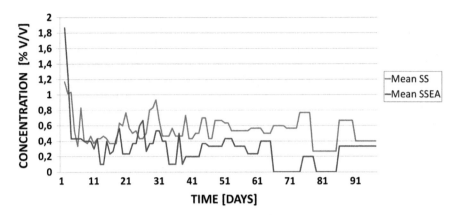

Fig. 9 Concentration of oxygen [%] (*SS* silage switchgrass, *SSEA* silage switchgrass exposed air)

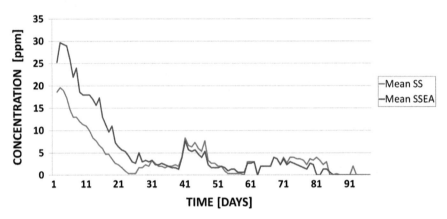

Fig. 10 Concentration of ammonia [ppm] (*SS* silage switchgrass, *SSEA* silage switchgrass exposed air)

4   Discussion

Chemical composition of the switchgrass silage is similar to roughage except of the concentrations of nitrogen free compounds, which is due to silage process. In this process bacteria utilize carbohydrates to produce short chain fatty acids which play main role in preserving silage (Bolsen et al. 1996). Switchgrass exposed to air was used as component in mixture to produce biogas in one variant of experiments. Exposing silages brings about losses in nutritional value. On the other hand, qualitative and quantitative composition of organic acids changes. Switchgrass is not commonly cultivated plant in Poland. High values of cumulative biogas production and average daily production in the first days of fermentation can be explained by preferential digestion of readily fermented chemical compounds like

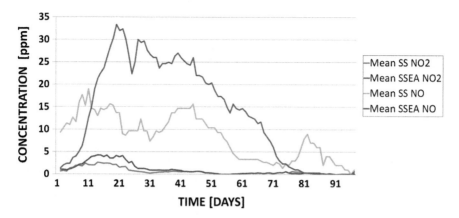

Fig. 11 NO and NO$_2$ concentration in biogases [ppm] (*SS NO2* silage switchgrass nitrogen dioxide concentration, *SSEA NO2* silage switchgrass exposed air nitrogen dioxide concentration, *SS NO* silage switchgrass nitric oxide concentration, *SSEA NO* silage switchgrass exposed air nitric oxide concentration)

carbohydrates. This corresponds to earlier observations from other scientists (Demirer and Chen 2008; Lu et al. 2007; Ahn et al. 2001).

Biogas is defined as mixture of gases. The content of methane decides about the caloric value of biogas. Literature data indicate the lowest, useful limit of methane between 45 and 55%. Stable emission of nitrogen oxides proved the steady process of methane fermentation with regards to biology. Primary higher emission of NO and NO$_2$ is caused by higher initial concentration of oxygen in fermenting mixture. This can suggest that in initial phase the fermentation can have oxygen character. Presence of NO$_x$ is undesired because of its harmful influence on the natural environment. The after mentioned carbon dioxide concentrations are within the standards of content in biogas. High concentration of CO$_2$ dilutes biogas and decreases its caloric value. High emission of ammonia, which is the result of protein compounds degradation, also proves the oxygenic character of reactions at the initial stage.

Decreasing emission of ammonia and steady emission of nitrogen oxides improved biogas content and point out that oxygen compounds were hydrolysed. Concentration of oxygen at the level of 0.1% in produced biogas is prospective in small scale agricultural systems of biogas production. Defined amount of oxygen is supplied to filters in some methods of decreasing sulphide hydrogen concentration at desulphurization installations located outside the fermentation chamber. The permissible concentration of sulphide hydrogen in biogas ranged between 18 and 20 ppm. Producers of current generators indicate that the concentration above this value may cause corrosion of engines. Ammonia and hydrogen sulphide are compounds with strong, unpleasant smell. Their emission negatively influences on the natural environment. There is a great interest in reducing odour pollution in rural areas. One a par with animal production, development of biogas production may be limited by emission of odour compounds and its insufficient controlling. On the other hand, anaerobic digestion helps to convert volatile organic compounds to less arduous compounds for environment (Mackie et al. 1998; McLaughlin and Walsh 1998).

References

Ahn HK, Smith MC, Konrad SL et al (2001) Evaluation of biogas production potential by dry anaerobic digestion of switchgrass—animal manure mixtures. Appl Biochem Biotechnol 160:965–975

Amon T (2007) Das Potenzial für Biogaserzeugung in Österreich. Beiratssitzung für Umwelt und soziale Verantwortung unserer Gesellschaft, Maria Enzersdorf• Imprint

AOAC (1995) Official methods of analysis, 16th edn. Association of Official Analytical Chemists, Washington, D.C.

Bolsen KK, Ashbell G, Weinberg Z (1996) Silage fermentation and silage additives—review. Asian-Australas J Anim Sci 9:483–493

Dach J, Zbytek Z, Pilarski K et al (2009) Badania efektywności wykorzystania odpadów z produkcji biopaliw jako substratu w biogazowni. Technika Rolnicza Ogrodnicza Leśna 6:7–9

Demirer GN, Chen S (2008) Anaerobic biogasification of undiluted dairy manure in leaching bed reactors. Waste Manage 28:112–119

DIN 38414 S8 (2012) German standard methods for the examination of water, waste water and sludge; sludge and sediments (group S); determination of the amenability to anaerobic digestion (S 8). DIN Deutches Institut fur Normung e.V., Berlin

Döhler H (red.)/KTBL (2009) Faustzahlen Biogas. 2. Auflage. Kuratorium für Technik und Bauwesen in der Landwirtschaft e.V. (KTBL), Darmstadt, Schauermanndruck GmbH, Gernsheim

Eder B, Schulz H (2007) Biogas Praxis. Okobuch Verlag und Versand GmbH, Staufen bei Freiburg

Fugol M, Szlachta J (2010) Zasadność używania kiszonki z kukurydzy i gnojowicy świńskiej do produkcji biogazu. Inżynieria Rolnicza 1(119):169–174

Görisch U, Helm M (2006) Biogasanlagen. Eugen Ulmer KG, Stuttgard (Hohenheim)

Jędrczak A (2007) Biologiczne przetwarzanie odpadów. Wydawnictwo Naukowe PWN, Warsaw

KTBL-Heft 84 (2009) Schwachstellen an Biogasanlagen verstehen und vermeiden. Kuratorium für Technik und Bauwesen in der Landwirtschaft e.V. (KTBL), Darmstadt, Druckerei Lokay, Reinheim

Lu S, Imai T, Ukita M et al (2007) Start-up performances of dry anaerobic mesophilic and thermophilic digestions of organic solid wastes. J Environ Sciences 19:416–420

Mackie RI, Stroot PG, Varel VH (1998) Biochemical identification and biological origin of key odor components in livestock waste. J Anim Sci 76(5):1331–1342

Majtkowski W, Majtkowska G (2000) Ocena możliwości wykorzystania w Polsce gatunków traw zgromadzonych w kolekcji ogrodu botanicznego IHAR w Bydgoszczy. In: Polsko-Niemiecka konferencja nt. Wykorzystania trzciny chińskiej Miscantus. 27–29 Sept, Połczyn Zdrój

Majtkowski W, Majtkowska G, Piłat J et al (2004) Przydatność do zakiszania zielonki traw C-4 w różnych fazach wegetacji. Biuletyn IHAR 234:219–225

McLaughlin SB, Kszos LA (2005) Development of switchgrass (Panicum virgatum) as a bioenergy feedstock in the United States. Biomass Bioenerg 28:515–535

McLaughlin SB, Walsh ME (1998) Evaluating environmental consequences of producing herbaceous crops for bioenergy. Biomass Bioenerg 14:317–324

Myczko A, Myczko R, Kołodziejczyk T et al (2011) Budowa i eksploatacja biogazowni rolniczych, Wyd. ITP Warszawa-Poznań

Parrish J, Fike JH (2005) The biology and agronomy of switchgrass for biofuels. CRC Crit Rev Plant Sci 24(5–6):423–459

Pisarek M, Śmigiewicz T, Wiśniewski G (2000) Możliwości wykorzystania biomasy do celów energetycznych w warunkach polskich. In: Polsko-Niemiecka konferencja nt. Wykorzystania trzciny chińskiej Miscantus. 27–29 Sept, Połczyn Zdrój

Polish Standard. PN-74/C-04540/00 Oznaczenie zasadowości. Wydawnictwo Normalizacyjne, Warsaw

Polish Standard. PN-75/C-04616/01 Oznaczanie suchej masy osadu i substancji organicznych. Woda i ścieki. Badania specjalne osadów. Oznaczanie zawartości wody, suchej masy, substancji organicznych i substancji mineralnych w osadach ściekowych. Wydawnictwo Normalizacyjne, Warszawa

Polish Standard. PN-75/C-04616/04 Oznaczenie lotnych kwasów tłuszczowych. Wydawnictwo Normalizacyjne, Warszawa

Polish Standard. PN-90 C-04540/01 Woda i ścieki. Badania pH, kwasowości i zasadowości. Oznaczanie pH wód i ścieków o przewodności elektrolitycznej właściwej 10 µS/cm i powyżej metodą elektrometryczną. Wydawnictwo Normalizacyjne, Warszawa

Sanderson MA, Reed RL, McLaughlin SB et al (1996) Switchgrass as a sustainable bioenergy crop. Bioresour Technol 56:83–93

Steppa M (1988) Biogazownie rolnicze. IBMER, Warsaw

Van Soest PJ, Robertson JB, Lewis BA (1991) Methods for dietary fiber, neutral detergent fiber, and non-starch polysaccharides in relation to animal nutrition. J Dairy Sci 74(10):3583–3597

Process Intensification in Practice: Ethylene Glycol Case Study

Magda H. Barecka, Mirko Skiborowski and Andrzej Górak

1 Process Intensification (PI): Potential for Innovation

The concept of PI as the development of breakthrough technologies in chemical engineering began to appear in the 1970s (Stankiewicz and Moulijn 2004). Numerous definitions for PI have been proposed, focusing on its potential to:

- "reduce the size of chemical plants" (Cross and Ramshaw 1986) and "combine multiple operation into fewer devices" (Tsouris and Porcelli 2003)
- "produce in a cleaner, more energy efficient and cheaper way" (Stankiewicz and Moulijn 2000; Gourdon et al. 2015)
- "target enhancement of involved phenomena at different scales to achieve a targeted benefit" (Lutze et al. 2010)
- use "alternative forms of energy supply" (Freund and Sundmacher 2010)
- "give each molecule the same processing experience" (Górak and Stankiewicz 2011).

Based on these proposed definitions, every novel process concept or equipment enabling targeted and drastic improvement can be categorized as PI. Hence, PI is often considered a "toolbox" that gathers examples of novel chemical engineering technologies. To provide a more systematic understanding of PI, van Gerven and Stankiewicz (2009) proposed the definition of fundamentals of intensification related to four domains: structure, energy, synergy and time.

M. H. Barecka (✉) · M. Skiborowski · A. Górak
Laboratory of Fluid Separations, TU Dortmund, Dortmund, Germany
e-mail: magda.barecka@tu-dortmund.de

M. H. Barecka · A. Górak
Faculty of Process and Environmental Engineering,
Lodz University of Technology, Lodz, Poland

© Springer International Publishing AG, part of Springer Nature 2018
M. Ochowiak et al. (eds.), *Practical Aspects of Chemical Engineering*,
Lecture Notes on Multidisciplinary Industrial Engineering,
https://doi.org/10.1007/978-3-319-73978-6_2

1.1 Barriers to Industrial Implementation of PI

The potential of PI technologies to improve chemical production processes has been widely demonstrated, but few PI examples can be found in industry. One of the main reasons for this reluctance is industrial trust in mature technologies. An equipment or method that has not proven its reliability over an extended period of time is always considered risky (Becht et al. 2009). Objections to the application of intensified technologies can also be attributed to high investment costs, usually related to PI. With few (or for some technologies even none) industrial examples of PI, it is difficult to prove that the technology will give a good return on investment. There are also concerns about the safety and control of a new process. As a result, the general risk associated with PI is high. Consequently, companies are not eager to take such an elevated risk, leading to a vicious circle that obstructs innovations (Adler 1998).

Another key aspect limiting PI application is the lack of general knowledge of how and where to apply such technologies. It is difficult to identify suitable PI technologies that will enhance the physical and chemical phenomena limiting the process of interest. Determination of most the promising equipment should be based on verified PI metrics, enabling benchmarking of various technologies (Curcio 2013). Such metrics would also serve as a basis to compare intensified versus classical equipment performance and systematically verify the benefit of PI for the considered technology. A lack of such metrics and approaches for determination of the most suitable PI technology results in the need for numerous experiments, adding to already high costs of technology implementation. Reassuming, numerous barriers for PI industrial implementation exist on the level of available knowledge, experience and costs related to intensified technologies (Fig. 1).

Fig. 1 Summary of barriers in industrial implementation of PI (Lutze et al. 2010; Becht et al. 2009; Stankiewicz and Moulijn 2000)

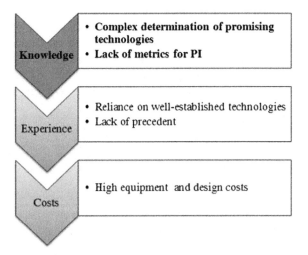

1.2 Retrofitting Approaches Supporting PI Implementation

Retrofitting targets the improvement of plant design by determination of main process limitations (bottlenecks) and subsequent changing or replacing of some unit operations to overcome this limitation (Lutze et al. 2010). Hence, process retrofitting can reduce the operational costs, waste generation or plant footprint. PI technologies have great potential for retrofitting, but it is complex to determine the right intensified technology for a given process. Systematic methodologies supporting process retrofitting by means of PI are consequently required. However, so far only a few approaches that consider PI technologies in the retrofit portfolio have been reported. Niu and Rangaiah (2016) proposed a heuristic-based methodology that enables base-case analysis and the generation of retrofit options, first without and next with additional investment. Proposed variants were evaluated by considering their potential to reduce overall manufacturing costs. The methodology was applied to a case study of isopropyl alcohol production. This approach enabled the generation of integrated process options, but only reactive and hybrid separations, which were achieved by combining existing operations in the process, could be determined. As a result, intensified equipment or the use of new driving forces or energy sources could not be systematically considered. Furthermore, due to a lack of quantitative metrics for the evaluation of different PI options, the possible process improvements had to be assessed by means of rigorous simulations.

The PI database-driven approach of Lutze (2011) was originally proposed for process synthesis problems. However, contrary to other methods dedicated to process synthesis, this approach also considered a set of indicators for determining the dominant process bottlenecks and can thus also be applied to retrofitting cases. This methodology undergoes a hierarchical procedure that first generates a significant number of intensified options from a knowledge-based tool and subsequently limits the number of options by considering various knock-out criteria. All options remaining after preselection must be evaluated by using derived models. Therefore, in the case of numerous options determined from the database screening, application of the method was time- and resource-consuming, which is undesirable in an industrial context. Recently, an extension of this method was reported, with additional integrated heuristics and general knowledge of process engineering for screening purposes (Benneker et al. 2016). However, the authors again addressed the difficulties in evaluating each retrofit option using tedious simulations.

Hence, all of the described methodologies face a major limitation: the proposed analyses are too complex to find successful implementation in terms of time- and cost-limited retrofitting projects. Due to the availability of a wide range of PI equipment, strategies enabling the systematic limitation of the number of options to the most promising ones are necessary. Moreover, such approaches should be capable of operating using only limited process data since detailed data are often not available during retrofit projects (ten Kate 2015). To overcome the limitations of existing methodologies, an alternative approach is presented that aims to determine the most promising retrofit options by obtaining insight into the phenomena limitation, resulting in the observed process bottleneck. By means of such an analysis, the options to be considered are drastically reduced at the very beginning of the analysis.

2 Proposed Methodology for Process Retrofitting

To support the industrial implementation of PI for process retrofitting, the proposed methodology targets the determination of the most promising retrofit options by using a simplified analysis. At first, the incentive for retrofitting is defined as, e.g., a minimization of operational costs. Next, the major process limitations (bottlenecks) are identified with respect to the defined goal of retrofitting. To this end, the process data are collected from either plant measurements or a flowsheet simulation. Further, bottlenecks are determined based on a sensitivity analysis of mass and energy efficiency factors calculated for each unit operation. Once the key bottlenecks are identified, retrofit options are generated from an extensive database of PI technologies. This knowledge-based tool is similar to the database of Lutze (2011) and contains over 150 technologies, characterized by unit operation, available phases, a range of operating parameters, bottlenecks that can be tackled and phenomena that are intensified in the given equipment. The PI database is screened stepwise. First, the unfeasible technologies are rejected by considering simple criteria, such as stream phases and process operating conditions. During the second step, a set of additional methods and tools (listed in Fig. 2) is used to determine the specific physical/chemical phenomena limitation responsible for the observed bottleneck. The space for improvement via the intensification of these phenomena is additionally quantified. Hence, metrics are available for benchmarking various PI options. The used methods are linked to the PI database and selected specifically to operate with the limited data and enable fast process analysis.

The preliminary design of the most promising intensified operations is performed in the third screening step. Since the database also contains reactive or hybrid separations, the determined options may require study of feasible separation processes that can be coupled to reactions or separations. The thermodynamic insight approach (Jaksland et al. 1995) is used for this purpose. Furthermore, if a solvent is required for the determined separation, computer-aided molecular design (CAMD) (Harper and Gani 2000) implemented in the ProCAMD tool is used to generate promising mass transfer agents. After the application of the set of additional tools, the most promising intensification options are determined. Finally, short-cut models are used to roughly estimate the equipment performance (fourth screening step). The most promising option is selected and integrated into the flowsheet, which is optimized with respect to the new operating conditions. The overall effect of retrofitting is evaluated based on simulation of the intensified process. Hence, the decision can be made if the improvement is significant enough to validate this option experimentally and introduce it in the chemical plant.

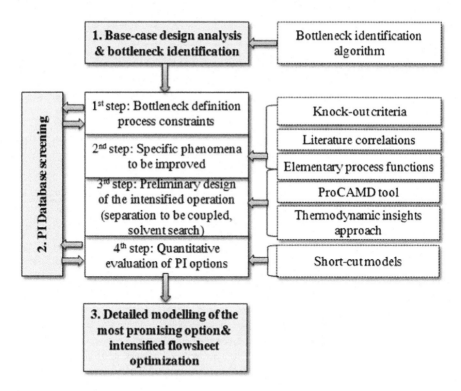

Fig. 2 Overall methodology structure (adapted from Barecka et al. 2017)

A further description of the methodology, algorithms used and related tools is given in the work of Barecka et al. (2017).

3 Ethylene Glycol Case Study

Monoethylene glycol (MEG) is an important raw material for industrial applications and is used mainly as an antifreeze agent and a starting material in the production of polyester fibers. Currently, this glycol is mainly produced by high-pressure and high-temperature hydrolysis of ethylene oxide (EO) (Rebsdat and Mayer 2010). The typical process flowsheet is depicted in Fig. 3 and is divided into three main sections: (1) MEG reactor, (2) multi-effect evaporation and (3) product refinement. The reaction system for MEG production (Fig. 4) consists of several consecutive reactions leading to the formation of higher glycols: diethylene glycol (DEG) and triethylene glycol (TEG). Although a wide variety of catalysts was previously reported, the uncatalyzed reaction system is still most commonly used in industry (Rebsdat and Mayer 2010). The improvement in reaction selectivity when using a catalyst was not sufficient to balance several disadvantages of such systems,

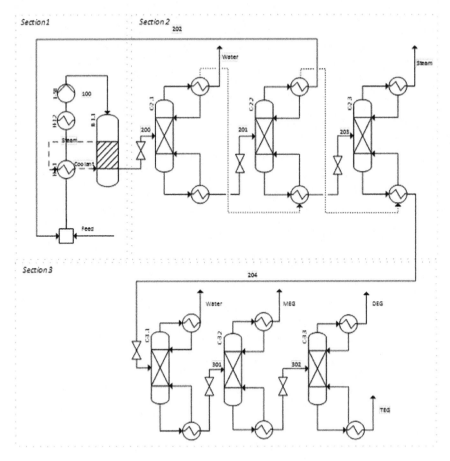

Fig. 3 Flowsheet of the well-established process for MEG production (Dye 2001)

$$\overset{+C_2H_4O}{H_2O \longrightarrow} \overset{+C_2H_4O}{C_2H_6O_2 \longrightarrow} \overset{+C_2H_4O}{C_4H_{10}O_3 \longrightarrow} C_6H_{14}O_4$$

Fig. 4 Reaction system for MEG production (Altiokka and Akyalçin 2009)

including the additional required purification steps and the risk of product contamination in case of insufficient catalyst separation.

The reaction system is highly exothermic and the heat produced by the reaction is used to pre-heat the cool feed stream. A 10- to 20-fold excess of water is usually used to limit the formation of by-products. Subsequently, the product is recovered from a diluted water stream. The purification starts with multi-effect evaporation, which enables the energy-efficient separation of water (Section 2). Most of the water is recycled back to the reactor. After water removal, the main product and by-products are separated by vacuum distillation (Section 3).

4 Base-Case Design Simulation

The well-established Siemens process for MEG production was selected as the base-case design and rigorously simulated in Aspen Plus. The process configuration was modelled as closely as possible to the reported flowsheet to give a good picture of the current state-of-the-art technology and was used for intensification bench-marking. The production rate was set to 220 kta (a typical capacity for MEG plants in Europe, Rebsdat and Mayer 2010), and the purity of the commercially available MEG was assumed (min. 99.8%). The byproducts DEG and TEG were as well separated as pure (min. 99.5%). The whole flowsheet was simulated using the Schwartzentruber and Renon (SR-Polar) property method, which enables accurate property prediction also for high-pressure and high-temperature operations (Schwartzentruber et al. 1989). An uncatalyzed reaction system for MEG produc-tion was selected due to the aforementioned predominance in industry and modelled with the kinetics reported by Altiokka and Akyalçin (2009). A multi-tubular reactor was used for MEG production. A water/EO molar ratio of 10/1 reported for the Siemens process was applied to represent the current technology. Reactants were mixed with excess water recovered from the second evaporation column (Fig. 3) and pre-heated. The reactor operated at 36 bar and was designed for ca. 4 min residence time, enabling the nearly total conversion of the key reactant-EO, similar to that reported for a typical MEG reactor operation (Dye 2001). The formation of MEG and higher glycols is highly exothermic; therefore, sufficient reactor cooling is a key issue for operational safety. The reaction mixture was maintained as a liquid through the reactor length using counter-current cooling with water. The steam produced from cooling was used for reactant pre-heating. The modelled reactor produced 27.6 tons of MEG per hour, with 79% selectivity.

Excess water was subsequently removed from the reactor outlet stream through a three-stage multi-effect evaporation chain. The first column operated at a pressure of 14 bars and consisted of 7 equilibrium trays. The pressure in the subsequent columns decreased step-wise (9 and 4 bars). In the first columns, the heat recovered from water condensation was subsequently used as a heating medium for the subsequent columns. The water recovered as the top product in the first column was used as recycled water in the plant. The water separated in the second column was recycled back to the reactor and re-used as excess reactant. In the third column, water was recovered as a useful medium (steam). This configuration optimized the energy requirement for water removal, and following Section 2, the water content was reduced to 40 wt%. The last section consisted of a chain of vacuum distillation columns, where the remaining water was removed and glycols were separated with the required purity.

4.1 Bottlenecks Identification

Because MEG is one of the most used bulk chemicals, minimization of the total operational cost (TOC) was selected as the retrofitting objective. Changes in the flowsheet configuration may require additional investments that must be considered before the final decision for the retrofit project. However, this investment is difficult to estimate reliably for novel technologies. Instead of providing a rough estimation of the investment costs, we focused on the determination of the operational cost reduction and resulting annual savings. Thus, quantified savings provide a basis to determine whether the retrofit design is promising enough to consider investment in the intensified equipment. At first, the operational cost was evaluated for the base-case design process (Eq. 1) based on a calculated consumption of reactants ($m_{raw,i}$, kg h^{-1}), utilities (E_i, kJ h^{-1}) and produced MEG flowrate (m_{prod}, kg h^{-1}). The price indicators for the cost of raw materials ($c_{raw,i}$, \$ kg^{-1}) and utilities ($c_{energy,i}$, \$ kJ^{-1}) were obtained from the literature (Peschel 2012; Barccka et al. 2017). No make-up chemicals were used in this process. The TOC was estimated as 610\$ t^{-1} tons of MEG, with the predominant contribution of the raw material costs (more than 99%). Due to the optimized energy recovery, the cost of the entire separation chain was responsible for only 0.02% of the TOC.

$$TOC = \frac{\sum c_{raw,i} \dot{m}_{raw,i} + \sum c_{energy,i} \dot{E}_i}{\dot{m}_{prod}} \qquad (1)$$

Subsequently, the mass and energy efficiency factors were calculated for each unit operation. The factors were further manipulated by some margin of uncertainty (e.g., ±5, 10%), and the objective function (operational costs) was again evaluated for the manipulated factors. The factors with the most significant impact on retrofitting were used to identify key process bottlenecks. By means of such analysis, the mass efficiency for the reaction (corresponding to reaction selectivity) was determined as the dominant bottleneck (Table 1). Mass and energy efficiency factors for the remaining unit operations had a minor influence on the TOC (far below 1%) and are not listed here.

Table 1 Mass efficiency factors for each piece of equipment with space for TOC improvement; n denotes component flowrate (kmol/hr)

Equipment/unit operation	Calculation formula	Value (%)	Space for TOC reduction for factor = 100% (%)
Reactor R-1.1	$m_{eff,R-1.1} = \frac{n_{C_2H_6O_2,S200}}{n_{C_2H_4O,S100} - n_{C_2H_4O,S200}}$	79.27	22

4.2 PI Database Screening: Tackling Reaction Selectivity Limitations

1st step: Bottleneck definition. The PI database was subsequently screened for technically feasible options for the reaction selectivity limitations bottleneck. The first screening step was performed based on data collected from the base-case design simulation and the resulting simple criteria characterizing the process:

- Type of unit operation: reaction
- Possible secondary operation: separation, mixing, heat exchange
- Inlet stream phase: liquid
- Outlet stream phase: liquid
- Reaction phase: liquid
- Catalyst: –
- Range of temperatures: 448–503 K
- Pressure: 36 bar
- Additional limitations: maintaining the current reaction pathway
- Bottleneck: reaction selectivity towards MEG production.

As a result of the screening procedure, 51 reactors and technologies were determined. These PI options address limitations related to phenomena:

- Mass transfer
- Heat transfer
- Product concentration
- Operating conditions
- Reactants concentration
- Chemical reaction system.

Consequently, it is necessary to verify which limitation has the most significant influence on reaction selectivity for MEG production. The effect of possible improvement of the listed phenomena is analyzed in the second screening step.

2nd step: Specific phenomena to be improved

Mass transfer The limitations of heterogeneous or homogenous mass transfer (mixing) can potentially result in limited selectivity. For the MEG production reaction, no heterogeneous catalyst was used, and all substrates were in the liquid phase. Therefore, there was no mass transfer between phases, and consequently, heterogeneous mass transfer limitations were not observed. The influence of possible mixing limitations on MEG reaction selectivity was verified using characteristic time analysis (Commenge and Falk 2014). The relevance of mixing phenomena was evaluated based on a comparison of the reaction time with the mixing time typically observed in the considered reactor. If the reaction time is close to the mixing time, poor mixing can potentially affect selectivity since the reaction begins before all reactants are uniformly distributed. If the reaction time clearly exceeds the mixing time, the mixing limitations are insignificant and can be neglected. The reaction time for a homogenous reaction (t_{hom}, Eq. 2) was estimated based on the available kinetic data as 70s (kinetic coefficient k, L mol^{-1} s^{-1}, key reactant concentration C_O, mol L^{-1}, reaction order n). The mixing time reported for a tubular reactor modelled in the base-case design was significantly lower (0.02–0.2 s). Hence, the reaction rate was not limited by the mixing phenomenon and PI solutions that targeted mixing enhancement were not considered in the following analysis steps.

$$t_{hom} = \frac{1}{k_n \cdot c_0^{n-1}} \tag{2}$$

Heat transfer, product and reactant concentration; chemical reaction system. The influence of these phenomena on reaction selectivity was evaluated using the concept of elementary process functions (EPF) introduced by Peschel et al. (2010). This method allows the determination of the maximal reaction selectivity under the optimal temperature/pressure/reactant and product concentration profiles. First, all technical limitations for the achievable profiles are neglected. Simplified equations describing the reactor are then used for the optimization, targeting selectivity enhancement. As a result, the margin of process improvement achieved by a perfect PI technology enhancing each phenomenon is quantified. For instance, the selectivity achieved for the operation under an optimal temperature profile evaluates the improvement with perfect heat transfer. By considering the selective removal of products, the potential for reactive separations can also be investigated. A rigorous application of the EPF approach necessitates a reaction kinetics model, but rough estimates can also be achieved based on correlations between product yield and temperature, pressure and reactant concentration from the plant data.

In the MEG case study, detailed kinetic data were available and used in the application of the described approach. At first, the production of MEG and its main by-products, DEG and TEG, was described as reported by Altiokka and Akyalçin (2009). To simplify the model, the total volumetric and molar flowrates were considered constant, and their values were taken as the inlet of the base-case design reactor. The rate of MEG production depended on temperature and water, EO and MEG molar concentrations. Therefore, the optimization targeted the determination of suitable temperature, water, EO and MEG profiles to reach maximal selectivity. Additionally, a constraint on the required MEG concentration at the reactor outlet (2.5 mol/L) was considered to guarantee sufficient conversion. Subsequently, the reactor was divided into segments, and the optimal values of each considered variables were determined in the segment.

First, the optimal temperature profile was obtained (Fig. 5). The temperature was slightly higher at the beginning of the reactor, thus enhancing the conversion. Further, as MEG concentration increased, the driving force for the consecutive side-reactions became more significant, and the temperature dropped to limit by-product formation. The selectivity calculated for optimal temperature was 84.3% (5% higher than in the base-case design).

Fig. 5 Optimal temperature profile for the MEG reactor, determined using the EPF approach

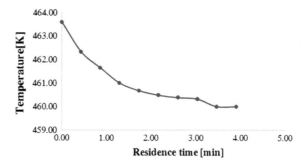

Subsequently, reactant concentration was optimized along the reactor. For water, the constraint on the maximal concentration was considered to avoid too high energy requirement for separation. The maximal water concentration corresponding to the conditions at the entrance of the base-case design reactor was assumed. EO concentration was not constrained. The determined optimal reactant concentration for both water and EO was constant throughout the reactor (water: 27.8 mol/L, EO 0.92 mol/L). Referring again to the reaction system, MEG production is driven by water concentration, in contrast to the rate of by-product formation. Consequently, a high concentration of this reactant enhanced selectivity, reaching 91.2% for the optimal case (11.9% improvement). On the contrary, EO accelerated both MEG and by-product formation; therefore, it was beneficial to maintain a constant low EO concentration. This enabled 91.4% selectivity towards MEG production (12.1% higher than for the base-case design). However, achieving such a constant reactant profile would naturally require additional reactant dosing streams.

Yet, the highest margin for selectivity improvement was related to MEG concentration optimization. In the optimal scenario, the MEG concentration should be kept at zero throughout the reactor, indicating that MEG should be immediately removed after its production. In such an idealized case, the selectivity reached 100% since there would be no substrate available for consecutive by-product formation. Consequently, controlling the product concentration via in situ removal showed the highest potential for selectivity improvement and was selected as the most promising intensification strategy. A feasible reactive separation process is determined and designed below.

3rd step: Preliminary design of the intensified operation. The control of the product concentration and consequent reactive separations were identified as the most promising strategy for improving reaction selectivity. Hence, a feasible separation technique to be coupled with MEG production reaction needed to be chosen. This part of the analysis was achieved using the thermodynamic insights approach (Jaksland et al. 1995). This method determines potentially feasible separation techniques for a given system based on mixture properties and differences in pure component properties. For each separation technique, the relevant component properties are specified, and the minimum value of relative component properties required for each separation is given as feasibility indices. For instance, distillation is expected to be feasible if no azeotrope is formed and if the boiling points ratio exceeds 1.01 (Holtbruegge et al. 2014; Jaksland et al. 1995).

First, a number of separation techniques were determined using the PI database (Table 2) for the given separation (liquid-liquid) and operating conditions for the MEG reactor (temperature 448–503 K, pressure: 36 bar). The feasibility of each separation was subsequently verified by analysis of key compound properties for each separation. The values of these properties for MEG, EO and water were collected from the literature, and the relative properties were calculated. Next, the calculated values were compared to the feasibility indices (Jaksland et al. 1995). The majority of separation techniques determined from the prescreening fulfilled the required feasibility indices. However, one key feasibility parameter for reactive microfiltration and reactive ultrafiltration (relative molecular diameter) was equal to one, resulting in a lack of driving force for the separation. Hence, those techniques were disregarded. In further screening steps, the commercial availability of membranes and solvent was verified, and additional heuristic rules were applied to determine the most promising separation.

Commercial availability screening. Literature was subsequently screened for a membrane that enabled the reactive separation of MEG from EO and water over a preferably wide range of concentrations. The number of membranes was previously

Table 2 Possible separation technologies determined for reactive separation for MEG production and summary of multistep screening for determination of feasible reactive separations. "X" denotes rejected separations

Separation technique from PI database	Feasibility index value	Commercial availability	Solvent availability	Heuristic screening
Reactive distillation				
Reaction extraction (solvent-based)			X	
Reactive distillation with gas sep. membrane		X		
Reactive distillation with pervaporation		X		
Reactive distillation dividing wall column				X
Membrane-assisted reactive divided wall column		X		
Reaction with in situ adsorption		X		
Slug flow reactor–extractor			X	
Reactive liquid membrane			X	
Reactive extractive distillation			X	
Reactive microfiltration	X			
Reactive pervaporation		X		
Reactive ultrafiltration	X			

reported for MEG separation, and a summary of experimental studies was published by Jafari et al. (2013), however, none of the reported membranes was used to separate a stream with any quantity of EO. Since EO is a highly reactive compound (Rebsdat and Mayer 2012), its presence, even in minor quantities, would result in problems with membrane stability. Therefore, due to a lack of data for the separation of MEG from both water and EO, separations including membranes were not considered further. Moreover, reactive adsorption was disregarded because no adsorbent was reported for MEG removal from a liquid stream.

Solvent screening. The remaining options for reactive separation included extraction-based separations. For a complete analysis of separation via extraction, a promising solvent was determined using computer-aided molecular design (Harper and Gani 2000) implemented using the ProCAMD software. The tool applies group contribution methods to generate feasible molecules that fulfill defined requirements in terms of component properties and separation performance. Hence, the separation type (liquid-liquid) was defined, as were the compounds to be separated (MEG from water and EO) and the process operating conditions. Additionally, a limit on the boiling point of the solvent was set to ensure that the solvent remained a liquid during operation. Based on the prescreening criteria, 5300 molecules were generated. Subsequently, an additional criterion of solvent minimum selectivity was added. The selectivity was defined as a ratio of the infinite dilution activity coefficient of the component not extracted to the activity coefficient of the solute. Since an appropriate solvent should always lower the activity coefficient of the solute (Seader et al. 2013), the selectivity should be higher than 1. None of the molecules met all of the requirements, which means that no solvent can ensure the selective separation of MEG from both the EO and water. Consequently, extraction, liquid membranes and reactive separations incorporating extraction were not considered.

Heuristic screening. Of the two remaining options, reactive distillation and reactive divided wall column, the latter was disregarded based on heuristic criteria, which are additionally available in the PI database. Divided wall columns were reported as economically interesting when the side stream was larger than the distillate and bottom (Kiss 2013), which was not the case for the analyzed case study process. As a result of the multistep screening, reactive distillation was the only remaining option for coupling MEG reaction with separation, as highlighted in the summary of screening steps (Table 2).

4.3 Modelling of the Intensified Process

Reactive distillation, which was determined to be the most promising option for tackling the selectivity limitation bottleneck, was used to replace the tubular reactor for MEG production. The column operation was rigorously simulated in Aspen Plus using the Radfrac model. The feed flowrate was the same as that used in the base-case design reactor. Water, characterized by a lower boiling temperature than the product (MEG), was fed at the top of column, following the general rules for reactive distillation design (Woods 2007). Operation at high water concentration was recommended; therefore, the column was designed to operate with a total reflux ratio. As a result, excess water was directly recycled back to the reactive zone.

Although the column was modelled with equilibrium stages, the liquid hold-up is the major design parameter that determines the extent of reaction; hence, an estimate of its value was provided. The liquid hold-up was adjusted to achieve the required residence time for the reaction (Harmsen 2007). The total liquid hold-up required was 5 m^3, divided into 5 reactive stages with 1 m^3 of hold-up per stage. The assumed hold-up fulfilled the limitation of a reasonable maximal liquid hold-up reported for bulk chemical production in reactive distillation columns (3 m^3 per stage, Huss et al. 2003). Subsequently, 5 stages for the stripping section were required to obtain a glycol mixture free of water. Another parameter with a key influence on reaction conversion was the operational pressure. Higher pressure enhanced conversion but increased the separation costs. An optimal pressure value, half of that in the base-case design, was selected (18 bar). The steady-state profiles for temperature and liquid composition are given in Fig. 6. Within the simulated column configuration, a 90% selectivity towards MEG was achieved (11% higher than in the base-case design).

Furthermore, due to the significant improvement of the reaction selectivity and separation of excess water already in the reactive distillation column, the separation chain needed to be refitted. The stream leaving the column was composed of glycols only; hence, further water removal was not necessary. The only remaining separation task was separation of the products MEG, TEG and DEG, which was achieved in a single column. Since the largest product stream is recovered at the top

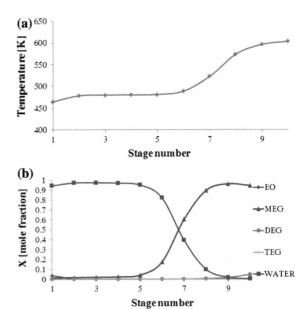

Fig. 6 Profiles for **a** temperature and **b** liquid composition for the proposed design of a reactive distillation column for MEG production

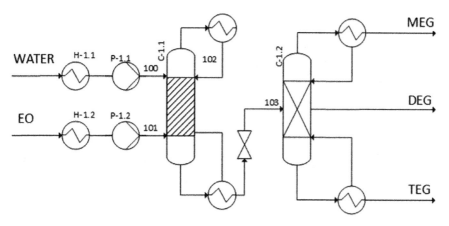

Fig. 7 Intensified process for MEG production using reactive distillation

of the column, a simple column with a side stream was economically more interesting than a divided wall column (Kiss 2013). The distillation column with side
stream removal used for product separation consisted of 30 stages, and the
side-stream recovery was located at the 17th stage. Operational pressure was very
low (0.1 bar), similar to product separation in the base-case design. A reflux ratio of
3 was used, and the product and by-products were recovered with the same purities
as in the base-case design. The intensified process is depicted in Fig. 7. Based on
the data for the retrofitted process and cost indicators used in the base-case design, a
new TOC was rigorously calculated. The TOC for the intensified design was
evaluated as 523$ t^{-1}, a 14.2% reduction with respect to the base-case design.
Considering the plant throughput (220 kta), this reduction in operational costs
would result in 19 M$ savings in operational costs per year.

5 Conclusions

The selectivity improvement, operational cost minimization and flowsheet simplification achieved in this case study highlight the benefit of retrofitting using PI
(Fig. 8). Those results would be even more valuable for the analysis of a plant that
is under the design process and has not yet been constructed. Using such a retrofit
study, not only the operational costs would be minimized, but as well the necessary
investment into equipment. The proposed structured approach enables the systematic consideration of a wide range of retrofit alternatives. Due to the availability
of quantitative metrics for determining the most promising options, analysis
required significantly lower modelling effort than do the approaches of Niu and
Rangaiah (2016) or Lutze (2011).

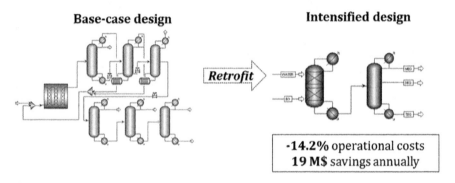

Fig. 8 Summary of results for retrofitting MEG production

Reactive distillation was determined as the most promising retrofit option for the MEG process. However, this option can also be interpreted as the most straight-forward intensification strategy, which could result from simple integration of the reactor with the following distillation column. Nevertheless, the proposed approach for determining the PI options has several advantages. Firstly, it was found out that considering the availability of equipment and various process constraints, other reactive separations cannot be used for MEG production. Moreover, we can compare MEG selectivity improvement for reactive distillation (11%) with the margin for selectivity enhancement for an ideal reactive separation (21%). This comparison highlights the potential benefit of a technique enabling to deal with limitations of distillation. Therefore, applying the retrofit methodology supports the identification of the technologies and equipment that should be further developed to achieve even more significant process improvement.

Is the determined intensified process by far the best option for MEG production?

The application of a retrofit methodology naturally raises the question of whether the new process design is optimal, given the currently available technologies. An EO hydrolysis reaction route for MEG processing was widely studied, and alternative technologies e.g., microreactors, and extraction-based separations have been discussed in the literature (Garcia Chavez 2013; Kockmann 2013). Based on extensive experimental studies, none of the reported technologies were promising for MEG production. However, by applying the proposed retrofit approach, all these unpromising options were rejected in just a few database screening steps, without any investment in unnecessary experiments. The determined option (re-active distillation) was previously reported as a case study to demonstrate different methodologies for the design of reactive distillation processes (Ciric and Gu 1994; Okasinski and Doherty 1998; Higler et al. 1999; Cardoso et al. 2000; Jackson and Grossmann 2001; Lima et al. 2006). Yet, the cited papers focused on the demon-stration of methods for column design, and optimization and general aspects of process retrofitting were not considered. As a result, the obtained results were not compared to any base-case design. No metrics were available to highlight the benefit of intensification and support the industrial implementation of a novel process variant that is doubtlessly more economical and efficient than the well-established technology. The present methodology enabled the quantification of the benefit of PI at the industrial application level. Therefore, the method can fill the gap in existing knowledge of retrofit design and can be successfully applied to a different range of chemical production processes.

References

Adler S (1998) Vision 2020: 1998 separations roadmap. Center for Waste Reduction Technologies, New York, NY

Altiokka MR, Akyalçin S (2009) Kinetics of the hydration of ethylene oxide in the presence of heterogeneous catalyst. Ind Eng Chem Res 48(24):10840–10844

Barecka MH, Skiborowski M, Górak A (2017) A novel approach for process retrofitting through process intensification: ethylene oxide case study. Chem Eng Res Des 123:295–316

Becht S, Franke R, Geißelmann A et al (2009) An industrial view of process intensification. Chem Eng Proc 48(1):329–332

Benneker AM, van der Ham Louis GJ et al (2016) Design and intensification of industrial DADPM process. Chem Eng Proc 109:39–50

Cardoso MF, Salcedo RL, Azevedo S de, Barbosa D (2000) Optimization of reactive distillation processes with simulated annealing. Chem Eng Sci 55(21):5059–5078

Ciric AR, Gu D (1994) Synthesis of nonequilibrium reactive distillation processes by MINLP optimization. AIChE J 40(9):1479–1487

Commenge J-M, Falk L (2014) Methodological framework for choice of intensified equipment and development of innovative technologies. Chem Eng Proc 84:109–127

Cross WT, Ramshaw C (1986) Process intensification: laminar flow heat transfer. Chem Eng Res Des 64(4):293–301

Curcio S (2013) Process intensification in the chemical industry: a review. In: Basile A, Piemonte V, de Falco M (eds) Sustainable development in chemical engineering: Innovative technologies. Wiley, Chichester, pp 95–118

Dye RF (2001) Ethylene glycols technology. Korean J Chem Eng 18(5):571–579

Freund H, Sundmacher K (2010) Process intensification, 1, fundamentals and molecular level. In: Ullmann's encyclopedia of industrial chemistry. Wiley, Weinheim

Garcia Chavez LY (2013) Designer solvents for the extraction of glycols and alcohols from aqueous streams. Technische Universiteit Eindhoven, Eindhoven

Górak A, Stankiewicz A (2011) Intensified reaction and separation systems. Annu Rev Chem Biomol Eng 2:431–251

Gourdon C, Elgue S, Prat L (2015) What are the needs for process intensification? Oil Gas Sci Technol—RevIFP Energies nouvelles 70(3):463–473

Harmsen GJ (2007) Reactive distillation: the front-runner of industrial process intensification. Chem Eng Proc 46(9):774–780

Harper PM, Gani R (2000) A multi-step and multi-level approach for computer aided molecular design. Comput Chem Eng 24(2–7):677–683

Higler AP, Taylor R, Krishna R (1999) The influence of mass transfer and mixing on the performance of a tray column for reactive distillation. Chem Eng Sci 54(13–14):2873–2881

Holtbruegge J, Kuhlmann H, Lutze P (2014) Conceptual design of flowsheet options based on thermodynamic insights for (reaction) separation processes applying process intensification. Ind Eng Chem Res 13412–13429

Huss RS, Chen F, Malone MF et al (2003) Reactive distillation for methyl acetate production. Comput Chem Eng 27(12):1855–1866

Jackson JR, Grossmann IE (2001) A disjunctive programming approach for the optimal design of reactive distillation columns. Comput Chem Eng 25(11–12):1661–1673

Jafari M, Bayat A, Mohammadi T et al (2013) Dehydration of ethylene glycol by pervaporation using gamma alumina/NaA zeolite composite membrane. Chem Eng Res Des 91:2412–2419

Jaksland CA, Gani R, Lien KM (1995) Separation process design and synthesis based on thermodynamic insights. Chem Eng Sci 50(3):511–530

Kiss AA (2013) Advanced distillation technologies: design, control and applications. Wiley, Chichester

Kockmann N (2013) Micro process engineering: fundamentals, devices, fabrication, and applications. Wiley-VCH, Weinheim

Lima RM, Salcedo RL, Barbosa D (2006) SIMOP: efficient reactive distillation optimization using stochastic optimizers. Chem Eng Sci 61(5):1718–1739

Lutze P (2011) An innovative synthesis methodology for process intensification. Ph.D. thesis, Department of Chemical and Biochemical Engineering Technical University of Denmark, Lyngby

Lutze P, Gani R, Woodley JM (2010) Process intensification: a perspective on process synthesis. Chem Eng Proc 49(6):547–558

Niu MW, Rangaiah GP (2016) Process retrofitting via intensification: a heuristic methodology and its application to isopropyl alcohol process. Ind Eng Chem Res 55(12):3614–3629

Okasinski MJ, Doherty MF (1998) Design method for kinetically controlled, staged reactive distillation columns. Ind Eng Chem Res 37(7):2821–2834

Peschel A (2012) Model-based design of optimal chemical reactors. Ph.D. thesis, Otto-von-Guericke-Universität Magdeburg, Magdeburg

Peschel A, Freund H, Sundmacher K (2010) Methodology for the design of optimal chemical reactors based on the concept of elementary process functions. Ind Eng Chem Res 49(21):10535–10548

Rebsdat S, Mayer D (2010) Ethylene glycol. In: Ullmann's encyclopedia of industrial chemistry. Wiley, Weinheim

Rebsdat S, Mayer D (2012) Ethylene oxide. In: Ullmann's encyclopedia of industrial chemistry, vol 13, pp 543–572

Schwartzentruber J, Renon H, Watanasiri S (1989) Development of a new cubic equation of state for phase equilibrium calculations. Fluid Phase Equilib 52:127–134

Seader JD, Henley EJ, Roper DK (2013) Separation process principles: chemical and biochemical operations

Stankiewicz A, Moulijn J (2000) Process intensification: transforming chemical engineering. Chem Eng Proc 96(1):22–34

Stankiewicz AI, Moulijn JA (2004) Re-engineering the chemical processing plant: process intensification. Marcel Dekker, New York

ten Kate A (2015) Industrially applied PSE for problem solving excellence. In: Gernaey KV, Huusom JK, Gani R (eds) 12th international symposium on process systems engineering and 25th European symposium on computer aided process engineering. Elsevier, pp 49–54

Tsouris C, Porcelli JV (2003) Process intensification—has its time finally come? Chem Eng Proc 99(10):50–55

van Gerven T, Stankiewicz A (2009) Structure, energy, synergy, time, the fundamentals of process intensification. Ind Eng Chem Res 48(5):2465–2474

Woods DR (2007) Rules of thumb in engineering practice. Wiley-VCH, Weinheim

Problems of Heat Transfer in Agitated Vessels

Magdalena Cudak, Marta Major-Godlewska and Joanna Karcz

1 Introduction

Agitation of liquids or heterogeneous systems is unit operation used frequently in chemical, biochemical and other processes. Agitated vessels operate as heat exchangers when heat transfer is limiting process affecting the productivity and then temperature of the fluid must be maintained within narrow limits. In practice, agitated vessels of different geometry are used (Fig. 1a).

They differ in position of the impeller shaft (central, eccentric, Fig. 1b) or side-entering (Fig. 1c), type and number of impellers (Fig. 3a) and baffles as well as type of the heat transfer surface area (outer jacket, Fig. 2a, b) or internal coil of different shape [helical (Fig. 2c) or vertical tubular (Figs. 2d, 3b, c, d)].

Up to now, problems of the heat transfer in agitated vessels have been considered in monographs (Nagata 1975; Oldshue 1983; Stręk 1981; Kurpiers 1985) and reviewed in papers (Poggeman et al. 1979; Steiff et al. 1980; Stręk and Karcz 1997; Mohan et al. 1992; Stręk 1963). An analysis of the literature data shows that commonly used standard geometrical parameters of the agitated vessel are not optimal dimensions from the point of view of the thermal processes which occur in such apparatuses.

In this chapter, problems of the heat transfer process intensification in the agitated vessels working as heat transfer exchangers are considered from the point of view the proper choice of the apparatus for a given technological task. Different factors affecting the enhancement of the heat transfer process, such as: the heat transfer surface area type, impeller and vessel types and geometry of the agitated

M. Cudak · M. Major-Godlewska · J. Karcz (✉)
Department of Chemical Engineering, West Pomeranian
University of Technology, Szczecin, Poland
e-mail: joanna.karcz@zut.edu.pl

© Springer International Publishing AG, part of Springer Nature 2018
M. Ochowiak et al. (eds.), *Practical Aspects of Chemical Engineering*,
Lecture Notes on Multidisciplinary Industrial Engineering,
https://doi.org/10.1007/978-3-319-73978-6_3

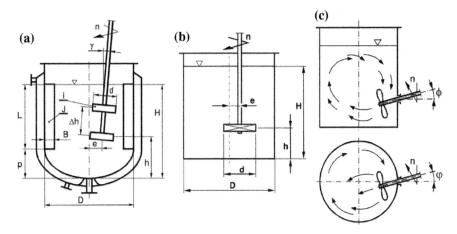

Fig. 1 **a** Geometrical parameters of an agitated vessel; **b** eccentric position of the shaft in a vessel; **c** side-entering position of the shaft in a vessel

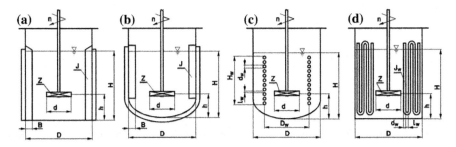

Fig. 2 Different types of the heat transfer surface area; **a, b** heating jackets; **c** helical coil; **d** vertical tubular coil

vessel are analyzed on the basis of our own experimental results and literature data. Mean and local values of the heat transfer coefficients as a function of the agitated vessel type are compared with regards to the fluids of different physical properties within the transitional and turbulent flow ranges.

2 Experimental

The measurements of the mean heat transfer coefficient for whole heat transfer surface area of the baffled jacketed agitated vessel were carried out using steady state thermal method in a vessel of inner diameter $D = 0.45$ m which was filled by Newtonian liquid up to the height $H = D$ ($0.5D < H < 1.9H$). Experimental set-up for this series of the measurements is presented in Fig. 4a.

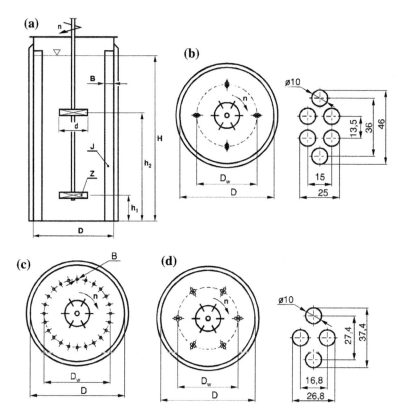

Fig. 3 **a** Jacketed, baffled agitated vessel with two impellers on the common shaft; **b** un-baffled agitated vessel with a tubular vertical coil G5 (k × m = 4 × 6); **c** tubular vertical coil G4 (k × m = 24 × 1); **d** tubular vertical coil G 6 (k × m = 6 × 4)

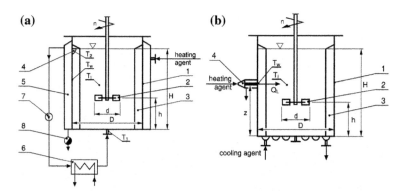

Fig. 4 Experimental set-up for the measurements of the: **a** mean heat transfer coefficient α_m; *1* vessel; *2* agitator; *3* baffle; *4* overflow; *5* jacket; *6* heat exchanger; *7* pump; *8* condenser pot; **b** local heat transfer coefficient α; *1* vessel; *2* agitator; *3* baffle; *4* local heat source

Both measuring methods were applied to determine local heat transfer coefficients on the wall of the baffled agitated vessel: thermal method with local heat source in a vessel of inner diameter of $D = 0.3$ m (Fig. 4b) and an electrochemical method in a vessel of inner diameter of $D = 0.288$ m (Fig. 5a). The measurements of the mean heat transfer coefficient for an un-baffled agitated vessel equipped with vertical tubular coils were performed using steady-state method in a vessel of inner diameter $D = 0.6$ m, filled by Newtonian or non-Newtonian liquid up to the height $H = D$ (Fig. 5b). The details of the measuring methods used in the study are given elsewhere [for mean heat transfer coefficients in a jacketed vessel in paper (Karcz and Stręk 1995) and in a vessel with tubular coil (Karcz and Major 2001), for local heat transfer coefficients measured by means of the local heat source in paper (Karcz 1999; Bielka et al. 2014) or an electrochemical method (Karcz et al. 2005; Bielka et al. 2014)]. Different high-speed impellers were tested in the study. Power consumption was measured by means of the strain gauge method (Karcz and Cudak 2002).

Mean heat transfer coefficient α_m for jacketed agitated vessel was calculated from the equation regarding the thermal resistance for cylindrical wall

$$\alpha_m = \frac{1}{\frac{1}{k} - \frac{1}{\alpha_o} \cdot \frac{D_{out}}{D} - \frac{D_{out}}{2\lambda_o} \ln\left(\frac{D_{out}}{D}\right)} \tag{1}$$

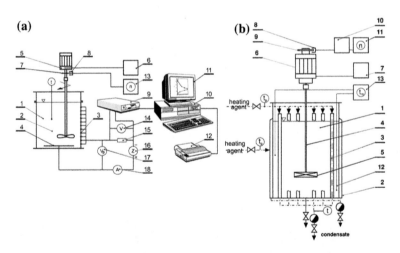

Fig. 5 Experimental set-up for the measurements of the: **a** local heat transfer coefficient α using an electrochemical method; *1* agitated vessel, *2* agitator, *3* cathode, *4* anode, *5* electric motor, *6* steering unit, *7* perforated disc, *8* photoelectric sensor, *9* A/D converter, *10* PC computer, *11* monitor, *12* printer, *13* electronic counter, *14,17* voltmeters, *15* resistor, *16* potential source, *18* ammeter; **b** mean heat transfer coefficient in a un-baffled vessel with vertical tubular coil; *1* agitated vessel, *2* jacket, *3* vertical tubular coil, *4* shaft, *5* agitator, *6* electric motor, *7* steering unit, *8* optic sensor, *9* measuring disc, *10* converter, *11* electronic counter, *12* thermocouple, *13* thermostat

where: k—overall heat transfer coefficient, α_o—heat transfer coefficient for heating agent, λ_o—conductivity for the vessel wall.

Local heat transfer coefficient α determined experimentally by thermal method was defined as follows

$$\alpha = \frac{dQ_l}{\Delta T_l dF} = \frac{q_w}{\Delta T_l} \tag{2}$$

where q_w—local heat flux, ΔT_l—driving difference of temperature, index w refers to the wall.

3   Results and Discussion

The results of the measurements of mean heat transfer coefficient α_m were described by means of the following equation

$$Nu = \frac{\alpha_m D}{\lambda} = CRe^A Pr^{0.33} Vi^{0.14} = C_o Re^A Pr^{0.33} Vi^{0.14} \Psi(i_1, \ldots, i_k) \tag{3}$$

where Reynolds and Prandtl numbers and viscosity simplex are defined as follows:

$$Re = \frac{nd^2\rho}{\eta}; \quad Pr = \frac{c_p\eta}{\lambda}; \quad Vi = \frac{\eta}{\eta_w} \tag{4}$$

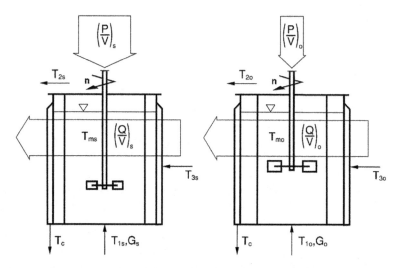

Fig. 6 Comparison of a standard (s) and optimal (o) geometry of the agitated vessel for heat transfer process

Out of the two compared geometries of the agitated vessels, these geometrical parameters are better for which lower agitation energy P/V is required to obtain the identical specific heat flux Q/V in both reactors (Fig. 6), i.e.

$$\left(\frac{Q}{V}\right)_s = \left(\frac{Q}{V}\right)_o \quad \text{and} \quad \left(\frac{P}{V}\right)_s > \left(\frac{P}{V}\right)_o \tag{5}$$

where s—standard, o—optimal geometries. Efficiency of the heat transfer process in our study was evaluated on the basis of the modified dimensionless heat transfer equation (Stręk et al. 1990; Stręk and Karcz 1991) which relates thermal effect with the agitation energy P/V

$$Nu = \frac{\alpha_m D}{\lambda} = K\left(\frac{\pi}{4}\right)^{A/3}\left(Re_{P,V}\right)^{A/3}Pr^{0.33}Vi^{0.14} \tag{6}$$

where modified Reynolds number $Re_{P,V}$ is defined as follows:

$$Re_{P,V} = \frac{(P/V)D^4\rho^2}{\eta^3} \tag{7}$$

and coefficient K is dependent on the geometrical parameters of the agitated vessel (for the turbulent range of the fluid flow):

$$K = \left(\frac{H}{D}\right)^{A/3}\left[\frac{C}{[Ne(D/d)]^{A/3}}\right] \tag{8}$$

where power number Ne:

$$Ne = \frac{P}{n^3 d^5 \rho} \tag{9}$$

In Eq. (3), term $\Psi(i_1, \ldots, i_k) = 1$ for standard geometry of the agitated vessel ($H/D = 1$; $J = 4$; $B/D = 0.1$; $d/D = 0.33$; $h/D = 0.33$). Experimentally determined values of the constants C_o from Eq. (3) obtained for statistically estimated exponent $A = 0.67$ and coefficients K from Eq. (5) are presented in Table 1 for jacketed baffled agitated vessel with standard impellers ($D = 0.45$ m; $H/D = 1$; $d/D = 0.33$; turbulent range of the liquid flow $10^4 < Re < 5 \times 10^4$).

Quantitative effects of the geometrical parameters of the vessel–impeller–baffle systems on the intensity of the heat transfer process in an jacketed agitated vessel are compared in Table 2 by means of the term $C = C_o\Psi(i_1, \ldots, i_k)$ in Eq. (3).

The results show clear effect of the impeller type [up-pumping pitched blade turbine (Karcz 1991), conical turbine (Stręk et al. 1990)], number of the impellers and connected with its higher value of the aspect ratio H/D [Fig. 3a, (Karcz and Stręk 1994)], geometry of the vessel and baffles (Stręk and Karcz 1985), position of the impeller shaft in a vessel [side entering, Fig. 1c, (Stręk et al. 1987) or eccentric,

Table 1 The values of the coefficients Co in Eq. (3) and K in Eq. (5) for jacketed baffled agitated vessel with standard impellers ($D = 0.45$ m; $H/D = 1$; $d/D = 0.33$; $104 < Re < 5 \times 10^4$)

No	Agitator	C_o	K	No	Agitator	C_o	K
1	Rushton turbine, Z = 6	0.73	0.382	5	Pitched blade turbine, Z = 6; β = 135°	0.43	0.308
2	Smith turbine (CD 6), Z = 6	0.62	0.382	6	Propeller, Z = 3; S/d = 1	0.30	0.295
3	Turbine, Z = 6; β = 90°	0.605	0.323	7	A 315, Z = 4	0.40	0.285
4	Pitched blade turbine, Z = 6; β = 45°	0.48	0.347	8	HE 3, Z = 3	0.28	0.269

Fig. 1b, (Karcz and Cudak 2002, 2006)] on the value of the mean heat transfer coefficient α_m.

Compared to Eq. (3), much more comprehensive assessment of the heat transfer process in the agitated vessel can be carried out based on the Eq. (6), where agitation energy is taken into account in the number $Re_{P,V}$. The dependences $Nu/Pr^{0.33}Vi^{0.14} = f(Re_{P,V})$ for jacketed vessel equipped with different types of impellers and baffles of different dimensions are compared in Fig. 7.

Within the turbulent range of the Newtonian liquid flow, the highest efficiency of the heat transfer process corresponds to disc turbine of diameter $d = 0.5D$ and $J = 8$ short baffles of length $L = 0.25H$ (line 9 in Fig. 7). The lowest one is ascribed to the standard agitated vessel with the propeller and standard baffles (line 1 in Fig. 7). Moreover, the results in Fig. 8 show that efficiency of the heat transfer process for the un-baffled ($J = 0$) agitated vessel with eccentrically located Rushton turbine (RT), propeller (P) or HE 3 impeller is almost the same as for that one equipped with the baffles ($J = 4$) and centrally located impeller.

For various tubular vertical coil—impeller systems immersed in a cylindrical un-baffled vessel, an efficiency of the heat transfer process can be evaluated on the basis of the criterion K from Eq. (6). The coefficients K calculated according to the definition (8) are compared in Figs. 9, 10, 11 and 12 for Newtonian and non-Newtonian (CMC solutions) liquids, as well as different types of the impellers ($d/D = 0.33$ or 0.5) and tubular vertical coils with the geometrical parameters described in Table 3. The values of the coefficients C and power numbers Ne are also given in Figs. 9, 10, 11 and 12 for comparative purposes. Level of the C and K values in Figs. 9 and 11 is different than that in Figs. 10 and 12 because of the varying values of the exponent A in Eq. (3). Dimensionless numbers for the non-Newtonian liquid in Eq. (3) were defined using apparent viscosity η_{ae}. Statistically evaluated mean value of the exponent A in Eq. (3) was equal to 0.8 within the transitional regime of the non-Newtonian liquid flow ($400 < Re < 1.9 \times 10^4$) and $A = 0.67$—within the turbulent range of the Newtonian liquid flow ($10^5 < Re < 4 \times 10^5$).

Table 2 The function $C = C_o \Psi (i_1, \ldots, i_k)$ in Eq. (3) for jacketed agitated vessel with different agitators (turbulent range of the liquid flow)

Agitator	$C = C_o \Psi (i_1, \ldots, i_k)$	Range
Disc turbine Stręk and Karcz (1985)	$C = 0.743\left(\dfrac{H}{D}\right)^{-0.44}\left(\dfrac{Z}{6}\right)^{0.21}\left(\dfrac{J}{4}\right)^{0.03}\left(\dfrac{10B}{D}\right)^{0.05}\left(\dfrac{3b}{D}\right)^{0.12}$	$D = 0.45$ m; $i = 1$; $0.73 \leq H/D \leq 1.9$; $0.25 \leq h/D \leq 0.58$; $2 \leq J \leq 10$; $0.067 \leq B/D \leq 0.2$; $a/d = 0.25$; $b/d = 0.2$; $4 \leq Z \leq 12$
Conical turbine Stręk et al. (1990)	$C = 0.48\left(\dfrac{3d}{D}\right)^{0.10}\left(\dfrac{Z}{6}\right)^{0.10}\left(\dfrac{5b}{d}\right)^{0.22}$	$D = 0.45$ m; $i = 1$; $H/D = 1$; $0.083 \leq h/D \leq 0.25$; $J = 4$; $B/D = 0.1$; $0.25 \leq d/D \leq 0.58$; $0.067 \leq b/d \leq 0.6$; $4 \leq Z \leq 12$
Pitched blade turbine PBT Karcz (1991)	$C = 0.605\left(\dfrac{3d}{D}\right)^{0.32}\left(\dfrac{Z}{6}\right)^{0.24}\left(\dfrac{5b}{d}\right)^{0.30}(sin\,\beta)^{0.58}$	$D = 0.45$ m; $i = 1$; $H/D = 1$; $h/D = 0.33$; $J = 4$; $B/D = 0.1$; $0.25 \leq d/D \leq 0.58$; $0.042 \leq b/D \leq 0.092$; $2 \leq Z \leq 15$; $30° \leq \beta < 90°$
Rushton turbine (i = 2) Karcz and Stręk (1994)	$C = -2.398 + 3.805\left(\dfrac{H}{D}\right) + 1.519\left(\dfrac{H}{D}\right)^2 + $ $- 3.148\left(\dfrac{H}{D}\right)^3 + 0.925\left(\dfrac{H}{D}\right)^4$	$D = 0.45$ m; $i = 2$; $1 \leq H/D \leq 1.9$; $h_1/H = 0.33$; $h_2/H = 0.67$; $J = 4$; $B/D = 0.1$; $d/D = 0.33$; $Z = 6$; $a/d = 0.25$; $b/d = 0.2$
Propeller side entering Stręk et al. (1987)	$C = 0.405(\cos \varphi)^{0.716}(\cos \phi)^{-0.036}\left(\dfrac{3l}{D}\right)^{-0.143}\psi_d \psi_s$ $\psi_d = -1.38 + 28.44\left(\dfrac{d}{D}\right) - 101.9\left(\dfrac{d}{D}\right)^2 + 114.55\left(\dfrac{d}{D}\right)^3$ $\psi_S = \dfrac{2.41\left(\frac{S}{d}\right)^2 + 4.133}{6.186 - 1.594\left(\frac{S}{d}\right) + 2.016\left(\frac{S}{d}\right)^2}$	$D = 0.9$ m; $i = 1$; $H/D = 1$; $0° < \varphi < 20°$; $-30° < \phi < 30°$; $D/6 < l < D/2$; $D/10 < d < D/3$
Propeller eccentric Karcz and Cudak (2002)	$C = 0.217\left(\dfrac{H}{D}\right)^{-0.16}\left[1.628\left(\dfrac{e}{R}\right)^2 - 0.189\left(\dfrac{e}{R}\right) + 1\right]$	$D = 0.45$ m; $i = 1$; $0.5 \leq H/D \leq 1.5$; $h/D = 0.33$; $J = 0$; $d/D = 0.33$; $0 \leq e/R \leq 0.53$; $Z = 3$
HE 3, eccentric Karcz and Cudak (2006)	$C = 0.235\left(\dfrac{H}{D}\right)^{-0.26}\left[1.126\left(\dfrac{e}{R}\right)^2 - 0.268\left(\dfrac{e}{R}\right) + 1\right]\left(\dfrac{3d}{D}\right)^{-0.084}$	$D = 0.45$ m; $i = 1$; $0.5 \leq H/D \leq 1.5$; $h/D = 0.33$; $J = 0$; $0 \leq e/R \leq 0.53$; $Z = 3$; $0.33 < d/D < 0.5$

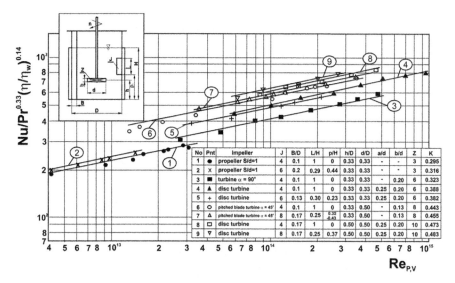

The table within Fig. 7:

No	Pnt	Impeller	J	B/D	L/H	p/H	h/D	d/D	a/d	b/d	z	K
1	●	propeller S/d=1	4	0.1	1	0	0.33	0.33	·	·	3	0.295
2	×	propeller S/d=1	6	0.2	0.29	0.44	0.33	0.33	·	·	3	0.316
3	■	turbine α = 90°	4	0.1	1	0	0.33	0.33	·	0.20	6	0.323
4	▲	disc turbine	4	0.1	1	0	0.33	0.33	0.25	0.20	6	0.388
5	+	disc turbine	6	0.13	0.30	0.23	0.33	0.33	0.25	0.20	6	0.382
6	○	pitched blade turbine α = 45°	4	0.1	1	0	0.33	0.50	·	0.13	8	0.443
7	△	pitched blade turbine α = 45°	8	0.17	0.25	0.32 -0.43	0.33	0.50	·	0.13	8	0.455
8	□	disc turbine	4	0.17	1	0	0.50	0.50	0.25	0.20	10	0.473
9	▽	disc turbine	8	0.17	0.25	0.37	0.50	0.50	0.25	0.20	10	0.483

Fig. 7 The relationship $Nu/Pr^{0.33}Vi^{0.14} = f(Re_{P,V})$ for different geometry of the jacketed agitated vessel; turbulent range of the Newtonian liquid flow

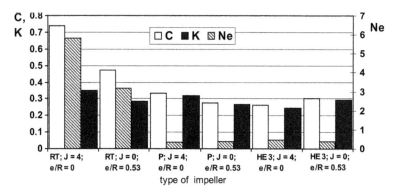

Fig. 8 Comparison of the dependence $K, C, Ne = f$ (type of the impeller) for the central ($e/R = 0$) and eccentric ($e/R \neq 0$) position of the shaft; $d = 0.33D$

Figures 9 and 10 illustrate the results of the comparative analysis for the given type of the vertical coil (G4, Fig. 3c) and different types of the impellers, whereas Figs. 11 and 12 show the results for the given type of the impeller and different vertical coils. The highest values of K in Figs. 9, 10, 11 and 12 correspond to the systems which enable to reach the most effective heat transfer in the agitated vessel with tubular vertical coil. As the data in Fig. 9 show, the system coil

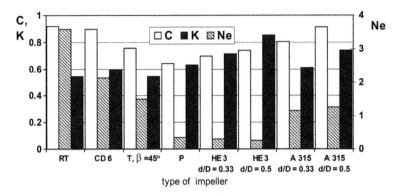

Fig. 9 The dependence K, C, Ne = f (type of the agitator); vertical coil G4; Newtonian liquid

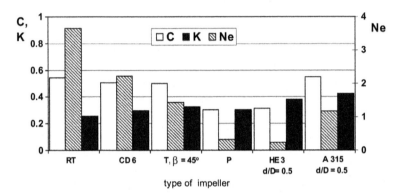

Fig. 10 The dependence K, C, Ne = f (type of the agitator); vertical coil G4; non-Newtonian liquid

G4—down-pumping HE 3 impeller of diameter $d = 0.5D$ can be recommended to agitate Newtonian liquid within the turbulent range of the flow. It is seen in Fig. 9, high values of the coefficients C correspond also to Rushton or Smith turbines or A 315 impeller of diameter $d = 0.5D$ operating with the vertical coil G4, however, high intensity of heat transfer process is reached at the high level of agitation energy, especially for the system coil G4—Rushton turbine. The highest values of the power number Ne at definition (8) give low values of the coefficient K. The lowest efficiency of the heat transfer process was also obtained for the system coil G4—Rushton turbine when non-Newtonian liquid is agitated within the range of the transitional liquid flow (Fig. 10).

The systems, A 315 impeller of diameter $d = 0.5D$—different tubular vertical coils, are compared in Fig. 11 for the turbulent Newtonian liquid flow in the

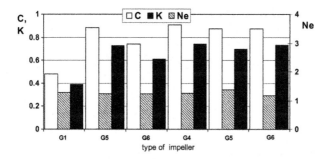

Fig. 11 The dependence K, C, Ne = f (type of the vertical coil); impeller A 315 (d/D = 0.5); Newtonian liquid

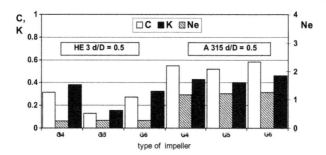

Fig. 12 The dependence K, C, Ne = f(type of the vertical coil); impeller A 315 or HE 3 (d/D = 0.5); non-Newtonian liquid

Table 3 Geometrical parameters of the tubular vertical coils used in the study (Major 2000; Michalska 2001)

Type	D_w/D	J	B/D	$k \times m$
G1	0.9	4	0.1	4×1
G2	0.7	16	2.5×10^{-2}	16×1
G3	0.64	16	2.5×10^{-2}	4×4
G4	0.7	24	1.67×10^{-2}	24×1
G5	0.64	24	1.67×10^{-2}	4×6
G6	0.65	24	1.67×10^{-2}	6×4

agitated vessel. Data in Fig. 11 show that coefficients K for the agitated vessels with the coil G2 or G4–G6 are maintained at the high level. Moreover, tubular vertical coils G4–G6 immersed in the vessel with the A 315 impeller of diameter $d = 0.5D$ enable to reach the best efficiency of the heat transfer process during the agitation of the non-Newtonian liquid (CMC solution) within the transitional range of the liquid flow (Fig. 12). In this case, heat transfer efficiency is greater than the one obtained for the vessel with the HE 3 impeller of diameter $d = 0.5D$ (Fig. 12).

Our experimental results prove that eccentric position of the impeller in the vessel affects the strong asymmetry of the thermal profiles on the heat transfer surface area. The effects of the impeller eccentricity e/R (where $R = D/2$) and the location of the measuring point (z/H, $\varphi /2\pi$) at the heat transfer surface, described by axial and angular coordinates on the local heat transfer coefficients were described for the system with the propeller, HE 3 ($d/D = 0.33$ or 0.5), Rushton turbine or A 315 impeller, using the following equation.

$$Nu = \frac{\alpha_l D}{\lambda} = CRe^{0.67}Pr^{0.33} = C_oRe^{0.67}Pr^{0.33}f_1\left(\frac{e}{R}\right)f_2\left(\frac{z}{H}\right)f_3\left(\frac{\varphi}{2\pi}\right) \qquad (10)$$

where functions f_1, f_2 and f_3 were approximated by means of the separated equations:

$$f_1 = x_1\left(\frac{e}{R}\right)^2 + x_2\left(\frac{e}{R}\right) + 1 \qquad (11)$$

$$f_2 = \left(x_3\left(\frac{z}{H}\right)^2 + x_4\left(\frac{z}{H}\right) + 1\right)\left(1 + x_5\left(\frac{z}{H}\right)^{x_6}\right)^{-1} \qquad (12)$$

$$f_3 = \left(x_7\left(\frac{\varphi}{2\pi}\right)^2 + x_8\left(\frac{\varphi}{2\pi}\right) + 1\right)\left(1 + x_9\left(\frac{\varphi}{2\pi}\right)^{x_{10}}\right)^{-1} \qquad (13)$$

The coefficients of the Eqs. (10–13) are collected in Table 4 for different high-speed impellers. The equations describe the results of the measurements within the range of the $Re \in < 2.3 \times 10^4; 7 \times 10^4 >$, $e/R \in < 0; 0.53 >$, $z/H \in (0; 1)$, $\varphi \in < 0; 2\pi >$.

The knowledge about intensity and efficiency of the heat transfer process determined on the basis of the mean heat transfer coefficients α_m is insufficient in order to take correct project decision because distributions of the coefficient α on the heat transfer surface area of the agitated vessel are observed. Figure 13 illustrates examples of the heat transfer coefficient distributions for the jacketed vessel, where the thermal profiles are compared for two high speed impellers on the common shaft and turbulent regime of the liquid flow (Fig. 13a), or transitional regime of the gas–liquid flow (Fig. 13b) or for a single A 315 impeller in a eccentric position in the agitated vessel (Fig. 13c). Distributions of the coefficient α strongly depend on the type and number of the impellers on the shaft, centric or eccentric position of the impeller shaft, as well as the presence of the dispersed gas phase in the agitated vessel.

Table 4 Coefficients C_o and parameters x_i in Eqs. (10–13)

No	Agitator	φ	C_o	x_1	x_2	x_3	x_4	x_5	References
1	Propeller	<0; π/2)	0.161	1.486	−0.116	−2.854	2.443	0	Karcz et al. (2005)
2		<π/2; π)	0.153	1.463	−0.108	−3.159	2.786	0	
3		<π; 3π/2)	0.249	1.638	−0.226	−3.944	3.574	0	
4		<3π/2; 2π)	0.46	1.584	−0.174	−3.784	3.358	0	
5	HE 3, d/D = 0.33	<0; π/2)	0.227	1.540	−0.4	−1.472	1.036	0	Karcz et al. (2005)
6		<π/2; π)	0.208	1.539	−0.375	−2.036	1.518	0	
7		<π; 3π/2)	0.095	1.537	−0.314	−1.551	0.905	0	
8		<3π/2; 2π)	0.186	1.064	−0.123	−0.072	−0.487	0	
9	HE 3, d/D = 0.5	<0; π/2)	0.233	2.395	−0.680	−0.283	0.162	0	
10		<π/2; π)	0.233	2.463	−0.651	−0.402	0.087	0	
11		<π; 3π/2)	0.313	2.026	−0.489	−0.854	0.493	0	
12		<3π/2; 2π)	0.397	2.512	−0.698	−0.366	0.117	0	
13	RT	<0; 2π)	0.027	−0.357	0.895	0.238	−0.762	−0.957	
14	A 315	<0; 2π)	0.029	−0.458	0.983	0.245	−0.553	−0.935	

No	Agitator	φ	x_6	x_7	x_8	x_9	x_{10}
1	propeller	<0; π/2)	0	0.992	−0.265	0	0
2		<π/2; π)	0	0.243	−0.116	0	0
3		<π; 3π/2)	0	1.170	−1.468	0	0
4		<π; 3π/2)	0	0.937	−1.621	0	0
5	HE 3, d/D = 0.33	<0; π/2)	0	−1.199	0.214	0	0
6		<π/2; π)	0	0.026	0.046	0	0
7		<π; 3π/2)	0	−4.193	5.222	0	0
8		<3π/2; 2π)	0	−1.206	1.962	0	0

(continued)

Table 4 (continued)

No	Agitator	φ	x_6	x_7	x_8	x_9	x_{10}	References
9	HE 3, d/D = 0.5	$<0; \pi/2)$	0	0.554	−0.105	0	0	Karcz and Cudak (2004)
10		$<\pi/2; \pi)$	0	−0.497	0.406	0	0	
11		$<\pi; 3\pi/2)$	0	0.746	−0.882	0	0	
12		$<3\pi/2; 2\pi)$	0	0.527	−0.906	0	0	
13	RT	$<0; 2\pi)$	6.175×10^{-3}	0.037	−0.887	−0.860	0.989	Karcz and Cudak (2008)
14	A 315	$<0; 2\pi)$	6.181×10^{-3}	−0.826	0.019	−0.819	2.035	Karcz and Cudak (2008)

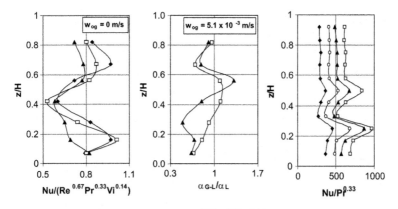

Fig. 13 Comparison of the profiles **a** $Nu/(Re^{0.67}Pr^{0.33}Vi^{0.14}) = f(z/H)$; distilled water; filled diamond—RT–RT, square—CD 6–RT, filled triangle—A 315–RT; **b** $\alpha_{G-L}/\alpha_L = f(z/H)$; $w_{og} = 0$, square—CD 6–RT, filled triangle—A 315–RT; air–solution of glycerol system $Re = 300$; **c** $Nu/Pr^{0.33} = f(z/H)$; A315 impeller; $\varphi = 0°$; $e/R = 0.53$ filled diamond—$Re = 2.3 \times 10^4$; circle—$Re = 3.8 \times 10^4$; filled triangle—$Re = 5.4 \times 10^4$; square—$Re = 6.9 \times 10^4$

4 Conclusions

Comparative analysis of the experimental results on the heat transfer process in the agitated vessel shows the following:

1. The choice of the heating/cooling surface area type is limited by the agitated vessel scale. The jackets are used in case of the vessels with smaller fluid volume. In case of the coils, it is easier to obtain surface area of the heat transfer required in the project.
2. Geometry of the vessel-impeller-baffles system strongly affects the heat transfer intensity. High values of the heat transfer coefficient can be obtained using radial flow impellers.
3. Estimation of the heat transfer quality on the basis of the efficiency criterion gives more information than simple heat transfer intensity because power consumption is taken additionally into account.
4. Full analysis of the heat transfer process should be completed by considerations on the heat transfer coefficient distributions as a function of the agitated vessel geometry.

References

Bielka I, Cudak M, Karcz J (2014) Local heat transfer process for a gas–liquid system in a wall region of an agitated vessel equipped with the system of CD 6-RT impellers. Ind Eng Chem Res 53:16539–16549

Cudak M, Karcz J (2008) Distribution of local heat transfer coefficient values in the wall region of an agitated vessel. Chem Pap 62(1):92–99

Karcz J (1991) Influence of geometry of the pitched blade turbine upon power consumption and intensity of heat transfer. Inż Chem Proc 12(1):47–61

Karcz J (1999) Studies of local heat transfer in a gas–liquid system agitated by double disc turbines in a slender vessel. Chem Eng J 72:217–227

Karcz J, Cudak M (2002) Efficiency of the heat transfer process in a jacketed agitated vessel equipped with an eccentrically located impeller. Chem Pap 56(6):382–386

Karcz J, Cudak M (2004) Eksperymentalna analiza procesów przenoszenia w mieszalniku z niecentrycznie zabudowanym mieszadłem. Inż Chem i Proc 25(3/2):1067–1073

Karcz J, Cudak M (2006) Analysis of transport processes near the wall region of a stirred tank with an eccentric impeller. Chem Process Eng 27:1469–1480

Karcz J, Major M (2001) Experimental studies of heat transfer in an agitated vessel equipped with vertical tubular coil. Inż Chem i Proc 22:445–459

Karcz J, Stręk F (1994) Experimental studies of heat transfer for a stirred tank equipped with two turbines. Inż Chem Proc 15(3):357–369 (in Polish)

Karcz J, Stręk F (1995) Heat transfer in jacketed agitated vessels equipped with non-standard baffles. Chem Eng J 58(2):135–143

Karcz J, Cudak M, Szoplik J (2005) Stirring of a liquid in a stirred tank with an eccentrically located impeller. Chem Eng Sci 60(8–9):2369–2380

Kurpiers P (1985) Waermeuebergang in Ein- und Merhphasenreaktoren. VCH, Weinheim

Major M (2000) Studies of heat transfer from vertical tubular coil into the pseudoplastic liquid mixed in an agitated vessel. In: Ph.D. Thesis, Technical University of Szczecin, Poland

Michalska M (2001) Heat transfer in an agitated vessel equipped with tubular vertical coil and high–speed impeller. In: Ph.D. Thesis, Technical University of Szczecin, Poland

Mohan P, Emery AN, Al-Hassan T (1992) Heat transfer to Newtonian fluids in mechanically agitated vessels. Exp Therm Fluid Sci 5:861–883

Nagata S (1975) Mixing. Principles and applications. Halsted Press, New York

Oldshue JY (1983) Fluid mixing technology. McGraw-Hill Publication, New York

Poggeman R, Steiff A, Weinspach PM (1979) Waermeuebergang in Ruehrkesseln mit einphasigen Fluessigkeiten. Chem Ing Techn 51(10):948–959

Steiff A, Poggeman R, Weinspach PM (1980) Waermeuebergang in Ruehrkesseln mit mehrphasigen Systemen. Chem Ing Techn 52(6):492–503

Stręk F (1963) Heat transfer in liquid mixers—study of a turbine agitator with six flat blades. Intern Chem Eng 3(3):533–556

Stręk F (1981) Mixing and Agitated Vessels. WNT, Warsaw

Stręk F, Karcz J (1985) Heat transfer in a baffled vessel agitated by a disc turbine impeller. Inż Chem Proc 6(1):133–146

Stręk F, Karcz J (1991) Experimental determination of the optimal geometry of baffles for heat transfer in an agitated vessel. Chem Eng Process 29(3):165–172

Stręk F, Karcz J (1997) Heat transfer to Newtonian fluid in a stirred tank—a comparative experimental study for vertical tubular coil and a jacket. Recents Prog en Genie des Procedes 51(11):105–112

Stręk F, Karcz J, Lacki H (1987) Problems of a scale—up of agitated vessels. Inż Chem Proc 8(4):601–615

Stręk F, Karcz J, Bujalski W (1990) Search for optimal geometry of a conical turbine agitator for heat transfer in stirred vessels. Chem Eng Technol 13(6):384–392

Non-invasive Measurement of Interfacial Surface States

Krystian Czernek and Małgorzata Płaczek

1 Introduction

Multi-phase flow is a common phenomenon in numerous engineering installations in industry and in various types of apparatus. The concurrent occurrence of gas and very viscous liquid phases in such systems results in many impediments to ensure adequate operating conditions for apparatus in which we have to do with the process of the heat transfer, momentum and mass (such as heat exchangers, evaporators, reactors). The basic reasons behind these problems are due to the largely stochastic characteristics of two-phase gas-liquid flow in which various flow patterns and forms are encountered—as the specific flow patterns are relative to the void fraction of the liquid and gas phases as well as physical properties of the components of the mixture. The various process conditions that correspond to these structures thereby constitute the basic obstacle in ensuring adequate flow regimes and, concurrently, the maintenance of required process parameters along the entire flow path. An important role is attributed to such solutions that secure the optimum operation of the process apparatus within maximum time ranges, which has to be accompanied by the elimination of adverse phenomena that can be associated with the course of the processes. The information found in the literature on the subject indicate that annular two-phase gas-liquid flow is applied more and more commonly, due to its efficiency with regards to processes of heat and mass transfer with the concurrent possibility of applying in these processes substances that vary in terms of physical properties.

Thin-film evaporators, film heat exchangers, film absorbers and thin-film condenser apparatus offer examples of apparatus that apply annular two-phase flow. To this date a large number of papers (Wojtan et al. 2005; Schmidt et al. 2006; Xu 2007; Czernek and Witczak 2013) have been disseminated as descriptions of almost

K. Czernek (✉) · M. Płaczek
Department of Process Engineering, Opole University of Technology, Opole, Poland
e-mail: k.czernek@po.opole.pl

© Springer International Publishing AG, part of Springer Nature 2018
M. Ochowiak et al. (eds.), *Practical Aspects of Chemical Engineering*,
Lecture Notes on Multidisciplinary Industrial Engineering,
https://doi.org/10.1007/978-3-319-73978-6_4

all types of flow patterns in multi-phase flow. The research covered in these papers includes the issues pertaining to identification and description of the forming flow patterns, areas of their occurrence (flow maps), determination of pressure drop and volume fraction of the specific components of the mixture as well as thickness of fluid films and characteristics of their wavy structures—all of which are standard issues put forward in the description of hydrodynamics of two-phase gas-liquid flow. However, regardless of the means in which two-phase gas-liquid flow is formed, in case of flow channels we have to do with various structures that originate as a consequence of the mutual configuration of the phases. The form and type of such structures are relative to volumetric flux of each phases, geometry of the channel, its configuration (vertical, horizontal, other layouts) as well as the physical and chemical properties of the components of two-phase mixture and, in particular, that of the liquid phase. These remarks are based on original experimental research by these authors (Troniewski et al. 2006; Czernek 2013; Czernek and Witczak 2013) conducted within liquid viscosity range up to 3500 mPa s, an example of which is found in Fig. 1.

The classification of flow patterns is usually undertaken on the basis of visual observation. For the cases when such observation is impossible, various visual systems are applied or unconventional observation techniques are applied, including:

- special photography systems: (Troniewski et al. 2006; Czernek and Witczak 2013)
- single- or multi-areas image capture in the cross-section of the stream: (Czernek 2013; Du et al. 2002)

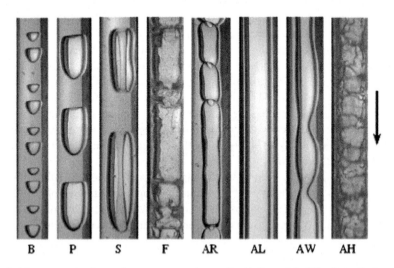

Fig. 1 Flow patterns of two-phase gas-very viscous liquid flow: *B* bubbly flow, *P* plug flow, *S* stalactite flow, *AR* annular core flow, *AL* annular smooth flow, *ALW* annular lightly wave flow, *AW* annular wave flow, *AH* annular hydraulic flow

- photography by means of a beam of X rays: (Saito et al. 2004; Stahl and Rohr 2004)
- optoelectronic cross-section image structure detectors: (Oriol et al. 2008)
- computer tomography: (Ikeda et al. 1983)
- holographic filming: (Lee and Kim 1986).

From the application viewpoint it is more practical to gather information regarding specific flow patterns in a manner that enables the user to predict the occurrence of a particular flow pattern. From the process perspective, such a course of research is well justified and even intentional since it can offer adequately high process results (increase of efficiency of heat exchange, improvement of flow rate, etc.). The areas of the occurrence of specific forms of two-phase flow are usually presented in charts that are called flow maps. They form a graphical interpretation of the flow conditions, through an indication of the areas of the occurrence of the particular flow patterns that are separated into boundary lines in such maps. In the literature on the subject, one comes across a large number of such studies. However, one can note that there is a lack of conformity between the authors with regards to the way in which relative co-ordinate systems are described. As a consequence, there is a deficiency in terms of universal flow maps that can be applied in various process conditions. These boundary lines are often in the form of conventional transfer areas with a large similarity between the changes occurring in them.

An example of the map that is universal for the annular flow that has been developed by (Czernek 2013), can be found in Fig. 2.

The universal characteristics of this map involve the application of dimensionless description of the co-ordinate system. It is important to note that it has been developed for two-phase flow for a fluid with the viscosity range of up to 3500 mPa s in channels with the diameter in the range of 12–40 mm.

A very relevant element that affects the course of heat and mass transfer process is the one connected with the surface of the exchange. For the co-current downward gas-very viscous liquid flow there is a possibility of disturbance in the interfacial surface in the form of waves. Such disturbance increases the surface of the contact between the phases. The knowledge of the flow parameters, for which the contact surface is possibly the largest and has no influence on the reduction of the flow rate or other complication, offers the possibility of conducting the process in an apparatus with virtually too small interfacial surface.

2 Experimental Testing

The development of engineering methods and improvements in measurement devices have made it possible to gain results that had been otherwise impossible previously. Optoelectronic devices form an example of a measurement system that is applied in the realization of experiments (Fig. 3).

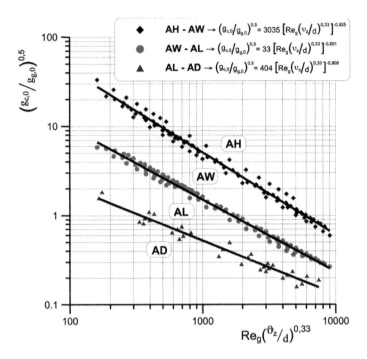

Fig. 2 Universal map of two-phase gas-very viscous liquid flow in a downward pipe

The volume flux of oil and gas were selected in such ratios that the resulting mixture, annular structure of two-phase gas-liquid flow was obtained. For this purpose, linear measurement system was applied (Fig. 4), which consisted of four single-axis measurement sensors situated along the cross section of the channel, which applied the principle of absorption of light beam passing through an oil film. The operating principle of this measurement system is illustrated in Fig. 4.

The results gained in this manner were averaged for local conditions, which made it possible to determine such parameters of downward film flow as:

- thickness as the value measured for local states and its averaged value
- wavy structures along the interfacial surface that was possible to map along the longitudinal surface of a given probe
- local phenomena on the boundaries of the liquid and gas phases resulting from this state, such as amplitude and speed of wave propagation as well as their height.

The averaging of the local conditions has consequently led to the determination of the mean interfacial surface for a given structure of annular gas-very viscous liquid flow. Examples of the measurements performed with the aid of a linear system are presented in Fig. 5. It was additionally indicated that the relative velocity of waves considerably decreases along with an increase of the equivalent Reynolds number. Whereas the above described point based system of optical

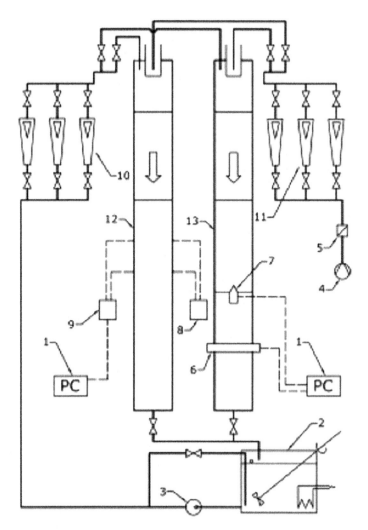

Fig. 3 Diagram of test stand: *1* PC, *2* oil tank with agitator and heater, *3* oil pump, *4* air compressor, *5* reducing valve, *6* laser knife, *7* camera, *8* laser illumination, *9* light detector, *10* oil rotameter assembly, *11* air rotameter assembly, *12* measurement channel with laser illumination, *13* measurement channel with laser knife

probes made it possible to capture the linear state of the configuration of gas-liquid flux, the other applied system—multi-channel system of object-oriented image analysis—enabled the authors to assess the state with regards to the entire cross-section of the channel—Fig. 6.

In its principle the system is formed by an optoelectronic image registration and analysis system based on a similar principle as a point-based one. However, in contrast to it, in this case a beam of light focused on the entire cross-section of the

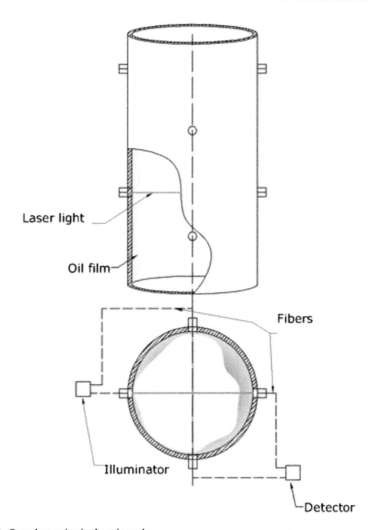

Fig. 4 Optoelectronic single-axis probe

channel was applied and the image was captured by means of an endoscope digital camera situated in the axis of the measuring channel. Overall, this system is based on the principle of a laser knife, which involves registration of an image that is visible in a vertical light beam that is focused in the cross-sectional plane of the channel with adjusting of the image resulting from the application of an optical system.

The application of optoelectronic measuring system in the form of laser knife offered supplementary output to the optoelectronic linear system and enabled authors to extend the range of identification with regards to geometrical parameters that are specific for the downward flow of liquid film.

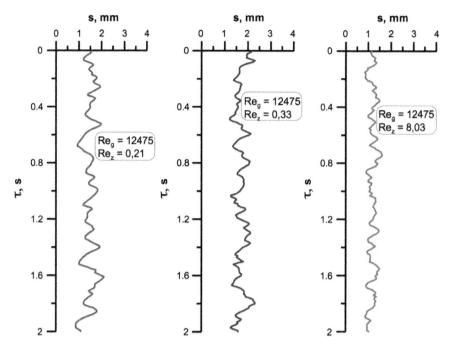

Fig. 5 Variable of liquid film thickness along the pipe: ($D = 22$ mm, $Re_g = 12,474$, $\eta_c = 1070$ mPa s)

In order to illustrate the changes occurring on the surface of the liquid film as a result of the increase of the velocity of co-current gas phase, Fig. 7 presents examples of planar images of this surface captured over the total circumference of the channel. These images concurrently reflect annular flow patterns and in quantitative framework (η = const.)—they indicate the change in downward liquid film referred to local conditions.

As one can note, an increase in the velocity of gas phase results in the occurrence of increasingly higher waves on the surface of liquid film. Beside the smooth and light wavy structures they often have irregular character and the appearance of wavy states on the boundaries between phases have spatially and temporally variable amplitude and length. The spectrum of these changes indicates that regardless of the initially smooth liquid films, within the Reynolds numbers (4000–10,000) of the gas there will always be wavy structures and the nature of the development of waves and their dimensions are considerably relative to the viscosity of the liquid (the more viscous liquid, the larger wave damping).

Exemplary results of calculations which illustrate the distribution of liquid film over the cross-section of the pipe in a given time interval that correspond to the conditions described in Fig. 7—are graphically illustrated in Figs. 8, 9, 10 and 11.

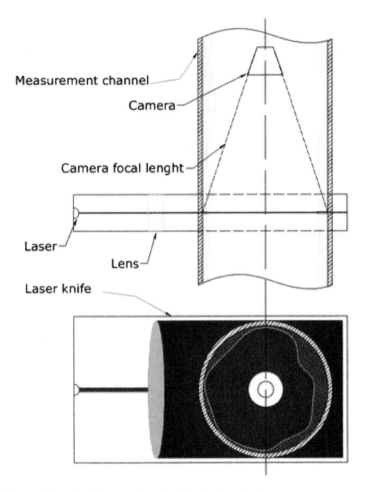

Fig. 6 Diagram illustrating the operating principle of a laser knife

These results confirm the previously stated remarks concerning the fact that the characteristics of two-phase gas-very viscous liquid flow form a considerably complex hydrodynamic phenomenon, which is accompanied by multidimensional asymmetry of the liquid film profile. This is further accompanied by stochastic characteristics of flow phenomena, scale of which is varied depending on the flow conditions. This is particularly discernible within the range of very wavy and hydraulic flow patterns, in which we have to do with very variable in time values of liquid film thickness. As a consequence, there is an occurrence of variable volume fraction of phases in the flow. In addition, some role is played by the variable interfacial surface in these conditions.

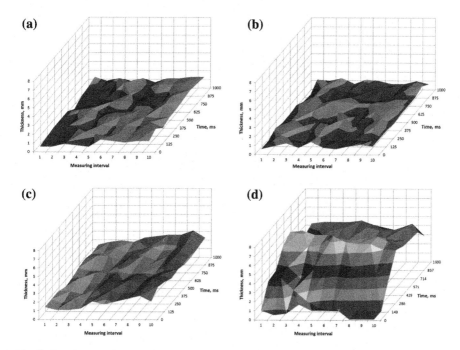

Fig. 7 Development of interfacial surface and distribution of oil film thickness in a pipe: **a** smooth flow ($Re_g = 1300$), **b** light wavy flow ($Re_g = 2250$), **c** wavy flow ($Re_g = 7800$), **d** hydraulic—high waves ($Re_g = 14,300$)

Exemplary results of calculations which illustrate the distribution of liquid film over the cross-section of the pipe in a given time interval that correspond to the conditions described in Fig. 7—are graphically illustrated in Figs. 8, 9, 10 and 11.

These results confirm the previously stated remarks concerning the fact that the characteristics of two-phase gas-very viscous liquid flow form a considerably complex hydrodynamic phenomenon, which is accompanied by multidimensional asymmetry of the liquid film profile. This is further accompanied by stochastic characteristics of flow phenomena, scale of which is varied depending on the flow conditions. This is particularly discernible within the range of very wavy and hydraulic flow patterns, in which we have to do with very variable in time values of liquid film thickness. As a consequence, there is an occurrence of variable volume fraction of phases in the flow. In addition, some role is played by the variable interfacial surface in these conditions.

In order to indicate the formation of interfacial surface for various values of flow parameters, Figs. 12 and 13 show a change of this surface related to specific internal surface of the channel.

The presented states in respect of the equivalent Reynolds number for liquid and Reynolds number for the gas indicate the variable distribution of this surface, in particular in the upper range of and Reynolds number for the gas. This is

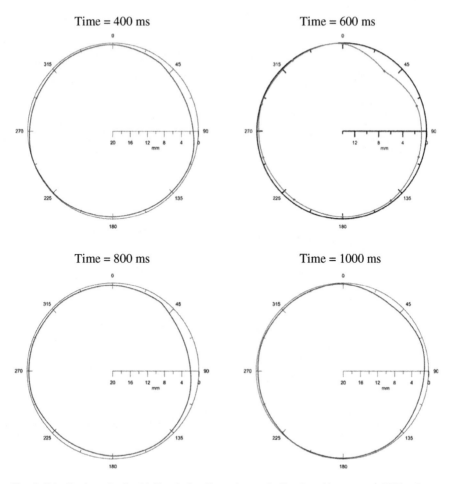

Fig. 8 Distribution of a liquid film during flow of smooth film $D = 40$ mm, $\eta_c = 2200$ mPa s

undoubtedly due to the variable wavy structures in this range which promotes the formation of additional capillary waves, which affects its size.

On the other hand, the higher dynamics of gas flow (in a certain range) promotes the development of interfacial surface, thus, as a consequence of these conditions brings an opposite effect for the ultimately even thinner and thinner hydraulic film in this range. This allows the statement that the size of interfacial surface is decided predominantly by the structures of the forming annular film but not film thickness which results from a given structure.

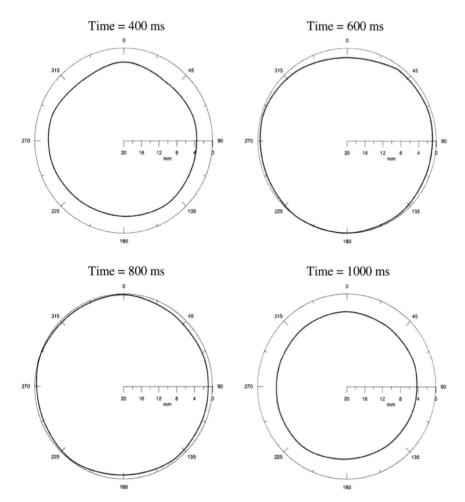

Fig. 9 Distribution of a wavy liquid film thickness $D = 40$ mm, $\eta_c = 2200$ mPa s

3 Summary

As a result of the analysis of the identified annular flow patterns of very viscous liquid flow, it was discovered that depending on the viscosity of the liquid and mutual relations of velocities of the two phases gas-liquid the annular flow patterns vary extremely—from a smooth liquid film until a highly dispersed hydraulic form.

By accounting for the similarity of the structures, three basic boundaries of the occurrence of annular forms of gas-very viscous liquid were identified. As a result, it was possible to develop a universal flow map—Fig. 2.

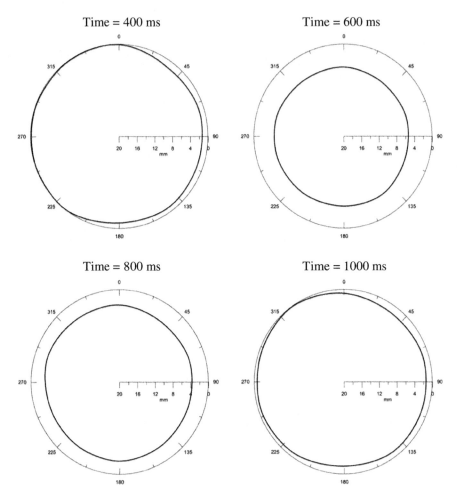

Fig. 10 Distribution of a high wave liquid film thickness with symmetrical distribution of waves $D = 40$ mm, $\eta_c = 2200$ mPa s

The presented optoelectronic measurement systems made it possible to identify such flow parameters as: liquid film thickness, velocity of waves dispersion on the surface as well as their dynamics.

The identification and assessment of the flow phenomena of the annular flow patterns were conducted on a wide scale. A considerable reduction of wave amplitude was noted on the interfacial surface as a result of damping the pulsations of annular flow by the very viscous liquid.

It was additionally indicated that an increase in the velocity of gas phase promotes the occurrence of higher and higher waves on the surface of a liquid film.

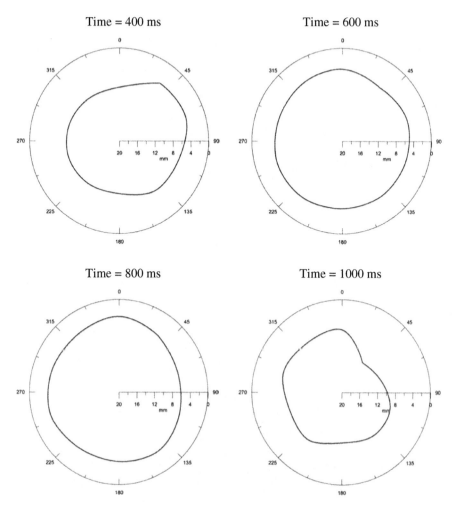

Fig. 11 Distribution of a hydraulic film thickness $D = 40$ mm, $\eta_c = 2200$ mPa s

With the exception of smooth films or light wavy ones these changes are irregular, which is manifested by the varied development of the interfacial surface. The occurrence of waves on the surface of phase separation is characterized by amplitude and length that are variable in time and space, which considerable complicated the quantitative description of the hydrodynamics of such phenomena.

From the analysis of experimental data one can conclude that within the entire range of variable process parameters all of them have a considerable effect on the formation of liquid film thickness. An increase in liquid viscosity always promotes an increase in the mean thickness of a liquid film. Concurrently, an increase in the

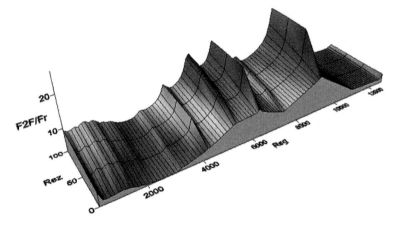

Fig. 12 Representation of interfacial surface in relation to the distribution in experimental points ($D = 40$ mm, $\eta_c = 2200$ mPa s)

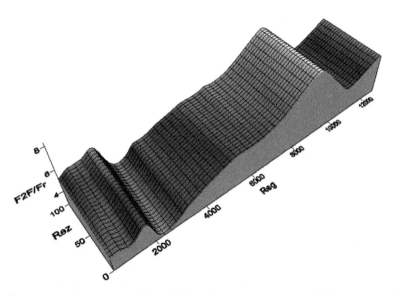

Fig. 13 Averaged representation of interfacial surface ($D = 40$ mm, $\eta_c = 2200$ mPa s)

volume flux of the gas phase for a constant equivalent Reynolds number for the liquid density leads to a thinner liquid film. These conditions are often accompanied by a phenomena of wave interference, which promotes their accumulation or results in a formation of additional capillary waves, which can lead to the occurrence of locally greater thickness.

References

Czernek K, Witczak S (2013) Non-invasive evaluation of wavy liquid film. Chem Proc Eng 34:241–252

Czernek K (2013) Hydrodynamic aspects design of thin film apparatuses for very viscous liquid. Opole University of Technology, Opole

Du XZ, Wang BX, Wu SR et al (2002) Energy analysis of evaporating thin falling film instability in vertical tube. Int J Heat Mass Tran 45(9):1889–1893

Ikeda T, Kotani K, Maeda Y et al (1983) Preliminary study on application of X-ray CT scanner to measurement of void fractions in steady-state two-phase flows. J Nucl Sci Technol 20(1):1–12

Lee YJ, Kim JH (1986) A review of holography applications in multiphase flow visualisation study. J Fluid Eng 108(3):279–289

Oriol J, Leclerc JP, Jallut C et al (2008) Characterization of the two-phase flow regimes and liquid dispersion in horizontal and vertical tubes by using coloured tracer and non-intrusive optical detector. Chem Eng Sci 63(1):24–34

Saito Y, Mishima K, Matsubayashi M (2004) Void fraction and velocity measurement of simulated bubble in a rotating disc using high frame rate neutron radiography. Appl Radiat Isot 61(4):667–674

Schmidt J, Giesbrecht H, van der Geld CWM (2006) Phase and velocity distributions in vertically upward high-viscosity two-phase flow. Int J Multiphas Flow 34(4):363–374

Stahl P, Rohr PR (2004) On the accuracy of void fraction measurements by single-beam gamma-densitometry for gas-liquid two-phase flows in pipes. Exp Therm Fluid Sci 28(6):533–544

Troniewski L, Witczak S, Czernek K (2006) Hydrodynamics and heat transfer during two-phase gas-high viscous liquid flow in film reactor. Chem Proc Eng 27:1341–1359

Wojtan L, Ursenbacher T, Thome JR (2005) Measurement of dynamic void fractions in stratified types of flow. Exp Therm Fluid Sci 29(3):383–392

Xu XX (2007) Study on oil-water two-phase flow in horizontal pipelines. J Petrol Sci Eng 59(1–2):43–58

The Influence of Rotating Magnetic Field on Biochemical Processing

Radosław Drozd, Agata Wasak, Maciej Konopacki, Marian Kordas and Rafał Rakoczy

1 Rotating Magnetic Field as a Tool in Chemical Engineering

The increase in productivity is one of the main aims in most biotechnological processes and may be achieved in several ways: using new or modified microorganisms strains; developing new bioreactors and optimizing the operation strategies; improving the separation processes; using efficient control systems and developing a more effective cell immobilization technique (Domingues et al. 2000). The quest for efficient processes involves non-convectional approaches for the stimulation of bioprocesses. The application of physical factors to achieve the most advantageous production performance has been studied for many years. Alternative methods of intensification of biotechnological processes are the use of various types of force fields (e.g., magnetic, electrical, ultrasound). A particularly interesting and potentially having high practical application may be the use of different types of magnetic fields (MFs) to stimulate bioprocesses.

The use of various types of force fields to support biological processes has been the subject of study for many years. The source of the MF acting in the magnetically assisted bioreactors may be generated in different ways. From the practical point of view, the generated field can be divided in two main categories: the axial MF (e.g. solenoids, Helmholtz pair, systems of coils) and the transverse MF (e.g. systems based on saddle coils generating a homogenous MF, cylindrical inductor

R. Drozd · A. Wasak
Faculty of Biotechnology and Animal Husbandry, West Pomeranian University of Technology, Szczecin, Poland

M. Konopacki · M. Kordas · R. Rakoczy (✉)
Faculty of Chemical Technology and Engineering, West Pomeranian University of Technology, Szczecin, Poland
e-mail: rafal.rakoczy@zut.edu.pl

© Springer International Publishing AG, part of Springer Nature 2018
M. Ochowiak et al. (eds.), *Practical Aspects of Chemical Engineering*,
Lecture Notes on Multidisciplinary Industrial Engineering,
https://doi.org/10.1007/978-3-319-73978-6_5

resembling the stator of three-phase asynchronous electrical engine producing a heterogeneous MF) (Kholoov 1974).

The magnetic fields can be divided into two main types: direct current magnetic field (DCMF) and alternating current magnetic field (ACMF). The DCMF does not change with time or changes very slowly. Such fields do not have the frequency (MF vector is constant in time and space). An example of MF of the constant current is the static magnetic field (SMF). In contrast to DCMF, ACMF varies with the frequency. An example of this type of MF is a pulsating magnetic field (PMF) characterized by an external MF vector which changes as a sine-wave with time at each point of the space. Vector of this type of field pulsates with the frequency of the current flowing through the coil. Alternatively, the superposition of three 120° out of phase PMFs is contributes to a rotating MF (RMF). This field has a constant intensity over time while it changes its direction continuously at any point of the domain (RMF is variable in space). Such a field is created for example in the stator and rotor windings as a result of supply of windings and makes the rotation of the engine possible. The rotating magnetic field is characterized by the fact that its axis rotates relative to the reference system (relative to the stator) and its return remains constant along the axis (Rakoczy and Masiuk 2011). This kind of a time-varying MF is commonly applied as electromagnetic stirrers (Molokov et al. 2007). It has been shown, that RMF can be a versatile option for enhancing molecular transport and diffusion in aqueous culture media (Hajiani and Larachi 2013). It should be noticed that the RMF can be modulated growth dynamics, cellular metabolic activity and is able to form biofilms by different species of bacteria (Rakoczy et al. 2016).

Static or alternating current magnetic fields may be used to augment the process intensity instead of the mechanical mixing (Hristov 2002; Hajiani and Larachi 2012). The alternating current magnetic field (ACMF), e.g. the rotating magnetic field (RMF), may be used as a non-intrusive stirring device and it can be engineered to provide any desired pattern of stirring (Moffat 1991). The application of this kind of magnetic field (MF) to augment the transport process intensity was discussed in the relevant literature (Rakoczy 2010; Rakoczy and Masiuk 2010, 2011; Rakoczy et al. 2017).

The rotating magnetic field (RMF) is a special case of electromagnetic field. This field is created due to an interaction between the force vectors of the electromagnetic fields generated by the coils situated in the circle every 120°. These coils are powered by the tri-phase alternate current, characterized by 120° electric phase shift. As the result of the superposition of the electromagnetic fields, a single vector of the Lorentz's force is created, which direction is rotating around the generator axis, around which the coils are situated. The position of the magnetic field in the generator core is constant, nevertheless, the direction of the magnetic force is changing depending on the phase of the current powered the windings (Rakoczy and Masiuk 2011). The dependency of the magnetic field force direction from the current phase is presented in a schematic figure (Fig. 1).

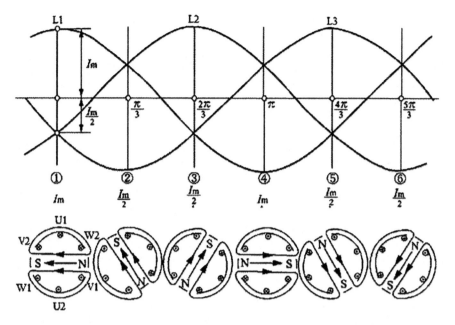

Fig. 1 Scheme of rotating magnetic field generation

The feature of the RMF is its ability to induce the time-averaged azimuthal force, which drives the flow of the electrical conducting fluid in the azimuthal direction. In this case the magnetic field lines, B, rotate horizontally with the rotation frequency of the magnetic field produced. Simultaneously, an electrical field, E, is produced perpendicularly to the magnetic field. It can be assumed, that the external magnetic field acting on the electrically conducting medium may induce virtual loop formation. The magnetic field passing through its enclosed area changes in time during one period of its revolution. According to the induction law, the density of electric current is induced along the virtual loop. The interaction between current density and the magnetic induction generates the electromagnetic force, F_{em}, which influences the fluid in the direction of the magnetic field rotation.

The impact of RMF over the formation of virtual loop causing movement of the liquid inside the generator of this kind of magnetic field is given in Fig. 2a. Figure 2b presents the effect of RMF on the exposed samples placed in the RMF generator. The medium in a sample is subjected to the impact of externally applied magnetic field, which induces the force causing movement of the fluid. The movement of the medium exposed to RMF can be explained on the basis of microlevel dynamo concept (Hristov 2010). The MF interacting with various charged particles, for example, ions, produces eddy currents in the culture medium (Gaafar et al. 2008; Anton-Leberre et al. 2010). The eddy currents may generate local MFs around the ions, which, in combination with an externally-applied MF, cause induction of their rotation and thus the movement of the liquid in accordance to the MF vector. As a consequence of this process the rotating ions create "dynamos" which cause the effect of micro-mixing. Micromixers utilizing magnetic forces represent important class of mixing possibility (Wang et al. 2008). These mixers use Lorentz force to agitate fluids and induce secondary complex chaotic flows (Ryu et al. 2004). Figure 2c shows the micro-mixing effect of the fluid in accordance with the dynamo concept.

Hunt et al. (2009), identified the class of bioreactors with the working volume entirely encircled by external magnetic systems and bioreactors with recirculation and external culture magnetization (this concept means a closed loop including the reactor and a chamber for magnetic exposure). The precursor of "magnetic support of bioprocesses" was Rosensweig, who proposed the concept of magnetic biore-actor (Rosensweig 1979). Reactor with the application of SMF was used for the production of enzymes (Moffat et al. 1994), cell cultures (Bramble et al. 1990) and the intensification of biomass growth (Sada et al. 1981). The concept of magnetic stirring reactor was also used to intensify the process of biochemical and enzymatic reactions (Sakai et al. 1989, 1990, 1992a, b, 1994, 1999; Gusakov et al. 1995, 1996; Sinitsyn et al. 1993; Bahar 2000). The application of MF in the bioreactor engi-neering is also presented in the relevant literature (Hristov and Ivanova 1999; Gogate et al. 2000). Different types of MFs were also implemented in many con-structions of bioreactors (Vangas et al. 1999).

The typical apparatus equipped with the RMF generator is presented in Fig. 3. This reactor contains an RMF generator, made of a three-phase stator of an induction squirrel cage motor. The RMF is generated by coils located around the cylinder and the axes are directed along the radius. When the alternating current supplies the windings, the generated magnetic field rotates around the cylinder axis

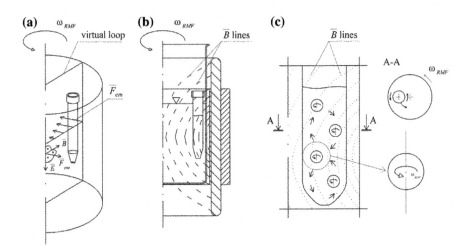

Fig. 2 The effect of the RMF on fluid: **a** effect of RMF on the fluid in the cylindrical container with the exposed samples; **b** effect of RMF on the medium in samples; **c** the concept of the micro-mixing effect in accordance with the dynamo concept

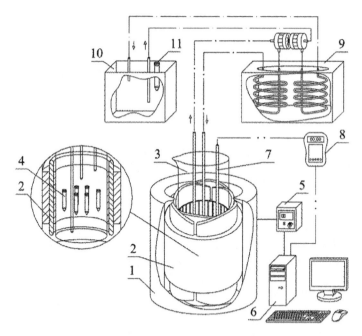

Fig. 3 The experimental set up: *1* cooling jacket, *2* RMF generator, *3* glass container, *4* sample, *5* A.C. transistorized inverter, *6* computer, *7* magnetic induction probe, *8* Hall-effect meter, *9* thermostat, *10* water bath, *11* control

with the constant angular frequency of RMF. In our experiments, the stator was supplied with 50 Hz three-phase alternating current. The ac transistorized inverter is used to change the frequency of the rotating magnetic field.

2 Influence of Magnetic Field on the Enzyme Catalyzed Reactions

One of the most well-known theories concerning the nature of magnetic field effects on enzyme catalyzed reactions is the radical pair recombination (Eichwald and Walleczek 1996; Grissom 1995; Steiner and Ulrich 1989; Taraban et al. 1997). Enzyme catalyzed reactions usually require less energy than the same chemical reactions in which other catalysts are used. This allows for the construction of an active enzyme center, allowing for a rapid transition state and the initiation of a reaction may take place with a relatively lower energy input. The catalytic center of enzymes possesses or gathers in one place all the necessary acceptors and donors and also creates the environment that forms the amino acids it builds and the cofactors are presented in it as some metal ions, nucleotides, vitamins etc., significantly increasing the likelihood of a reaction (Szefczyk et al. 2004). An external magnetic field can modify the likelihood of recombinant pairs of free radicals even when it is much weaker than the locally produced magnetic field due to the magnetic effects of unpaired electrons. During radical catalyzed reactions, free radicals may be formed during electron transfer in oxidoreductase reactions such as cytochrome C oxidase (Blank and Soo 2001; Woodward 2002). Another example of reaction in which free radical formation may occur is bond homolysis, where example may be some of the lyase as B12 ethanolamine ammonia lyase (Jones et al. 2007). During generation of free radicals pair due to spin conservation, the unpaired electron spins are initially correlated in either a singlet (S), an anti-parallel or triplet (T), a parallel configuration. In case when unpaired spin possesses different local magnetic field, the S and T radicals spin-states will interconvert (Messiha et al. 2014). Depending on the value of the density of the flux of the external magnetic field in the free radical reactions triplet and singlet yield is different. Under conditions of low density magnetic flux the T yield decreasing (Eichwald and Walleczek 1996). The influence of magnetic field on the enzymes activity that is important in the chemical engineering with a potential to be used in industry was reported several times. For horse radish peroxidase RPM was postulated as a reason for altering its activity under MF exposure (Taraban et al. 1997). The results of the exposition of the fungal laccase to RMF also revealed some significance in enzyme activity (Fig. 4 The influence of rotating). This enzyme is also oxidoreductase which contains in active center four cooper inions that are important in electron transfer during catalytic act, which can be altered by external magnetic field (Giardina et al. 2010).

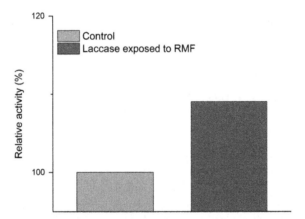

Fig. 4 The influence of rotating magnetic field on residual activity of laccase from Trametes versicolor. Laccase was exposed to RMF at 50 Hz frequency with 18 mT magnetic induction. The reaction was conducted at 30 °C in 50 mM sodium acetate buffer solution pH 4.0 with 0.5 mM 2,2′-azino-bis(3-ethylbenzothiazoline-6-sulphonic acid as substrate

3 Influence of Magnetic Field on the Enzyme Structure

However, not in all enzymatic reactions transient complexes are formed by the mechanism of radical pair recombination and reports indicate the possibility of changing their activity due to the effect of external magnetic field.

Enzymatic proteins consist of twenty amino acids linked together by a peptide bond formed by the amino and carboxy groups of the main amino acid chain. At this level we are talking about the backbone of a globular protein capable of forming regular secondary structures stabilized by the hydrogen bonds that arise between the carbonyl and amino of the peptide bond. Hydrogen bonds, despite their low strength but owing to the high number, correspond to the stabilization of this level of organization of proteins (Sheu et al. 2003). At this level, we recognize beta and alpha structures (Fig. 5). At the next level of organization, which is the tertiary structure of proteins, salt bridges formed by carboxyl groups amino acids such as

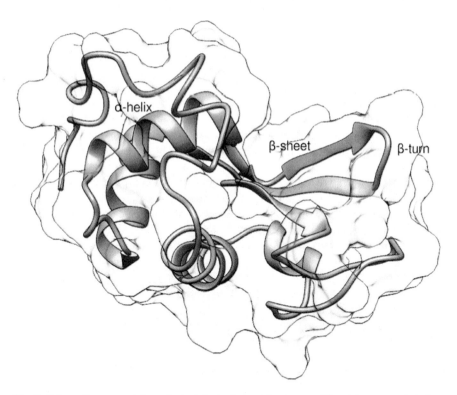

Fig. 5 Schematic representation of principles in the protein structure. The picture presents tertiary structure of hen egg lysozyme (Protein data bank entry 1lyz, Diamont 1974). The flat arrows indicate the β-strands and the twisted a α-helix

aspartate, glutamate and amino from lysine and arginine play a key role in stabilization of enzyme structure against high temperature (Van den Burg et al. 1998).

Magnetic field can probably induce changes in the structure of enzyme proteins.
The exposition of bovine serum albumin and human hemoglobin to EMF, the
source of which was an electromagnet, resulted in changes in intensity of the amide
bands in the IR spectra and the β-sheet to α-helix ratio. These changes were
reversed to the control level at the end of the EMF exposure. However, the
relaxation time for the analyzed proteins was different, possibly related to the
different molecular structure of albumin and hemoglobin. A similar effect was also
demonstrated for the exposition of albumin to SMF (200 mT), where changes in the
secondary structure of this protein were observed. So far, the impact of EMF on the
structural changes of a significant number of enzymes has been studied in detail
only for a few enzymes and mostly for lysozyme. As a result of exposure to
ultra-low electromagnetic field of 180 µT, it was found that the incubation of this
enzyme in EMF lasting for 4 h has a negative effect on its structure. The response
of biomolecules as proteins and including enzymes to the magnetic field is connected with their diamagnetic anisotropy and coherent electric polar vibrations
(Magazù and Calabrò 2011; Calabrò and Magazù 2012).

In the study regarding the effect of SMF (1–9 T) on the Epidermal Growth
Factor Kinase Domain Receptor, the influence of this factor on the level of the
second- and third-order structure induced changes may be influenced by the
appropriate conformation of the monomeric subunits of this domain required to
form functional dimer. However, based on the value of the EGFR dipole moment
and the direction of the magnetic field flux, it has been found that the magnetic field
can significantly influence the proper positioning of the EGFR monomers, making
it difficult or impossible for them to mutually assimilate to the proper functioning of
the dimer. Many enzymatic proteins, in order to reach complete catalytic activity
require the formation of homo or heteroligomeric structures (Mei et al. 2005;
Rumfeldt et al. 2008). Typically, the structure of the oligomer subunits is already
prepared to properly connect with another subunit.

One of the factors affecting the speed and specificity of mutual recognition of the
correct orientation is the electrostatic potential on the molecular surface of the
subunits (Campbell et al. 2014; Sinha and Smith 2002) (Fig. 6 Schematic representation of electrostatic potential lines lysozyme).

Thus, the ability to influence the distribution of electrostatic field (EMF) on the
molecular surface of enzymes can cause changes in their activity and at the same
time a way to regulate the efficiency of enzyme catalyzed reactions in the process
where in most cases the enzymatic reactions take place in buffers whose main
component is water and the solution also contains various types of metal ions or
organic compounds whose properties can be altered by the external magnetic field.
For example phosphate buffer solution exposed to SMF (250 mT) exhibited
increasing in its conductivity and also deionized water exhibits after exposition on
the MF increasing in values of this parameter. The change in the conductivity of
buffers can subsequently result in the weakening the hydration shell around ions
occurring in solution, which can indirectly influence the structure and the activity of

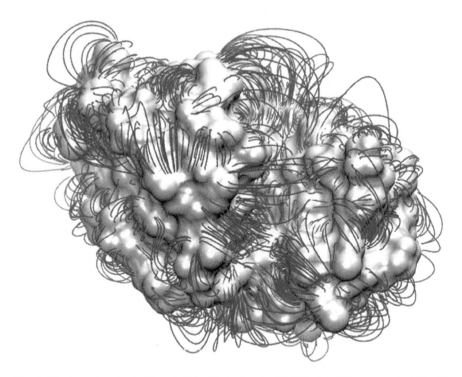

Fig. 6 Electrostatic potential lines (−0.5 kT/e (red) and +0.5 kT/e (blue)) lysozyme, in 150 mM salt at pH 7.0. The structure used was hen egg lysozyme. The electric field was calculated with using a Adaptive Posion Boltzman Solver and visualized by using VMD software

enzymes (Holysz et al. 2007; Szcześ et al. 2011; Li et al. 2014; Myśliwiec et al. 2016).

In technological processes, enzymes are used both in free and immobilized form. Immobilization of enzymes of various types has many advantages in enhancing their stability to reaction conditions, ease of separation from the reaction medium, and reusability, which significantly reduces the cost of obtaining enzymes (Mohamad et al. 2015). Various soluble and solid organic carriers of natural origin (e.g. chitosan, cellulose), artificially synthesized polymers (e.g. polyacrylamide, polyutherane) and inorganic carriers (e.g. silica, porous glass, clay minerals) can be used to immobilize enzymes (Dong-Hao et al. 2013; Sheldon and van Pelt 2013; Mehta et al. 2016). These media are often modified by ferromagnetic or paramagnetic materials that are sensitive to external magnetic fields. The advantage of this type of connection is the substantial advantage of facilitating the separation of such media from the reaction medium with the use of a normal magnet. In addition to combining magnetic nanoparticles with other substances, they are often subjected to a variety of modifications such as amination, silanization or carboxylation to increase their potential use in the industry (Vaghari et al. 2016). In addition to the obvious advantages, the use of MNP also has the potential to reduce the activity of enzymes as a result of conformational changes during immobilization and to obstruct mass transfer but not to the extent of high porosity media (Singh et al. 2013). One of the important problems in the immobilization of enzymes is the frequent interruption of the mass exchange process (Xiu et al. 2001).

This is a twofold difference. Firstly, the enzyme for optimum activity requires the proper concentration of the starting substrate and secondly reaction products often become competitive inhibitors. In many cases, a significant increase in the K_m constant of an enzyme immobilized relative to its free form is observed. This is due to the difficulty of diffusion of the substrate in the microenvironment formed by the carrier towards the catalytic region of the enzyme and the locally increasing concentration of reaction products whose diffusion beyond the matrix is also impeded (Berendsen et al. 2008). In one of the first works on the use of PM in enhancing the efficiency of immobilized enzymes on MNP-containing carriers, the urease was packed in a polyacrylamide matrix encrusted with MNP. In this work Sada et al. (1981), was one of the first to use a set of electromagnets surrounding the reaction vessel generating a variable magnetic field. The efficiency of the system under review has increased significantly with respect to the traditional mechanical stirrer reactor, probably as a result of vertical and revolutionary movements of the immobilized-enzyme beads. The ability to oscillate and rotate the media by exposure to variable magnetic fields also has a beneficial effect on starch hydrolysis by immobilized glucoamylase (Yang et al. 2010). Zheng et al. (2013), reported a significant increase in reaction in bi-enzymatic system consisting of glucose dehydrogenase degradation, glutamate dehydrogenase and co-factor as NADH. Variable magnetic field, depending on its strength and frequency, modified the probability of collision of both substrates and co-factors with analyzed enzymes.

Mizuki et al. (2010, 2013) analyzed the effect of rotational magnetic field (RMF) on lipase and chitinase and α-amylase that were immobilized on modified MNPs. Immobilization alone did not have a positive effect on the activity of the enzymes tested, which was probably due to conformational changes in the structure of the enzymes after adsorption on the support. However, the exposure to RMF significantly increased the activity of immobilized lipase and chitinase and α-amylase, what was not observed for free enzyme forms. The activity of immobilized enzymes correlated also closely with the RMF frequency, the reasons for its highest increase were observed at 5 Hz ($\approx$12 mT). Another studied immobilized enzyme like laccase, when exposed to RMF showed slight altering of its activity compared to immobilized enzyme not exposed to rotating magnetic field (Fig. 7).

It can be assumed that the increase in activity was the result of a reduction in mass transfer reduction due to the oscillating and rotational movements of the carrier exposed to the variable magnetic field (Webb et al. 1996).

Klyachko et al. (2012), observed a significant decrease in immobilized α-chymotrypsin and immobilized β-galactosidase activity on MNP-modified polymers subjected to pulsatile exposure to alternating current magnetic field (AC-MF). MNP radio-frequency (RF) alternating current (AC) magnetic fields can induce the heating of single-domain. However, in the case of AC magnetic field, no significant elevation of the carrier temperature was observed, which excludes this factor as a

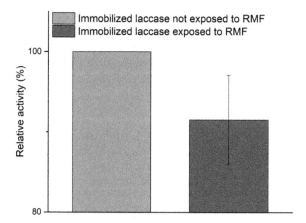

Fig. 7 The effect of rotating magnetic field on residual activity of laccase from *Trametes versicolor* immobilized on polyethyleneimine modified ferromagnetic particles. Immobilized laccase was exposed to RMF at 50 Hz frequency with 18 mT magnetic induction. The reaction was conducted at 30 °C in 50 mM sodium acetate buffer solution pH 4.0 with 0.5 mM 2,2'-azino-bis(3-ethylbenzothiazoline-6-sulphonic acid as substrate

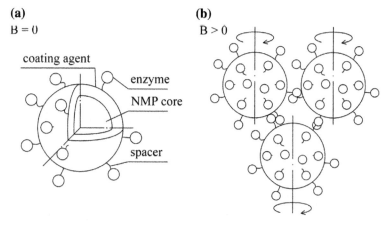

Fig. 8 Schematic representation for the consequence of SMNP movement under MF exposure. The attached enzymes molecules are on negative interaction due to possibility of collision with other SMNP (Klyachko et al. 2012)

reason for the decrease in mobilized enzyme activity. The results of structural analysis using circular dichroism indicated the induction of changes in the secondary structure of immobilized enzymes, which may be the reason for the decrease in activity of the enzymes tested. Probably the structural changes in the carrier resulting from exposure to AC MF have destabilized immobilized enzymes (Fig. 8).

Golovin et al. (2014), have shown that SMNPs subjected to ESEs can significantly alter the structure of their associated macromolecules—enzymes. SMNP movements, the possibility of collisions and rotation around one's own axis, affect the probability of modifying chemical bonds and further secondary structure of enzymes, changing the conformation of the catalytic region.

4 Magnetic Field Modulation of Enzymes Activity in Cellular System

The electromagnetic field is one of the factors causing the environmental stress of the microorganisms. The action of the external electromagnetic field influences the expression of the genes in answers on the environmental stress. This effect causes the modulation of the metabolic processes that taking place inside the cells occurs and influence of microorganisms growth kinetic (Fig. 9).

The conducted research showed the possibility of application of the rotating magnetic field to stimulate the metabolic paths responsible for the creation of the desirable secondary metabolites, in an example the ethanol in the fermentation process. It was shown, that under the action of the rotating magnetic field the yeast cells produced 28% more quantity of the ethanol during 24 h of the continuous exposition in comparison to the control process (Figs. 9 and 10).

According to Bialek et al. (1989), the electromagnetic field causing the transitional changes in the enzymatic reactions increases the metabolic activity of the

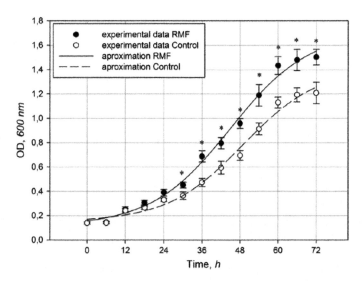

Fig. 9 The influence of the rotating magnetic field on yeast cells growth. The culture was conducted in static conditions in yeast extract Peptone Dextrose (glucose) broth at 28 °C

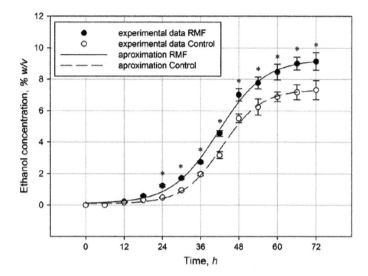

Fig. 10 The influence of the rotating magnetic field on the ethanol production by the yeast cells. The culture was conducted in static conditions in yeast extract Peptone Dextrose (glucose) broth at 28 °C

cells. The rotating magnetic field probably influences the charged particles inside the cell, inducing changes in the ions transfer through the cell wall and following in supplying the substrates for the enzymatic reactions and the removal of the product. The changes in the transfer of these components could stimulate or inhibit the enzymatic reactions rates, which are often combined together in the metabolic paths on the basis of the feedback (Ledakowicz 2011). Pringle et al. (1981), based on the conducted research suggested that the magnetic field influences the cell cycle duration due to activation of the enzymes, making it shorter than the time of the cell division.

5 Conclusion

Enzymes, due to their obvious advantages and wide application in many industries, also become important tools in chemical engineering. The application of enzymes is connected with the selection of the proper conditions that ensure the stability of enzyme structure and the speed of action. The rotating magnetic field may be applied as a tool of chemical engineering in order to the control the catalyzed reactions involving free and immobilized enzymes. Moreover, this kind of magnetic field is affected by the direction and the rate of the biochemical processes in the living cells.

Acknowledgements The authors are grateful for the financial support of the National Science Centre Poland within the PRELUDIUM 11 Programme (Grant No. 2016/21/N/ST8/02343).

References

Anton-Leberre V, Haanappel E, Marasaund N et al (2010) Expsoure to high static or pulsed magnetic fields does not affect cellular processes in the yeast *Saccharomyces cerevisiae*. Bioelectromagnetics 31:28–38

Bahar T, Çelebi SS (2000) Performance of immobilized glucoamylase in a magnetically stabilized fluidized bed reactor (MSFBR). Enzyme Microb Tech 26:28–33

Berendsen WR, Lapin A, Reuss M (2008) Investigations of reaction kinetics for immobilized enzymes–identification of parameters in the presence of diffusion limitation. Biotechnol Prog 22:1305–1312

Bialek W, Bruno WJ, Joseph J et al (1989) Quantum and classical dynamics in biochemical reactions. Photosynth Res 22(1):15–27

Blank M, Soo L (2001) Optimal frequencies for magnetic acceleration of cytochrome oxidase and Na, K-ATPase Reactions. Bioelectrochemistry 53(2):171–174

Bramble JL, Graves DJ, Brodelius P (1990) Plant cell culture using a novel bioreactor: the magnetically stabilized fluidized bed. Biotechnol Prog 6:452–457

Calabrò E, Magazù S (2012) Electromagnetic fields effects on the secondary structure of lysozyme and bioprotective effectiveness of trehalose. Adv Phys Chem 2012, Article ID 970369

Campbell B, Petukh M, Alexov E et al (2014) On the electrostatic properties of homodimeric proteins. J Theor Comput Chem 13(3):1440007

Diamond R (1974) Real–space refinement of the structure of hen egg–white lysozyme. J Mol Biol 82:371–391

Domingues L, Vicente AA, Lima N et al (2000) Applications of yeast flocculation in biotechnological processes. Biotechnol Bioprocess Eng 5:288–305

Dong-Hao Z, Li-Xia Y, Li-Juan P (2013) Parameters affecting the performance of immobilized enzyme. J Chem 2013, Article ID 946248

Eichwald C, Walleczek J (1996) Model for magnetic field effects on radical pair recombination in enzyme kinetics. Biophys J 71:623–631

Gaafar ESA, Hanafy MS, Tohamy EY et al (2008) The effect electromagnetic field on protein molecular structure of *E. coli* and its pathogenesis. Rom J Biophys 18:145–169

Giardina P, Faraco V, Pezzella C et al (2010) Laccases: a never-ending story. Cell Mol Life Sci 67 (3):369–385

Gogate PR, Beenackers AACM, Pandit AB (2000) Multiple-impeller systems with a special emphasis on bioreactors: a critical review. Biochem Eng J 6:109–144

Golovin YI, Gribanovskii SL, Golovin DY et al (2014) Single-domain magnetic nanoparticles in an alternating magnetic field as mediators of local deformation of the surrounding macromolecules. Phys Solid State 56(7):1342

Grissom CB (1995) Magnetic field effects in biology: a survey of possible mechanisms with emphasis on radical-pair recombination. Chem Rev 95(1):3–24

Gusakov AV, Sinitsyn AP, Davydkin IYOV et al (1995) Use of a bioreactor with intense mass transfer for enzymatic hydrolysis of cellulose-containing materials. Appl Biochem Micro 31:310–314

Gusakov AV, Sinitsyn AP, Davydkin IY et al (1996) Enhancement of enzymatic cellulose hydrolysis using a novel type of bioreactor with intensive stirring. Appl Biochem Micro 56:141–153

Hajiani P, Larachi F (2012) Reducing Taylor dispersion in capillary laminar flows using magnetically excited nanoparticles: nanomixing mechanism for micro/nanoscale applications. Chem Eng J 203:492–498

Hajiani P, Larachi F (2013) Giant effective liquid-self diffusion in stagnant liquids by magnetic nanomixing. Chem Eng Process 71:77–82

Holysz L, Szcześ A, Chibowski E (2007) Effects of a static magnetic field on water and electrolyte solutions. J Colloid Interface Sci 316(2):996–1002

Hristov J (2002) Magnetic field assisted fluidization—a unified approach. Part 1: fundamentals and relevant hydrodynamics. Rev Chem Eng 18:295–509

Hristov J (2010) Magnetic field assisted fluidization—a unified approach. Part 8: mass transfer: magnetically assisted bioprocess. Rev Chem Eng 26:55–128

Hristov JY, Ivanova V (1999) Magnetic field assisted bioreactors. Recent Res Dev Ferment Bioeng 2:41–95

Hunt RW, Zavalin A, Bhatnagar A et al (2009) Electromagnetic biostimulation of living cultures for biotechnology, biofuel and bioenergy applications. Int J Mol Sci 10:4515–4558

Jones AR, Hay S, Woodward JR et al (2007) Magnetic field effect studies indicate reduced geminate recombination of the radical pair in substrate-bound adenosylcobalamin-dependent ethanolamine ammonia lyase. J Am Chem Soc 129(50):15718–15727

Kholoov Y (ed) (1974) Influence of magnetic field on biological objects. U.S. Joint Publications Research Service, Arlington, VA

Klyachko NL, Sokolsky-Papkov M, Pothayee N et al (2012) Changing the enzyme reaction rate in magnetic nanosuspensions by a non-heating magnetic field. Angew Chem Int Ed 51:12016–12019

Ledakowicz S (2011) Inżynieria biochemiczna. Wydawnictwo Naukowo Techniczne, Warszawa

Li JY, Wang AJ, Ren NQ et al (2014) Effects of static magnetic field on phosphate buffer solution. Adv Mat Res 953–954:1293–1296

Magazù S, Calabrò E (2011) Studying the electromagnetic-induced changes of the secondary structure of bovine serum albumin and the bioprotective effectiveness of trehalose by Fourier transform infrared spectroscopy. J Phys Chem B 115(21):6818–6826

Mehta J, Bhardwaj N, Bhardwaj SK et al (2016) Recent advances in enzyme immobilization techniques: metal-organic frameworks as novel substrates. Coord Chem Rev 322:30–40

Mei G, Di Venere A, Rosato N et al (2005) The importance of being dimeric. FEBS J 272(1):16–27

Messiha HL, Wongnate T, Chaiyen P et al (2014) Magnetic field effects as a result of the radical pair mechanism are unlikely in redox enzymes. J R Soc Interface 12:20141155

Mizuki T, Watanabe N, Nagaoka Y et al (2010) Activity of an enzyme immobilized on superparamagnetic particles in a rotational magnetic field. Biochem Biophys Res Commun 393 (4):779–782

Mizuki T, Sawai M, Nagaoka Y et al (2013) Activity of lipase and chitinase immobilized on superparamagnetic particles in a rotational magnetic field. PLoS One 8(6):e66528

Moffat HK (1991) Electromagnetic stirring. Phys Fluids A 3:1336–1343

Moffat G, Williams RA, Webb C et al (1994) Selective separation in environmental and industrial processes using magnetic carrier technology. Miner Eng 7:1039–1056

Mohamad NR, Marzuki NH, Buang NA et al (2015) An overview of technologies for immobilization of enzymes and surface analysis techniques for immobilized enzymes. Biotechnol Biotechnol Equip 29(2):205–220

Molokov S, Moreau R, Moffat HK (2007) Magnetohydrodynaics. In: Molokov S, Moreau S, Moffatt R, Keith H (eds) Historical evolution and trends. Springer, The Netherlands

Myśliwiec D, Szcześ A, Chibowski S (2016) Influence of static magnetic field on the kinetics of calcium carbonate formation. J Ind Eng Chem 35:400–407

Pringle JR (1981) The Saccharomyces cerevisiae cell cycle. The molecular biology of yeast Saccharomyces: life cycle and inheritance, pp 97–142

Rakoczy R (2010) Enhancement of solid dissolution process under the influence of rotating magnetic field. Chem Eng Process 49:42–50

Rakoczy R, Masiuk S (2010) Influence of transverse rotating magnetic field on enhancement of dissolution process. AIChE J 56:1416–1433

Rakoczy R, Masiuk S (2011) Studies of mixing process induced by a transverse rotating magnetic field. Chem Eng Sci 66:2298–2308

Rakoczy R, Konopacki M, Fijałkowski K (2016) The influence of a ferrofluid in the presence of an external rotating magnetic field on the growth rate and cell metabolic activity of a wine yeast strain. Biochem Eng J 109:43–50

Rakoczy R, Lechowska J, Kordas M et al (2017) Effects of a rotating magnetic field on gas–liquid mass transfer coeffcient. Chem Eng J 327:608–617

Rosensweig RE (1979) Fluidization: hydrodynamics stabilization with a magnetic field. Science 204:57–60

Rumfeldt JA, Galvagnion C, Vassall KA et al (2008) Conformational stability and folding mechanisms of dimeric proteins. Prog Biophys Mol Biol 1:61–84

Ryu KS, Shaikh K, Goluch E et al (2004) Micro magnet stir-bar mixer integrated with parylene microfluidic channels. Lab Chip 4:608–613

Sada E, Katoh S, Shiozawa M et al (1981) Performance of fluidized-bed reactors utilizing magnetic-fields. Biotechnol Bioeng 23:2561–2567

Sakai Y, Taguchi H, Takahashi F (1989) The effect of alternative magnetic field on the pigment ejection from magnetic anisotropic gel beads. B Chem Soc Jpn 62:3207–3210

Sakai Y, Kuwahata M, Takahashi F (1990) The effect of alternating magnetic field on the magnetic anisotropic gel beads immobilized catalase. B Chem Soc Jpn 63:2358–2362

Sakai Y, Kuwahata M, Takahashi F (1992a) Numerical formulation of pigment release from magnetically anisotropic gel beads with respect to the magnetic moment in an alternating magnetic field. B Chem Soc Jpn 65:396–399

Sakai Y, Osada K, Takahashi F et al (1992b) Preparation and properties of immobilized glucoamylase on a magnetically anisotropic carrier comprising a ferromagnetic powder coated by albumin. B Chem Soc Jpn 65:3430–3433

Sakai Y, Tamiya Y, Takahashi F (1994) Enhancement of ethanol formation by immobilized yeast containing iron powder or Ba-ferrite due to eddy current or hysteresis. J Ferment Bioeng 77:169–172

Sakai Y, Oishi A, Takahashi F (1999) Enhancement of enzyme reaction of magnetically anisotropic polyacrylamide gel rods immobilized with ferromagnetic powder in an alternating magnetic field. Biotechnol Bioeng 62:363–367

Sheldon RA, van Pelt S (2013) Enzyme immobilisation in biocatalysis: why, what and how. Chem Soc Rev 42:6223–6235

Sheu SY, Yang DY, Selzle HL, Schlag EW (2003) Energetics of hydrogen bonds in peptides. Proc Natl Acad Sci USA 100(22):12683-12687. 28 October 2003

Singh RK, Tiwari MK, Singh R et al (2013) From protein engineering to immobilization: promising strategies for the upgrade of industrial enzymes. Int J Mol Sci 14:1232–1277

Sinha N, Smith-Gill SJ (2002) Electrostatics in protein binding and function. Curr Protein Pept Sci 3(6):601–614

Sinitsyn AP, Gusakov AV, Davydkin IY et al (1993) A hyperefficient process for enzymatic cellulose hydrolysis in the intensive mass transfer reactor. Biotechnol Lett 15:283–288

Steiner UE, Ulrich T (1989) Magnetic field effects in chemical kinetics and related phenomena. Chem Rev 89:51–147

Szcześ A, Chibowski E, Hołysz L et al (2011) Effects of static magnetic field on electrolyte solutions under kinetic condition. J Phys Chem A 115(21):5449–5452

Szefczyk B, Mulholland AJ, Ranaghan KE et al (2004) Differential transition-state stabilization in enzyme catalysis: quantum chemical analysis of interactions in the chorismate mutase reaction and prediction of the optimal catalytic field. J Am Chem Soc 126(49):16148–16159

Taraban MB, Leshina TV, Anderson MA et al (1997) Magnetic field dependence of electron transfer and the role of electron spin in heme enzymes: horseradish peroxidase. J Am Chem Soc 119:5768–5769

Vaghari H, Jafarizadeh-Malmiri H, Mohammadlou MS et al (2016) Application of in smart enzyme immobilization. Biotechnol Lett 38:223–233

Van den Burg B, Vriend G, Veltman OR et al (1998) Engineering an enzyme to resist boiling. Proc Natl Acad Sci USA

Vangas J, Viesturs U, Fort I (1999) Mixing intensity studies in a pilot plant stirred bioreactor with an electromagnetic drive. Biochem Eng J 3:25–33

Wang Y, Zhe J, Chung BTF et al (2008) A rapid magnetic particle driven micromixer. Microfluid Nanofluid 4:375–389

Webb C, Kang H, Moffat G et al (1996) The magnetically stabilized fluidized bed bioreactor: a tool for improved mass transfer in immobilized enzyme systems? Chem Eng J 61:241–246

Woodward JR (2002) Radical pairs in solution. Prog React Kinet Mech 27:165–207

Xiu GH, Jiang L, Li P (2001) Mass-transfer limitations for immobilized enzyme-catalyzed kinetic resolution of racemate in a fixed-bed reactor. Biotechnol Bioeng 74:29–39

Yang K, Xu NS, Su WW (2010) Co-immobilized enzymes in magnetic chitosan beads for improved hydrolysis of macromolecular substrates under a time-varying magnetic field. J Biotechnol 148(2-3):119-127. 20 July 2010

Zheng M, Su Z, Ji X, Ma G, Wang P, Zhang S (2013) Magnetic field intensified bi-enzyme system with in situ cofactor regeneration supported by magnetic nanoparticles. J Biotechnol 168 (2):212-217. 20 October 2013

Chemical Processing of Switchgrass (*Panicum virgatum*) and Grass Mixtures in Terms of Biogas Yield in Poland

Karol Durczak, Mariusz Adamski, Piotr Tomasz Mitkowski, Waldemar Szaferski, Piotr Gulewicz and Włodzimierz Majtkowski

1 Introduction

In recent years there has been an increase in the interest in the use of plant biomass in many areas of renewable energy production (Parish and Fike 2005; Sanderson et al. 1996). Mainstreams of the investigations in its utilization as energy source are: the combustion, the thermochemical conversion, the production the ethanol or the production of the biogas (McLaughlin and Kszos 2005; Ahn et al. 2001). The development of waste biomass from agriculture and food processing is an opportunity to significantly reduce the amount of biodegradable waste in the country and also allows to acquire organic energy as well as, above all, renewable energy (Bullard and Metcalfe 2001). Recycling of waste biomass in oxygen and aerobic stabilization processes provides the possibility of implementing the European Union's recommendations on waste management and on the share of renewable energy in the country's energy balance. In case of inadequate organic substrates for anaerobic conversion, plant substrates are sought. We are looking for plants with a low cost of cultivation in relation to the yield. What aroused the researchers'

K. Durczak · M. Adamski (✉)
Institute of Biosystems Engineering, Faculty of Agriculture and Bioengineering, Poznan University of Life Sciences, Poznań, Poland
e-mail: mariusz.adamski@up.poznan.pl

P. T. Mitkowski · W. Szaferski
Institute of Chemical Technology and Engineering, Faculty of Chemical Technology, Poznan University of Technology, Poznań, Poland

P. Gulewicz
Department of Animal Nutrition and Feed Management, Bydgoszcz University of Technology and Life Sciences, Bydgoszcz, Poland

W. Majtkowski
Plant Breeding and Acclimatization Institute, Botanical Garden, Bydgoszcz, Poland

© Springer International Publishing AG, part of Springer Nature 2018
M. Ochowiak et al. (eds.), *Practical Aspects of Chemical Engineering*,
Lecture Notes on Multidisciplinary Industrial Engineering,
https://doi.org/10.1007/978-3-319-73978-6_6

interest is switchgrass and the growth of meadows and pastures (Pisarek et al. 2000; Majtkowski and Majtkowska 2000).

Grasses belonging to C_4 plants, in comparison to native grasses of C_3 carbon fixation pathway are better adapted to light soils conditions and can be the nutritious fodder and the substratum for the biogas production (Majtkowski et al. 2004). The switchgrass belongs to C_4 plants and is one of oldest cultivated plants. The significance of the switchgrass decreases along with the development of the agriculture.

From the previous state of knowledge, alkaline pretreatment, pre-treatment with strongly alkaline solution, can be used to control the production of biogas from maize straw by anaerobic digestion of organic matter. Various concentrations of sodium hydroxide [1, 2.5, 5.0 and 7.5% (m/m)] are used (Jiying et al. 2010). Prototype processing methods give a satisfactory result in the delivery of different forms of energy during the initial transformation of biomass (Hu et al. 2008). NaOH concentrations, as indicated in the literature, were tested for samples that contained maize straw. Degradation (decomposition) of lignin during pre-treatment, increased from 9.1 to 46.2% when NaOH concentration increased from 1.0 to 7.5%. Pre-prepared maize straw with NaOH was fermented and seeded with liquid effluent as a graft and a nitrogen source. Preparation of straw with a 1% NaOH concentration did not significantly improve the yield of biogas.

The highest biogas yield of 372.4 $Ndm^3 \cdot kg^{-1}$ of dry matter content was obtained at 5% concentration (content) of NaOH in pretreated maize straw which was 37.0% higher than that of unprocessed maize straw. However, a higher NaOH dose of 7.5% m/m resulted in faster production of volatile fatty acids during the hydrolysis and acidogenesis steps that inhibited methanogenesis. Simultaneous use of NaOH and anaerobic digestion did not significantly improve biogas production ($P > 0.05$) (Greenfield and Smith 1974).

Switchgrass (*Panicum virgatum*) has been devoted much attention in recent years. The cereal beetle has been the subject of bioenergetic research for over ten years. It is a very efficient North American plant, perennial, seasonal warm grass, suitable to grow on marginal, poor soils. As such, it has a great potential as a durable and bioenergetic plant (Bullard and Metcalfe 2001). Most of the research done so far focuses on the use of switchgrass by burning, thermochemical conversion or the production of cellulosic ethanol.

The biogas productivity of the switchgrass is evaluated at 187–212 $Ndm^3 \cdot kg^{-1}$ of dry organic matter (Gorisch and Helm 2006; Lemus et al. 2002). In the literature, the biogas productivity of switchgrass, treated by leaching with sodium hydroxide (NaOH), after conversion increased by 8–10%, depending on the sodium hydroxide concentration (Bullard and Metcalf 2001; Greenfiel and Smith 1974). It should be noted that these results correspond to the values of lye concentration only above 5% (Hu et all 2008; Lemus et all 2002).

It should be pointed out that optimization of biogas production takes place according to a group of technological and economic criteria. From an environmental point of view, it is also advisable to use waste materials to limit the share of potential feed and food products. Considering the above mentioned criteria, it is appropriate to use plant biomass derived from fallow, uncultivated or waste soils. This kind of solution requires the development of guidelines for the biogasization process.

From the point of view of development of grass for energy purposes, anaerobic method allows the use of fresh substrate or conditioning. It should be noted that the results of research on the chemical processing of plant substrates do not include blends, but only concern monosubstituted solid fractions. Minimizing the dosage of chemicals, the effects of the improved efficiency of biogas per unit mass of the mixture is a major objective of the research for many years (KTBL Heft 84 2009).

In Poland, there is an increasing interest in biogas installations. At the same time, biogas plants are intensively searching for new waste and vegetable substrates that can be used in terms of cost and technology.

2 Materials and Methods

2.1 Plant Material

Switchgrass (*Panicum virgatum*) was cultivated in Botanic Garden of the Plant Breeding and Acclimatization Institute in Bydgoszcz. Grass mixture consists of 60% of Perennial Ryegrass (*Lolium perenne*), 30% of white clover and 10% flavor herbs.

2.2 The Methane Fermentation

Research into the production of biogas was performed on the test stand, using eudiometric tanks (Fig. 1).

The biogas yield tests were carried out in accordance with DIN 38 414-S8 in a multi-chamber fermentation station (Fig. 2), based on an eudiometric system that stores the biogas generated on a 1 dm^3 fermentation tank (KTBL-Heft-84 2009). A measurement station for methane, carbon dioxide, hydrogen sulphide, oxygen, ammonia, nitric oxide and nitrogen dioxide was used for the biogas gas concentration test. For the preparation of inoculum a methanogenic thermostated biostat with a capacity of 1650 ml was used.

The fermentation station was equipped with a thermostatic tank, keeping the set thermal parameters of the process, fermentation tanks of 1 and 2 dm^3 and tanks for the storage of biogas with a capacity of 1200 ml. Biogas tanks are equipped with valves and connectors which allow to the remove the stored gas and the injection into the gas route equipped with biogas gas concentration analyzers (DIN 38414 s.8 2012).

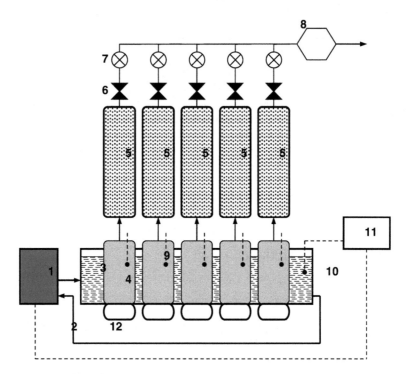

Fig. 1 Diagram of an eudiometric system for the research of biogas productivity of substrates. **1** water heater with temperature regulator, **2** insulated tubes for heating fluid, **3** water jacket with temperature control, **4** biofermentor with a capacity of 1 or 2 dm³, **5** biogas tank, **6** shut-off valves, **7** gas flowmeters, **8** gas analyzers (CH_4, CO_2, NH_3, H_2S, O_2, NO_X), **9** pH sensors, **10** temperature sensor, **11** registration control unit, **12** magnetic stirrers of contents

Measurements of concentration and volume of secreted gas were carried out at 24-hour intervals. A mixture of identical composition was placed in two biofermentors in order to improve the result correctness. MG-72 and MG-73 series measuring heads have been used for measurement of the biogas composition produced within the measuring ranges 0–100% of the volume and the measuring resolution in the order of 0.1 ppm to 1% volume.

Based on our own research and literature analysis (Jędrczak 2007; Myczko et al. 2011; Steppa 1988), these factors have been identified to characterize the fermentation pulp. Factors that may have a significant impact on the biogas production process include, but are not limited to, the dry substance content, organic matter content, batch weight, reaction rate, percentage of ingredients in the fermenting mix, and time from the start of the experiment. The following standards were used: PN-74/C-04540/00, PN-75/C-04616/01—04, PN-90 C-04540/01, AOAC 1995.

The parameter to be evaluated is the volume of generated biogas and the cumulative value (Dach et al. 2009). During the study the process temperature was 36 °C (AOAC 1995).

Fig. 2 Research stand for the study of biogas productivity of substrates according to DIN 38414 s.8 (left), inoculum station for quasi continuous fermentation work (right)

The object of the study was a mixture of solid and liquid substrates, subjected to anaerobic degradation. Cattle slurry and inoculum were also used for the study. The content of dry matter was set over at 6–8% m/m for the introduction into the process an increased dose of substrate representing the lignocellulose complex (Fugol and Szlachta 2010).

The mixture (approximately 6.28–7.44% of the dry matter content) of loose bovine slurry and switchgrass or grass fresh forage, treated and non-treated with NaOH, were substrates for methane fermentation. Before preparing mixture, silage of the switchgrass was chopped into the 3 cm pieces. The mixture was inoculated with post-fermentative pulp from biogas plant in Liszkowo (Amon 2007). Content of mixtures used for frementation is presented in Tables 1, 2, 3 and 4.

The methane fermentation was carried out in a water-jacked biofermentor with the thermostat at temperature 36°C with 1 min mixing every 2 h. Biogas was collected in eudiometrical containers filled with neutral displaced liquid. The level

Table 1 Parameters of mixture with switchgrass fresh forage ($N = 3$)

Constituent	Dry mass of constituent [%]	Contribution in mixture [%]	Contribution in mixture dry mass [g]	Contribution in mixture fresh mass [g]	pH
Slurry	2.9	69	39.5415	1363.5	
Inoculate	5.8	18	20.3174	350.3	Cond
Switchgrass	34	13	85.748	252.2	15.20 mS
Mixture	7.4		145.6069	1966.0	6.28

Table 2 Parameters of mixture with switchgrass fresh forge treated with NaOH ($N = 3$)

Constituent	Dry mass of constituent [%]	Contribution in mixture [%]	Contribution in mixture dry mass [g]	Contribution in mixture fresh mass [g]	pH
Slurry	2.9	69	39.5966	1365.4	
Inoculate	5.8	18	20.416	352.0	Cond
Switchgrass	34	13	85.816	252.4	18.45 mS
Mixture	7.4		145.8286	1969.8	7.26

Table 3 Parameters of mixture with grass fresh forge ($N = 3$)

Constituent	Dry mass of constituent [%]	Contribution in mixture [%]	Contribution in mixture dry mass [g]	Contribution in mixture fresh mass [g]	pH
Slurry	2.9	69	39.4052	1358.8	
Inoculate	5.8	18	20.3522	350.9	Cond
Switchgrass	29	13	72.674	250.6	12.22 mS
Mixture	6.8		132.4314	1960.3	6.81

Table 4 Parameters of mixture with grass fresh forge treated with NaOH ($N = 3$)

Constituent	Dry mass of constituent [%]	Contribution in mixture [%]	Contribution in mixture dry mass [g]	Contribution in mixture fresh mass [g]	pH
Slurry	2.9	69	39.5125	1362.5	
Inoculate	5.8	18	20.4798	353.1	Cond
Switchgrass	29	13	73.4425	253.3	22.95 mS
Mixture	6.8		133.4348	1968.9	7.44

of the liquid decreased with the increasing volume of collected biogas. Each container was connected with a set of gas analyzers (methane, ammonia, carbon dioxide and hydrogen sulphide detectors). Experiments were made in three replications.

2.3   Treating with NaOH

Glass containers (2 dm^3) were filled with grass mixture and switchgrass. 3.5% and 2% m/m of 1.25 M NaOH was added respectively to grass mixture and switchgrass and containers were tightly closed. Studies of biogas productivity began after 24 h.

2.4 Statistical Analysis

Data were expressed as the mean ± standard deviation of three indenpendent replicates. Data were subjected to mulifactor analysis of variance (ANOVA) using last-squares differences (LSD) test with the Statistica 8.0 Program.

3 Results

Content of cellulose, lignin and hemicellulose in switchgrass and grass mixture fresh forage is presented in Table 5. Switchgrass contains 42.19% of cellulose, 22.84% of lignin and 40.21 of hemicellulose. Content of these components differs slightly from that in grass mixture. Cumulative biogas production is presented on the Fig. 3. In the all cases increase of production biogas was increasing fast to 18–21 day and after these day dropped down.

Average daily production of biogas is depicted in Fig. 4. Production of biogas from mixture with switchgrass fresh forage stongly increased from the beging of the experiment. On the ninth day, the production achieved its peak at 2.10 dm^3. Subsequently, the biogas production dramatically decreased to 0.88 dm^3. Next, the gas production day by day went out irregurally achieving the value of 0.07 dm^3 on the 71st day. The curve of average daily gas production from mixture with mixture grass forage shows that in this case production of biogas had similar course. Production achieved its peak at the level of 1.86 dm^3 on the ninth day and decreased irregulary but more steadly achieving the value of 0.06 dm^3 in the 71st day. Treating switchgrass with NaOH results in longer and more stable average day production of biogas to the 20th day. The achieved peak of biogas production was lower than in the case of untreated forages—1.78 on the 7th day and 1.88 on the 10th day respectively for SNa and GNa.

Methane concentration in biogases produced in examined cases is depicted by Fig. 5. Concentration of methane in biogas from mixtures with switchgrass and grass forage exceeded 60% on the 4th day of experiment and maintained above this value to the end of the experiment. The content of methane in SNa and GNa biogases exceeding 60% was achieved respectively on the 7th and 6th day and also maintained above this value to the end of the experiment. The concentrations of methane reached their peaks on the days 47–52 (70%), 4 (65%), 25–26 (71%) and 24–26 (72%) respectively for S, SNa, G and GNa biogases. Cumulative methane

Table 5 Content of cellulose, lignin and hemicellulose in switchgrass and grass mixture fresh forage

Material	Cellulose [%]	Lignin [%]	Hemicellulose [%]
Switchgrass	42.19	22.84	40.21
Grass mixture	44.83	25.35	39.40

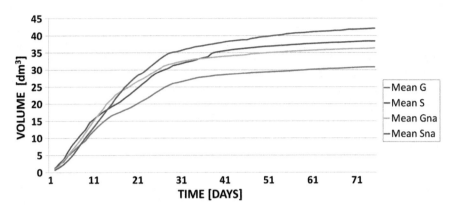

Fig. 3 Cumulative biogas production on 1 kg of mixture (*G* Grass, *S* Switchgrass, *GNa* mixed grass processed with NaOH, *SNa* mixed switchgrass processed with NaOH)

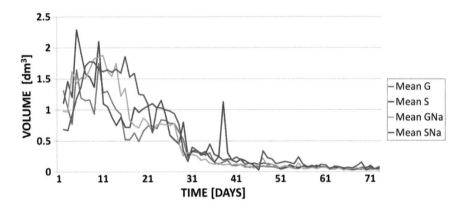

Fig. 4 Average daily production of biogas (*G* grass, *S* switchgrass, *GNa* mixed grass processed with NaOH, *SNa* mixed switchgrass processed with NaOH)

production is presented in Fig. 6. The correlation between cumulative biogas production and cumulative methane production can be observed—correlation coefficients 0.98; 0.99; 0.98; 0.98 for S, G, SNa and GNa respectively. Total methane volumes produced during fermentation was 25.84 and 17.95 dm^3 respectively for S and G biogases. Treating forages with NaOH resulted in the rise of these values to 25.50 and 21.67 dm^3.

Concentration of carbon dioxide in biogases is depicted in Fig. 7. From the 2nd day of fermentation, the concentration of carbon dioxide ranged from 14 to 45% and from 19 to 49% in S biogas and G, respectively. In case of forages treated with NaOH the ranges of CO_2 in biogas were following 13–41% and 12–45% for SNa and GNa biogases.

Figure 8 presents the concentrations of hydrogen sulphide. The highest concentrations of H_2S were observed in the first two weeks of fermentation. In case of S

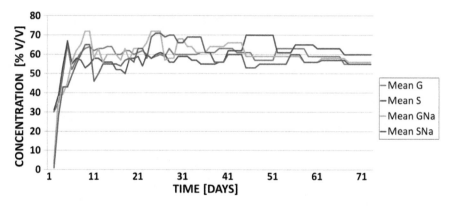

Fig. 5 Methane concentration in biogases [%] (*G* grass, *S* switchgrass, *GNa* mixed grass processed with NaOH, *SNa* mixed switchgrass processed with NaOH)

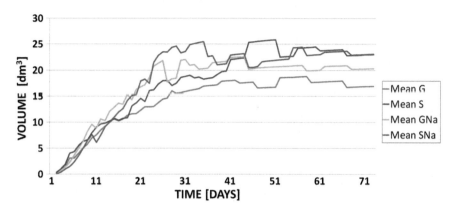

Fig. 6 Cumulative methane production [dm³] (*G* grass, *S* switchgrass, *GNa* mixed grass processed with NaOH, *SNa* mixed switchgrass processed with NaOH)

biogas the concentration achieved the value of 500 ppm and for G biogas 565 ppm. Treating forages with NaOH resulted in the growth of maximum concentration of H_2S to 525 and 625 ppm, respectively, in SNa and GNa biogas.

Concentrations of oxygen in biogases (Fig. 9) were the highest on the first day—2.9 and 2.3% respectively for S and G biogases. Afterwards, the concentration dropped reaching the concentration between 0.3 and 0.7% in these both biogases. In case of biogas produced for forages treated with NaOH dropping of O_2 was observed on the 3rd day. The minimum concentration was at the same level, 0.3%, for both biogases.

Figure 10 shows the concentration of ammonia in biogases. The concentration of ammonia in biogases was not stable and varied during the fermentation. The maximum concentration was observed in case of G biogas—9 ppm and the

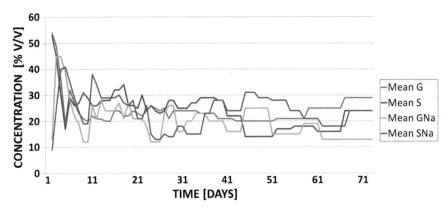

Fig. 7 Concentration of carbon dioxide in biogases [%] (*G* grass, *S* switchgrass, *GNa* mixed grass processed with NaOH, *SNa* mixed switchgrass processed with NaOH)

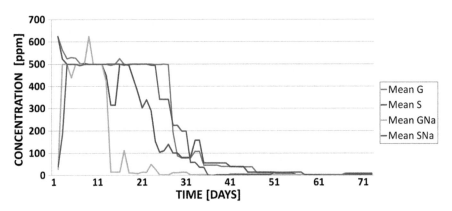

Fig. 8 Concentration of hydrogen sulphide [ppm] (*G* grass, *S* switchgrass, *GNa* mixed grass processed with NaOH, *SNa* mixed switchgrass processed with NaOH)

minimum one in SNa biogas. Treating with NaOH caused the maximum concentration of NH_3 to decrease.

NO_2 and NO concentrations in the biogases produced in all cases are shown in Figs. 11 and 12. Maximum contents of NO_2 were 5.82; 2.8 2.9; 2.7 ppm respectively for S, G, SNa and GNa biogases. The minimum was at the same level for all variants—0.1 ppm. The concentration of NO achieved its maximum (17 ppm). Treating with NaOH results in the decreasing maximum value of 9 ppm in SNa biogas. In case of G biogas the maximum concentration of 11 ppm was higher than in GNa biogas.

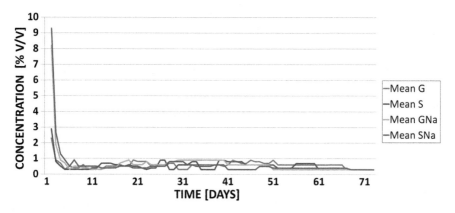

Fig. 9 Concentration of oxygen [%] (*G* grass, *S* switchgrass, *GNa* mixed grass processed with NaOH, *SNa* mixed switchgrass processed with NaOH)

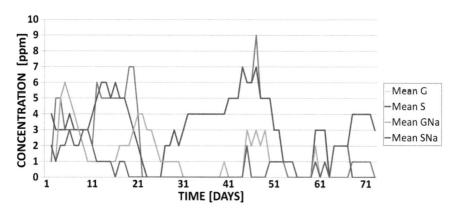

Fig. 10 Concentration of ammonia [ppm] (*G* grass, *S* switchgrass, *GNa* mixed grass processed with NaOH, *SNa* mixed switchgrass processed with NaOH)

Productivity of biogas and methane is presented in Table 6. The highest biogas and methane productivity expressed as amount of m^3 per 1 Mg of dry mass is observed in case of switchgrass treated with NaOH—290.27 and 208.99. Treatment with NaOH caused the productivity to increase in comparison to switchgrass forage —263.79 m^3 of biogas/Mg of dry mass and 195.20 of methane/Mg of dry mass. Similar rise of productivity was observed after treating grass mixture with NaOH. Biogas and methane productivity increased respectively from 23,265 m^3/Mg of dry mass to 273.47 m^3/Mg of dry mass and from 172.16 m^3/Mg of dry mass to 199.63 m^3/Mg of dry mass.

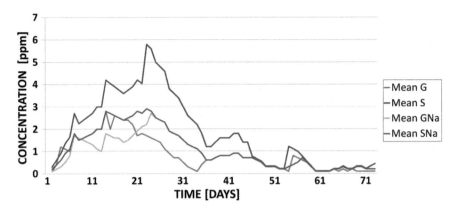

Fig. 11 NO$_2$ concentration in biogases [ppm] (*G* grass, *S* switchgrass, *GNa* mixed grass processed with NaOH, *SNa* mixed switchgrass processed with NaOH)

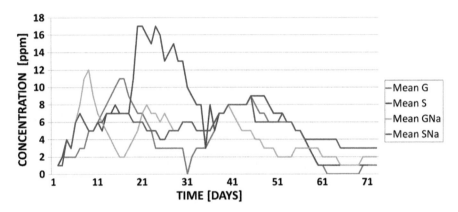

Fig. 12 NO concentration in biogases [ppm] (*G* grass, *S* switchgrass, *GNa* mixed grass processed with NaOH, *SNa* mixed switchgrass processed with NaOH)

4 Discussion

Cellulose, lignin and hemicellulose content in switchgrass is similar to grass mixture. For methane fermentation, forages of switchgrass and grass mixture were used. Using forages treated with NaOH was also examined. Hydrolyzing with NaOH resulted in enhancement of methanogenesis as literature data shows (Van Soest et al. 1991). High values of cumulative biogas production and average daily production on the first days of fermentation can be explained by preferential digestion of readily fermented chemical compounds like carbohydrates. This corresponds to earlier observations made by other scientists (Demirer and Chen 2008; Lu et al. 2007; Ahn et al. 2001; Hu et al. 2008). Longer persistence of higher values of average biogas production was observed after NaOH treating of both plant materials.

Table 6 Productivity of biogas and methane

Material unit		Productivity [m³/Mg]	
		Biogas	Methane
Switchgrass forage	Dry mass	263.79	195.20
	Fresh mass	19.54	14.46
	Dry organic mass	332.49	246.04
Swichgrass NaOH	Dry mass	290.27	208.99
	Fresh mass	21.49	15.47
	Dry organic mass	366.43	263.83
Grass mixture	Dry mass	232.65	172.16
	Fresh mass	15.72	11.63
	Dry organic mass	297.17	219.91
Grass mixture NaOH	Dry mass	273.47	
	Fresh mass	18.53	
	Dry organic mass	349.05	

Biogas is definied as mixture of gases. The content of methane decides about the biogas caloric value. Literature data indicate the lowest, useful limit of methane between 45 and 55%. In all examined groups, the maximum concentration of methane achieved or exceeded the value of 80%. High concentration of CO_2 dilutes biogas, decreasing its caloric value (Eder and Schulz 2007; Bolsen et al. 1996; Döhler 2009).

High emission of ammonia, which is a result of protein compounds degradation, proves also an oxygenic character of reactions at the initial stage. In case of plant material untreated with NaOH, there appears the second peak of ammonia concentration between the 40th and 50th day of fermentation in comparison to treated material. Treating the plant material with NaOH caused decreasing concentration of ammonia, especially between the 25th and the 51st day of fermentation.

Stable emission of nitrogen oxides proved the steady process of methane fermentation in regard to biology. Primary higher emission of NO and NO_2 is caused by higher initial concentration of oxygen in fermenting mixture. This can suggest that in an initial phase the fermentation can have oxygen character. Presence of NO_x is undesired because of harmful influence on the natural environment. Stated carbon dioxide concentrations are within the standards of content in biogas.

Strongly variable emission of ammonia and nitrogen oxides indicates the intensification of biomass decomposition processes also at the end of the fermentation process. Concentration of oxygen at the level of 0.1 to 1% V/V in produced biogas is prospective in the small scale agricultural systems of biogas production. A defined amount of oxygen is supplied to filters in some methods of decreasing the sulphide hydrogen concentration of desulphurisation installations located outside the fermentation chamber. The permissible concentration of sulphide hydrogen in biogas ranged between 18 and 20 ppm. Producers of current generators indicate that the concentration above this value may cause corrosion of engines. Ammonia and hydrogen sulphide are compounds with strong, unpleasant smell. Their emissions negatively influence the natural environment. There is a great interest in reducing odour pollution in rural area. Likewise, in animal production, the development of biogas production may be limited by emission of odour compounds and its insufficient controlling. On the other hand, anaerobic digestion helps to convert volatile organic compounds into less arduous compounds for environment (Mackie et al. 1998; McLaughlin and Walsh 1998). Shapes of obtained characteristics of biogas productivity showed the correct course of the methane process. The total biogas productivity of switchgrass as a result of processing with sodium hydroxide gave better effect than with similarly processed grass mixtures. Cumulative biogas productivity of switchgrass mixtures increased by about 10% v/v, as a result of chemical processing. Cumulative biogas productivity of grass mixtures, consequently to chemical processing, increased by approximately 17.5% v/v.

Studies have shown that NaOH processing allows biogas yields at low reagent concentrations (2–3.5% m/m) to improve. Alkaline conversion of biomass makes the fermentation process cause the increased emission of hydrogen sulphide between the 10th and 30th day of the process. Increased biogas productivity and biogas emission levels indicate an improvement in fibber digestibility without the need for additional thermal interference in the processing of plant substrates (Hu et al. 2008; Ahn et al. 2001; Lu et al. 2007; Jiying 2010).

References

Ahn HK, Smith MC, Konrad SL et al (2001) Evaluation of biogas production potential by dry anaerobic digestion of switchgrass—animal manure mixtures. Appl Biochem Biotechnol 160:965–975

Amon T (2007) Das Potenzial für Biogaserzeugung in Österreich. Beiratssitzung für Umwelt und soziale Verantwortung unserer Gesellschaft, Maria Enzersdorf Imprint

AOAC (1995) Official methods of analysis, 16th edn. Association of Official Analytical Chemists, Washington D.C

Bolsen KK, Ashbell G, Weinberg Z (1996) Silage fermentation and silage additives—review. Asian-Australas J Anim Sci 9(5):483–493

Bullard M, Metcalfe P (2001) Estimating the energy requirements and CO_2 emissions from production of the perennial grasses miscanthus, switchgrass and reed canary grass. ADAS Consulting Ltd

Dach J, Zbytek Z, Pilarski K et al (2009) Badania efektywności wykorzystania odpadów z produkcji biopaliw jako substratu w biogazowni. Technika Rolnicza Ogrodnicza Leśna 6:7–9

Demirer GN, Chen S (2008) Anaerobic biogasification of undiluted dairy manure in leaching bed reactors. Waste Manag 28:112–119

DIN 38414 s.8 (2012) German standard methods for the examination of water, waste water and sludge; sludge and sediments (group S); determination of the amenability to anaerobic digestion (S 8). DIN Deutches Institut fur Normung e.V., Berlin

Döhler H (red.)/KTBL 2009: Faustzahlen Biogas. 2. Auflage. Kuratorium für Technik und Bauwesen in der Landwirtschaft e.V. (KTBL), Darmstadt, Schauermanndruck GmbH, Gernsheim

Eder B, Schulz H (2007) Biogas Praxis. Okobuch Verlag und Versand GmbH, Staufen bei Freiburg

Fugol M, Szlachta J (2010) Zasadność używania kiszonki z kukurydzy i gnojowicy świńskiej do produkcji biogazu. Inżynieria Rolnicza 1(119):169–174

Görisch U, Helm M (2006) Biogasanlagen. Eugen Ulmer KG, Stuttgard (Hohenheim)

Greenfield SB, Smith D (1974) Diurnal variations of nonstructural carbohydrates in the individual parts of switchgrass shoots at anthesis. J Range Manag 27(6):466–469

Hu ZH, Wang YF, Wen ZY (2008) Alkali (NaOH) pretreatment of switchgrass by radio frequency-based dielectric heating. Appl Biochem Biotechnol 148(1–3):71–81

Jędrczak A (2007) Biologiczne przetwarzanie odpadów. Wydawnictwo Naukowe PWN, Warsaw

Jiying Z, Caixia W, Yebo L (2010) Enhanced solid-state anaerobic digestion of corn stover by alkaline pretreatment. Bioresour Technol 101(19):7523–7528

KTBL-Heft 84 (2009) Schwachstellen an Biogasanlagen verstehen und vermeiden. Kuratorium für Technik und Bauwesen in der Landwirtschaft e.V. (KTBL), Darmstadt, Druckerei Lokay, Reinheim

Lemus R, Brummer EC, Moore KJ et al (2002) Biomass yield and quality of 20 switchgrass populations in southern Iowa USA. Biomass Bioenerg 23(6):433–442

Lu S, Imai T, Ukita M et al (2007) Start-up performances of dry anaerobic mesophilic and thermophilic digestions of organic solid wastes. J Environ Sci 19:416–420

Mackie RI, Stroot PG, Varel VH (1998) Biochemical identification and biological origin of key odor components in livestock waste. J Anim Sci 76(5):1331–1342

Majtkowski W, Majtkowska G (2000) Ocena możliwości wykorzystania w Polsce gatunków traw zgromadzonych w kolekcji ogrodu botanicznego IHAR w Bydgoszczy. In: Polsko-Niemiecka konferencja nt. Wykorzystania trzciny chińskiej Miscantus. 27–29 September, Połczyn Zdrój

Majtkowski W, Majtkowska G, Piłat J et al (2004) Przydatność do zakiszania zielonki traw C-4 w różnych fazach wegetacji. Biuletyn IHAR 234:219–225

McLaughlin SB, Kszos LA (2005) Development of switchgrass (*Panicum virgatum*) as a bioenergy feedstock in the United States. Biomass Bioenerg 28:515–535

McLaughlin SB, Walsh ME (1998) Evaluating environmental consequences of producing herbaceous crops for bioenergy. Biomass Bioenerg 14:317–324

Myczko A, Myczko R, Kołodziejczyk T et al (2011) Budowa i eksploatacja biogazowni rolniczych. Wyd, ITP Warszawa-Poznań

Parrish J, Fike JH (2005) The biology and agronomy of switchgrass for biofuels. CRC Crit Rev Plant Sci 24(5–6):423–459

Pisarek M, Śmigiewicz T, Wiśniewski G (2000) Możliwości wykorzystania biomasy do celów energetycznych w warunkach polskich. In: Polsko-Niemiecka konferencja nt. Wykorzystania trzciny chińskiej Miscantus. 27–29 September, Połczyn Zdrój

Polish Standard. PN-74/C-04540/00 Oznaczenie zasadowości. Wydawnictwo Normalizacyjne, Warsaw

Polish Standard. PN-75/C-04616/01 Oznaczanie suchej masy osadu i substancji organicznych. Woda i ścieki. Badania specjalne osadów. Oznaczanie zawartości wody, suchej masy, substancji organicznych i substancji mineralnych w osadach ściekowych. Wydawnictwo Normalizacyjne, Warszawa

Polish Standard. PN-75/C-04616/04 Oznaczenie lotnych kwasów tłuszczowych. Wydawnictwo Normalizacyjne, Warszawa

Polish Standard. PN-90 C-04540/01 Woda i ścieki. Badania pH, kwasowości i zasadowości. Oznaczanie pH wód i ścieków o przewodności elektrolitycznej właściwej 10 μS/cm i powyżej metodą elektrometryczną. Wydawnictwo Normalizacyjne, Warszawa

Sanderson MA, Reed RL, McLaughlin SB et al (1996) Switchgrass as a sustainable bioenergy crop. Bioresour Technol 56:83–93

Steppa M (1988) Biogazownie rolnicze. IBMER, Warsaw

Van Soest PJ, Robertson JB, Lewis BA (1991) Methods for dietary fiber, neutral detergent fiber, and non-starch polysaccharides in relation to animal nutrition. J Dairy Sci 74(10):3583–3597

Analysis of Flow Through the Entry Region of a Channel with Metal Foam Packing

Roman Dyga, Małgorzata Płaczek, Stanisław Witczak
and Krystian Czernek

1 Introduction

One of many types of structural packing utilized in the design of process apparatus is the one based on the application of open-cell metal foams. Such foams have a structure formed by compact, empty, multi-wall blocks. These blocks are interconnected by open walls, while the edges of the cells form a skeleton of foam. The size of the ligaments is several times smaller from cell dimensions. Such a design provides high porosity of the foams, usually exceeding 90%, which is accompanied by the maintenance of a large specific surface. The high thermal conductivity of the foam skeleton and continuous layout of the skeleton ligaments lead to the limitation of the thermal resistance. The technology of open-cell foam manufacture provides adequate solutions needed to obtain this material from a variety of metal alloys. As a result, it is possible to apply foams even in an extreme range of operating conditions, in which high thermal and chemical resistance properties are require. In addition, the technology of applying catalysts on the surface of the cellular framework is also achieved to a satisfactory standard.

The above characteristics of open-cell metal foams lead to the recognition of these materials as an alternative to more common structural packing and porous ones applied in heat exchangers (Cookson et al. 2006; Ozmat et al. 2004), thermal energy storage systems and heat regenerators (Wang et al. 2007; Vadwala 2011; Tian and Zhao 2013) as well as chemical reactors, including catalytic reactors (Incera Garrido et al. 2008; Lévêque 2009; Tschentscher et al. 2011).

Metal foams to a relatively small extent also contribute to laminarization of the flow. This characteristic can be applied in reactors with short-channel packing.

This type of apparatus takes advantage of the fact that the intensity of mass and heat exchange assumes the highest intensity at the inlet of the packing, where the

R. Dyga (✉) · M. Płaczek · S. Witczak · K. Czernek
Faculty of Mechanical Engineering, Opole University of Technology, Opole, Poland
e-mail: r.dyga@po.opole.pl

© Springer International Publishing AG, part of Springer Nature 2018
M. Ochowiak et al. (eds.), *Practical Aspects of Chemical Engineering*,
Lecture Notes on Multidisciplinary Industrial Engineering,
https://doi.org/10.1007/978-3-319-73978-6_7

flow is more turbulent than after it assumes a steady state at a certain distance from the inlet of the channel.

Due to the scarcity of comprehensive descriptions of the flow behaviour at the entry region of the channels and apparatus with metal foam packing, the study was undertaken in this respect.

2 Simulation of the Flow in the Entry Region

The description of the phenomena occurring at the entry region of the channel is associated with the need to get to know the local values of the temperatures of the foam skeleton and fluid as well as fluid velocity and pressure. The small dimensions and irregular shape of the foam result in severe difficulties during the experimental determination of these quantities. The results of simulations were conducted side by side with the experiments conducted with regards to non-adiabatic fluid flow through channels with metal foam packing (Dyga and Troniewski 2015). Although the experiments did not directly involve the description of the phenomena occurring in the entry region of the channel, they proved satisfactory for the purpose of determination of the conditions for numerical simulations and validation of their results.

Both types of research (numerical and experimental) involved flow of air, water and Velol-9Q machine oil through a horizontal channel with the internal diameter of 20 mm. The channel was packed with open-cell metal foam with the pore density of 20 PPI (pores per inch), porosity $\varepsilon = 0.934$, cell diameter $d_c = 3.452$ mm and pore diameter $d_p = 1.094$ mm. The diameters of the pores and cells were established graphically on the basis of the analysis of microscopic images of the foam skeleton (Fig. 1b), conducted in accordance with the methodology described by Kamath et al. (2011).

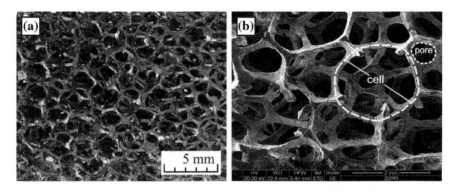

Fig. 1 AlSi7Mg alloy foam, **a** actual view, **b** microscopic view of foam skeleton

The channel was heated from outside by means of resistance heaters. The heating power was set so that the potential stepwise increase of the fluid temperature could be equal to at least 10 K. The experiments were realized in such range of fluids velocities change that ensured the occurrence of laminar and turbulent flow (Table 1). The Reynolds number Re_f describing the flow was defined by the equation,

$$Re_f = \frac{w_f d_p \rho_f}{(1 - \varepsilon)\mu_f}, \quad f \equiv a, ol, w \tag{1}$$

where: w_f denotes the mean velocity in the transverse cross-section of the empty channel (without the foam).

The Velol-9Q oil applied in the experiments was characterized by the following properties (at the temperature of 20 °C): viscosity $\mu_{ol} = 0.0086$ Pa s, density $\rho_{ol} = 859.8$ kg/m^3, specific heat $c_{ol} = 1848.8$ J/(kg K), thermal conductivity coefficient $k_{ol} = 0.128$ W/(m K). Concurrently, for the foam material, the following parameters were adopted: $c_s = 1848.8$ J/(kg K) and $k_s = 150.4$ W/(m K).

Due to the complex structure of the foam skeleton, numerical simulations were performed by application of a simplified geometrical foam model in the shape of Kelvin structure. This was formed by adjacent regular tetrakaidecahedron-shaped cells. The results reported by Boomsma et al. (2003), Bai and Chung (2011), and Dyga et al. (2013) confirm the possibility to achieve satisfactory results of numerical analysis conducted by application of a simplified model of metal foams. In addition, these models offer the considerable reduction of the computing power necessary to perform the necessary calculations.

The dimensions of tetrakaidecahedron were taken to be equal to the cell diameter (Fig. 2). The diameter of the skeleton fibre was selected so that the porosity of the Kelvin structure was equal to the foam porosity.

The length of the entry region subjected to the analysis was established on the basis of the insights from literature. Mostafida (2007) states that the unit pressure drop during flow through foams with various thickness assumes a constant value if the foam thickness is in the excess of 30 mm. This implies that such a value is sufficient to ensure hydrodynamic flow development. Dukhan and Suleiman (2014) indicate that the hydrodynamic development of flow occurs along a length of six cells. In the consideration of this fact, the length of the channel subjected to analysis was adopted to the 34.52 mm, which is equal to 10 times the size of a foam cell.

Table 1 Conditions of the experiment

Fluid f	Velocity w_f, m/s	Reynolds number Re_f,	Temperature of fluid t_f, °C
Air, a	0.028–9.88	23–13,244	21–95
Oil, ol	0.003–0.167	3–293	19–93
Water, w	0.003–0.270	30–4509	24–88

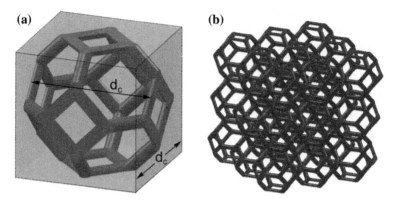

Fig. 2 Simplified geometric model of foam, **a** single cell, **b** Kelvin structure

The repeatability and symmetry of the cells in the Kelvin structure resulted in the possibility of additionally reducing the modelled region by application of two symmetry planes—XY plane and a one that is inclined at an angle 45° with relation to the first one (Fig. 3a). Some of the numerical calculations were realized in the periodic system layout, which comprised a section of the channel with the length of a single foam cell (Fig. 3b). In this case, the objective of the simulation involved the determination of the distribution of fluid velocities for developed flow. This distribution was later applied at the stage of interpretation of the velocity variations along the entry region.

The numerical calculations were conducted for the steady flow. Water and oil were considered as incompressible fluids, whereas air as an ideal gas. The fluid properties as well as the thermal conductivity and specific heat of the foam were taken to be relative to the temperature changes. Steady mass flux (m = const.) and constant fluid temperature (t = const.) were taken to represent the boundary conditions at the entry to the analyzed region. Fluid was discharged into an area with a given pressure value. In the periodic system, the external areas that were normal in relation to the flow direction played a dual role—as the fluid inlet and outlet—with the components of the velocity vector repeatable at the intervals given by d_c. An assumption regarding lack of fluid slip along the channel surface area and foam skeleton was assumed (components of the fluid velocity $u_x = u_y = u_z = 0$). For the symmetry planes, the Neumann condition was taken $\left(\frac{\partial u_x}{\partial n} = \frac{\partial u_y}{\partial n} = \frac{\partial u_z}{\partial n} = 0\right)$. Along the external channel wall surface, a constant value of the heat flux was adopted, q_h = const. Along the contact surface between the channel and foam with the fluid, an assumption regarding the continuity of the temperature and heat flux density q was adopted. The values taken as the initial and boundary conditions were observed during the course of the experimental research.

The results of the experiments indicate that the flow demonstrates laminar or transient characteristics along a wide range of the variable fluid velocities. Due to this, the numerical analysis was conducted on the basis of the turbulence model

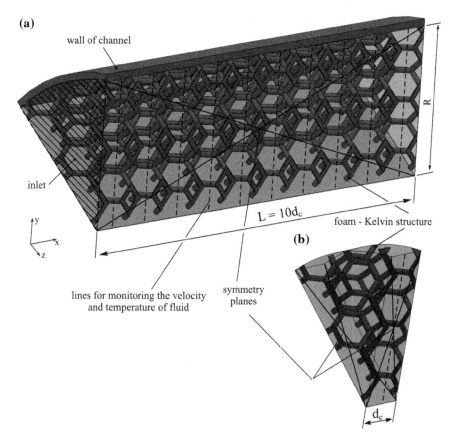

Fig. 3 Simulation region, **a** entry region, **b** periodic region (elementary flow section)

named "realizable $k - \varepsilon$", taking into account near-wall phenomena. Such a turbulence model serves to adequately describe the hydrodynamic phenomena accompanying flow with a small degree of turbulence and for the cases when the medium recirculates and fluid stream detaches from a surface with a considerable curvature. Such phenomena can occur near the ligaments in the foam skeleton. The basic relations of the "realizable $k - \varepsilon$" model are transport equations describing the kinetic energy of the turbulence k,

$$\frac{\partial}{\partial \tau}(\rho k) + \frac{\partial}{\partial x_j}(\rho k u_j) = \frac{\partial}{\partial x_j}\left[\left(\mu + \frac{\mu_t}{\sigma_k}\rho k u_j\right)\frac{\partial k}{\partial x_j}\right] + G_k + G_b - \rho\varepsilon - Y_M + S_k \quad (2)$$

and its dissipation rate ε,

$$\frac{\partial}{\partial \tau}(\rho \varepsilon) + \frac{\partial}{\partial x_j}(\rho \varepsilon u_j) = \frac{\partial}{\partial x_j}\left[\left(\mu + \frac{\mu_t}{\sigma_\varepsilon}\rho k u_j\right)\frac{\partial \varepsilon}{\partial x_j}\right] + \rho C_1 S\varepsilon + \rho C_2 \frac{\varepsilon^2}{k + \sqrt{v\varepsilon}}$$
$$+ C_{1\varepsilon}\frac{\varepsilon}{k}C_{3\varepsilon}G_b + S_\varepsilon \tag{3}$$

For the constants occurring in Eqs. (2) and (3), the following values were adopted: $\sigma_k = 1.0$; $\sigma_\varepsilon = 1.2$, $C_1 = 1.44$; $C_2 = 1.9$; whereas $C_1 = \max[0, 43; \eta/(\eta + 5)]$. The η parameter is derived from the relation,

$$\eta = \frac{k}{\varepsilon}S \tag{4}$$

$$S = \sqrt{2S_{ij}S_{ij}} \tag{5}$$

The G_k and G_b parameters represent the generation of turbulence kinetic energy due to the mean velocity and buoyancy and are defined by the relations,

$$G_k = \mu_t S^2 \tag{6}$$

$$G_b = b g_i \frac{\mu_t}{\text{Pr}_t}\frac{\partial t}{\partial x_i} \tag{7}$$

The Y_M parameter represents the contribution of the fluctuating dilatation in compressible turbulence to the overall dissipation rate. The turbulent viscosity μ_t, which complements the description of the energy dissipation, is defined by the relation,

$$\mu_t = \rho C_\mu \frac{k^2}{\varepsilon} \tag{8}$$

In the "realizable $k - \varepsilon$" model, a complex means of determination of a dimensionless quantity C_μ, is applied. The way in which C_μ, is derived along with many other parameters of the turbulence model applied in this paper (which explanation is not provided here), are more exhaustively presented in the ANSYS Fluent Theory Guide, Ansys Inc., 17.2.

The mathematical description of the flow is complemented by energy equations for fluids,

$$\rho c_f \frac{\partial}{\partial x_i}(u_i t_f) = \frac{\partial}{\partial x_j}\left(k_f \frac{\partial t_f}{\partial x_j}\right) \tag{9}$$

and for the foam skeleton and channel wall,

$$\frac{\partial}{\partial x_j}\left(k_s \frac{\partial t_s}{\partial x_j}\right) = 0 \tag{10}$$

In these equations, k_f and k_s denote, respectively, the thermal conductivity of the fluid and solid. For the case of turbulent flow, k_f is substituted by the effective thermal conductivity.

3 Analysis of the Numerical Research Results

The substitution of the metal foam by its simplified geometrical model results in the inevitable differences between the results of the data derived from simulations and the measurement corresponding to them. The pressure drop determined numerically and measured in experiment differ by a maximum of several per cent (the mean error is equal to 11.3%). The variability characteristics of the pressure drops derived from numerical calculations are the same as for the case of the measured values (Fig. 4). Due to the fact that the objective of the research was to determine the length of the entry region, it was assumed that the maintenance of the qualitative similarity between the results of the numerical analysis and experimental data is sufficient for the assessment of the flow. The region of developed flow is characterized by the stabilization of such parameters as fluid velocity, unit pressure drop and heat transfer coefficient.

On the periodic system, the distribution of the fluid velocity was obtained for the case of the developed flow (i.e. away from the entry region). Figure 5 represents the distribution of the local fluid velocity $u_f(y/R)$ along the central axis of the cells, which is marked in Fig. 3b in the x-y symmetry plane. The extremes of the fluid velocity are associated with the variable cross-sections of the foam cells.

The distribution of the local velocity obtained for the periodic system forms a reference applied for the assessment of hydrodynamic flow development in the

Fig. 4 Comparison between calculated and measured value of the pressure drop

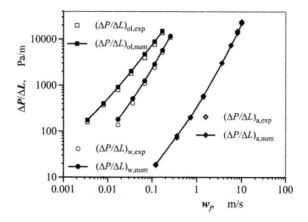

Fig. 5 Distribution of the local fluid velocity in vertical axis of the channel determined in the periodic system

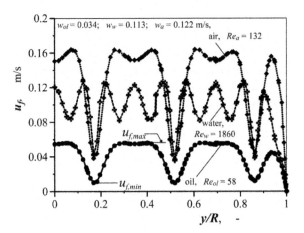

entry region of the channel. An assumption was adopted that along a given distance from the inlet to the channel, the radial distribution of the local fluid velocity takes a shape that was gained for the distribution in the periodic system.

The other criterion applied for the assessment of the length of the flow development region was associated with the analysis of the variability of the local fluid velocity along the central axis of the channel (x axis). The u_f velocity cyclically varies due to the occurrence of the regular variations of the cell cross-sections. Nevertheless, we can note a trend associated with the increase of the local velocity along with the greater distance from the inlet to the channel ($x/d_c = 0$—Fig. 6). The distance from the inlet for which the velocity in the channel axis no longer increases was adopted to represent the length of the hydrodynamic flow development region. For the case of the laminar flow, flow is developed in the third foam cell, i.e. for $x/d_c \approx 2.5$ (Fig. 6a).

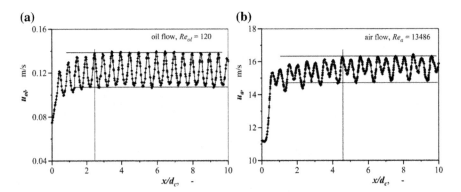

Fig. 6 Development of the local fluid velocity along the channel length, **a** laminar oil flow, **b** turbulent air flow

The conformity of the radial distribution of the velocity $u_f(y/R)$ with the one that was gained during the simulation of the flow through the periodic system has been noted at the same distance from the inlet. The increase of the flow turbulence leads to the greater length of the channel needed for flow development, nevertheless the stabilization of flow occurs relatively fast. For example in turbulent air flow ($Re_a = 13{,}486$), the velocity profile is set in the fifth foam cell ($x/d_c \approx 4.5$), which corresponds to a length of 15.5 mm. Hence, even for the case of the turbulent flow, the channel section needed for hydrodynamic flow development is shorter than the cross-section of the channel (20 mm).

The distribution of the temperature varies in a characteristic way in the entry region. In the vicinity of the channel axis, there is no temperature increase along the channel. The fluid heats up only near the channel wall. In this region, the fluid gains a greater temperature along the channel length (in the function of x/d_c). In the channel core, the temperature is the lowest and it is nearly constant. In the laminar flow, the fluid does not heat up in the area enclosed by the channel $y/R = 0$–0.4. For the case of the turbulent flow, this area occupies an even greater space, $y/R = 0$–0.7 (Fig. 7).

The small intensity of heat transfer from the channel wall to the laminar fluid results in a greater conducted heat through the foam skeleton. As a consequence, the foam temperature increases in the channel core and the temperature of the fluid flowing around the foam increases as well. In addition, the participation of the heat transfer through the foam increases along with the decrease in the thermal conductivity of the fluid.

The smaller ratio of the thermal conductivity of the fluid and foam material k_f/k_s and the smaller turbulence of the flow (Re_f), the greater contribution of the heat transfer through the foam. The ratio of the heat transfer to the fluid through the foam skeleton Q_{mf} to the heat delivered by the channel wall Q_k is visualized in Fig. 8. For the case when air is heated, the heat flux to the fluid through the foam is over a dozen times greater from the heat flux delivered by the wall of the channel despite the fact that the surface of the foam skeleton is only around 3.5 times greater

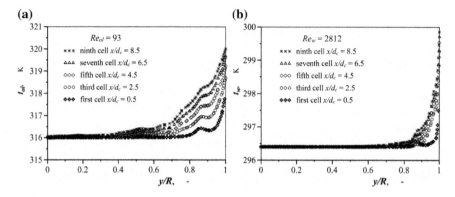

Fig. 7 Distribution of the fluid temperature along the section of the channel for various distances from the inlet (x/d_c): **a** laminar oil flow, **b** turbulent water flow

Fig. 8 Foam ratio in heat
transfer

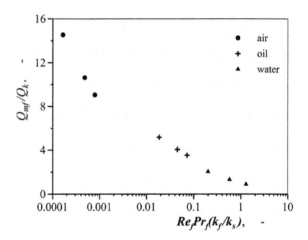

from the surface of the channel itself. For the case of water, the role of foam in the
heat transfer is considerably smaller and the Q_{mf}/Q_s ratio does not exceed the value
of 2 (under the experiment conditions).

The thermal development in the entry region was assessed on the basis of the
variability of the heat transfer coefficient h_f determined according to equation

$$h_f = \frac{q}{\bar{t}_s - \bar{t}_f} \qquad (11)$$

where: q denotes the heat flux density transferred from the internal channel surface
and foam skeleton to the fluid, t_f is the mean fluid temperature, and t_s is the mean
temperature of the channel surface and foam skeleton.

The heat transfer coefficient was derived independently for each of the ten cell
layers forming the total length of the entry region analysed. The h_f values determined numerically range from 60 to 220% greater in the entry region in comparison
to the ones that were determined during the experiment for the fully developed
flow. A considerably higher intensity of heat transfer at the inlet to the channel
forms a standard condition which is attributable to the flow development in the
entry region. The value of the heat transfer coefficient decreases monotonously
along with the greater distance from the channel inlet. In the analyzed region it does
not assume a constant level (Fig. 9). It means that in the flow there is no thermal
development.

The results gained during the experiment do not permit the statement regarding
the distance from the inlet to the channel at which the flow is thermally developed.
We can only state that the thermal development of the flow requires a considerable
length of the channel in comparison to the hydrodynamic development. The results
of further numerical research suggest that for the case of non-isothermal flow, the
entry region tends to be 4–5 times longer than for the case when the flow occurs
without heat transfer.

Fig. 9 Variations of heat transfer coefficient in the entry region

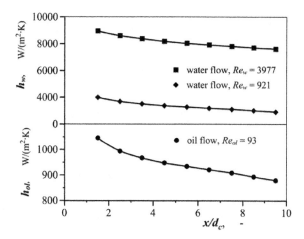

4   Summary

In a channel with open-cell metal foam, the hydrodynamic flow development occurs relatively quickly near the channel inlet. For the case of the turbulent flow, the distribution of the velocity is formed along a section with the length corresponding to five times the size of a foam cell. During laminar flow, this section is even shorter. The thermal development requires a several times greater length of the channel section. The length of the entry region for the non-isothermal flow has not been clearly defined to this date. The heat exchange in the entry region occurs directly between the channel wall and the heated fluid. Heat transfer occurs to a smaller extent through the foam itself. The role of the foam in heat transfer from the heated surface through the fluid is relative to the turbulence of the flow and thermal properties of the fluids.

The local fluid velocity demonstrates considerable variability due to the periodic changes in the cross-section of the foam cells. If the velocity changes in the areas occupied by single cells are to be neglected, we can state that the velocity and temperature of a fluid considered in relation to the channel dimensions are constant in the middle part of the channel. The variations of these values are only recorded near the channel wall. The impact of the wall over the velocity distribution is visible only at the distance not exceeding the diameter of a cell. The fluctuations of the fluid temperature are only discernible at the distance equal to three times the cell size, i.e. they reach as far as the half of the radius of the channel.

References

Bai M, Chung JN (2011) Analytical and numerical prediction of heat transfer and pressure drop in open-cell metal foams. Int J Thermal Sci 50:869–880

Boomsma K, Poulikakos D, Ventikos Y (2003) Simulations of flow through open cell metal foams using an idealized periodic cell structure. Int J Heat Fluid Flow 24:825–834

Cookson EJ, Floyd DE, Shih AJ (2006) Design, manufacture, and analysis of metal foam electrical resistance heather. Int J Mechanical Sciences 48:1314–1322

Dukhan N, Suleiman AS (2014) Simulation of entry-region flow in open-cell metal foam and experimental validation. Transp Porous Med 101:229–246

Dyga R, Witczak S, Filipczak G (2013) Symulacja przepływu cieczy przez piane metalowa FEC z wykorzystaniem periodycznego modelu geometrycznego piany. Inż Ap Chem 5:410–411

Dyga R, Troniewski L (2015) Convective heat transfer for fluids passing through aluminum foams. Arch Thermod 36(1):139–156

Incera Garrido G, Patcas FC, Lang S et al (2008) Mass transfer and pressure drop in ceramic foams: a description for different pore sizes and porosities. Chem Eng Sci 63:5202–5217

Kamath PM, Balaji C, Venkateshan SP (2011) Experimental investigation of flow assisted mixed convection in high porosity foams in vertical channels. Int J Heat Mass Transfer 54:5231–5241

Lévêque J, Rouzineau D, Prévost M et al (2009) Hydrodynamic and mass transfer efficiency of ceramic foam packing applied to distillation. Chem Eng Sci 64:2607–2616

Mostafid AM (2007) Entrance and exit effects on flow through metallic foams, Master Thesis, Department of Mechanical and Industrial Engineering, Concordia University, Montreal, Canada

Ozmat B, Leyda B, Benson B (2004) Thermal applications of open cell metal foams. Character Manufact Proc 19(5):839–862

Tian Y, Zhao CY (2013) Thermal and exergetic analysis of metal foam-enhanced cascaded thermal energy storage (MF-CTES). Int J Heat Mass Transfer 58:86–96

Tschentscher R, Schubert M, Bieberle A et al (2011) Tomography measurements of gas holdup in rotating foam reactors with Newtonian, non-Newtonian and foaming liquids. Chem Eng Sci 66:3317–3327

Vadwala PH (2011) Thermal energy storage in copper foams filled with paraffin wax, Master of Applied Science. Mechanical and Industrial Engineering University of Toronto, Toronto

Wang K, Ju YL, Lu XS et al (2007) On the performance of copper foaming metal in the heat exchangers of pulse tube refrigerator. Cryogenics 47:19–24

CFD Modelling of Liquid-Liquid Multiphase Slug Flow with Reaction

Jakub Dzierla, Maciej Staszak and Krzysztof Alejski

1 Microstructured Reactors

Slug flow as a flow of two immiscible liquid phases is characterized by high mass and heat transfer due to its double mechanism. High surface-to-volume ratio allows to obtain fast transfer by diffusion between two consecutive slugs, while shearing motion between slug and capillary wall creates internal circulation which enhances convection inside an individual slug (Kashid et al. 2007). A typical approach to generate slug flow is using a microstructured reactor (MSR), which has been demonstrated to be useful both for kinetic research and process intensification instruments (Kashid and Kiwi-Minsker 2009). Great transfer performances, uniform flow patterns in combination with precise control of residence time and small volume of potentially hazardous materials give possibility to obtain optimum yield and selectivity (Wang et al. 2013a, b; Dummann et al. 2003; Burns and Ramshaw 2001; Burns and Ramshaw 1999). MSRs are also considered to be a much more energy efficient solution in comparison to the traditional equipment, such as: stirred tanks, bubble, packed, plate columns, straight or coiled tubular reactors and static mixer reactors (Kashid et al. 2011). Efficiency of systems utilizing slug flow depends on various parameters: physical properties of both phases and differences between them (e.g. surface tension, density), parameters of reaction (such as kinetics, thermal condition, stoichiometry), post-reaction separation, mass and heat-transfer, slug length, residence time, reactor diameter and shape. In addition to that, slug length depends on flow velocity, geometry of channel, surface tension between two phases and material of channel (Kashid and Kiwi-Minsker 2009;

J. Dzierla · M. Staszak (✉) · K. Alejski
Institute of Technology and Chemical Engineering,
Poznan University of Technology, Poznań, Poland
e-mail: maciej.staszak@put.poznan.pl

© Springer International Publishing AG, part of Springer Nature 2018
M. Ochowiak et al. (eds.), *Practical Aspects of Chemical Engineering*,
Lecture Notes on Multidisciplinary Industrial Engineering,
https://doi.org/10.1007/978-3-319-73978-6_8

Burns and Ramshaw 2001). Some of these aspects are inherent and can be controlled only to a certain point, wherever the other ones create the complex relationships which make designing of such a system even more complicated. Optimization of microreactor systems can involve long and expensive experimental work. In order to attain the faster and cheaper one, there is a need for a model which would be able to predict both flow field and mass-heat transfer (Harries et al. 2003). Computational fluid dynamics (CFD) can be used to predict flow field and heat transfer with high precision for liquid-liquid systems such as MSRs (Harries et al. 2003).

MSRs have received increasingly more attention since 1990's due to their advantages in comparison to classical systems. Their advantages include: smaller size and reduced length of pipes, low energy input and higher energy efficiency, minimized waste and byproducts, low hold-up and precise control over thermal condition (Kashid and Kiwi-Minsker 2009). The main disadvantage of typical MSR is small cross section area of the channel which generates high pressure drop across the reactor. In consequence, a significant portion of required energy is used to counteract these effects. One way to solve this issue may be using the Y-shaped instead of T-shaped mixing element which provides less friction losses (Watts and Wiles 2007, 2012; Wiles and Watts 2011; Kashid et al. 2010). Another flaw of a small cross section area is the impossibility of using a solid catalyst covering the inner walls of channel (similar to chromatographic techniques), which limits the potential applications. To expand the possible applications more work on the channels with higher diameter (up to a few mm, where slug flow can be still observed) is needed, including both experimental and numerical investigations. Slightly bigger reactors can be also cheaper and easier to produce and work with, while maintaining their advantages over conventional reactors.

Microstructured reactors and segmented/slug flow in general have many potential applications. Table 1 presents several selected applications of liquid-liquid segmented flow with comparison to the parameters of microreactors used. There is also a possibility to use MSR for gas-liquid systems and for parallel flow. Overall, MSR can be used to enhance such processes as: gas absorption into liquid (Sobieszuk et al. 2010; Irandoust et al. 1992), non-reactive mixing (van Baten and Krishna 2004; Günter et al. 2004; Berčič and Pintar 1997), titration reaction (Dessimoz et al. 2008; Burns and Ramshaw 2001), online analysis (Antes et al. 2003), enzymatic reaction (Maruyama et al. 2003), phase transfer catalysis (Ahmed-Omer et al. 2007; Ueno et al. 2003), polymerization (Kiparissides 1996; Nakashima et al. 1991), fluorination reaction (de Mas et al. 2003; Chambers et al. 2001; Jahnisch et al. 2000), chlorination reaction (Ehrich et al. 2002), nitration (Jähnisch et al. 2004; Burns et al. 2002; Doku et al. 2001), oxidation reaction (Chambers et al. 2003), sulfonation reaction (Müller et al. 2005), extraction (Smirnova et al. 2006), transesterification, hydrolysis (Ahmed-Omer et al. 2006, 2007).

Table 1 Applications of MSR in liquid-liquid systems

Type of MSR	Application	Publications
Cross-shape glass microreactor with 380 μm width rectangular channel	Titration reaction	Burns and Ramshaw (2001)
Y- and T-junction glass microreactor with 269 and 400 μm width rectangular channel	Titration reaction	Dessimoz et al. (2008)
Silicon microreactor with nine parallel rectangular channel (250 μm width) with G-shaped mixing elements	Online analysis of nitration reaction	Antes et al. (2003)
Two sets of Y-junction chips with 100 and 40 μm width rectangular channels	Extraction and detection of carbaryl derivative	Smirnova et al. (2006)
T-junction borosilicate glass reactor with 200 μm width and 100 μm deep channel	Nitration of benzene	Doku et al. (2001)
Y-junction reactor with 500 μm width channel	Nitration of benzene	Burns et al. (2002)
Y-junction reactor with 500 μm width channel	Nitration of toluene	Burns et al. (2002)
Thirty two rectangular, parallel channels with 900 μm width and 60 μm depth	Vitamin precursor synthesis	Löwe and Ehrfeld (1991)
T-junction PMMA rectangular reactor (300 × 300 μm) and PTFE cylindrical reactor (300 μm internal diameter)	Hydrolysis of p-nitrophenol acetate	Ahmed-Omer et al. (2006)
T-junction PMMA rectangular reactor (300 × 300 μm) and Y-junction PMMA rectangular reactor (width 150 μm and depth 300 μm)	Hydrolysis of p-nitrophenol acetate	Ahmed-Omer et al. (2007)
Y-junction glass reactor with 100 μm width and 25 μm depth	Enzymatic reactions	Maruyama et al. (2003)
Y-junction Teflon cylindrical reactor with 200 μm internal diameter	Phase transfer catalysis	Ueno et al. (2003)
T-junction PMMA rectangular reactor (300 × 300 μm) and Y-junction PMMA rectangular reactor (width 150 μm and depth 300 μm)	Hydrolysis of p-nitrophenol acetate (1)	Ahmed-Omer et al. (2007)
Y-junction Teflon cylindrical reactor with 200 μm internal diameter	Benzylation reaction	Ueno et al. (2003)

2   Interfaces in Computational Fluid Dynamics

In the field of computational fluid dynamics the interface is treated as a surface geometrical manifold of an infinitesimal thickness which is located in the area of sudden change of physical and chemical parameters. The force field is affected by forces originating from the interface due to the molecular forces imbalance near and at its surface. The importance of the shape of the interface was realized and analyzed by Gibbs (1906). It was assumed that interface is of infinitesimal thickness and its curvature is related to pressure jump according to Laplace formula (2). The other convention is to treat the transition region as a composite of three volumes, namely two bulk phases and third volume of the interface itself. Such convention proposed by Guggenheim (1940) assumes that there is no step change in any property from phase to another phase but they are changing gradually. Despite the simpler convention of infinitesimal interface thickness in CFD technique, the influence of the interface reaches far more than the nearest neighbouring area, reconstructing very realistic image of the nature of the interface. From a purely thermodynamic point of view, the interface is described using the concept of surface free energy and thermodynamic excess functions. In case of equilibrium state and assuming constant temperature, pressure in both phases and amount of substance, the interfacial tension is defined as a Gibbs free energy derivative per unit area of the interface:

$$\gamma = \left. \frac{\partial G}{\partial A} \right|_{T,P,N} \tag{1}$$

The Gibbs free energy is defined as an excessive quantity which arises as a consequence of the interface existence. A general thermodynamical description of an evolving interface in the systems far from equilibrium was given by Gurtin and Voorhees (1996). This theory does not assume a linear relationship between fluxes and forces and does not limit to small departures from equilibrium.

On the other hand, mechanical treatment, which is also based on thermodynamic considerations, expresses the thermodynamic parameters in terms of pressure. Deep analysis and derivation of the mechanical expressions for interfacial parameters from their thermodynamic definition was given by Oversteegen et al. (1999). It is demonstrated that their approach is thermodynamically consistent and remains in accordance with generalized Laplace equation of capillarity. They derive the generalized Laplace equation based on the definition of thermodynamic grand potential defined as pressure volume integral over space of two phases containing interface. It was stated that the grand potential of a two-phase system is not a function of the position of the Gibbs dividing plane (Gibbs 1878) but of all its contributions. For CFD technique the interface description is much simpler and involves the standard form of Young-Laplace Eq. (2). The forces produced by the molecular movements of the molecules at the interface (surface tension forces) are expressed as a pressure jump across the interface. This pressure jump can be related to the surface tension and the curvature of the interface:

$$\Delta P = \gamma \left(\frac{1}{R_1} + \frac{1}{R_2} \right) \tag{2}$$

This equation, which describes the mechanical equilibrium of the interface, has proved to be very useful when modelling the behaviour of the interface surface. The interface curvature defined as a sum of reciprocals of the principal radii is, in fact, a simplification which is accurate for small curvatures. In fluid dynamics, the equation is used for force source term computation when estimating the velocity field in the system. The formulation of the CFD technique is based on several assumptions where the one of the most fundamental is that the volume of the simulated system is divided into a set of smaller volumes where the size of the small volumes constitutes the accuracy of the solution. Generally, such a technique known as finite element method, becomes, in the case of volumetric elements in the CFD case, the finite volume method.

The classical Navier-Stokes set of equations, which originate from Newton's Law of Motion, constitutes the basis of the method.

$$\frac{\partial(\rho\vec{v})}{\partial t} + \nabla \cdot (\rho\vec{v}\vec{v}) = -\nabla p + \nabla \cdot (\bar{\bar{\tau}}) + \rho g + \vec{F} \tag{3}$$

The mass conservation (or continuity equation) is used for total mass balance:

$$\frac{\partial\rho}{\partial t} + \nabla \cdot (\rho\vec{v}) = \Delta m \tag{4}$$

This differential equations set is valid for the entire fluid volume and when being applied onto the volumetric elements, the equations are discretized according to the chosen discretization scheme. For any scalar quantity φ, the discretization of the governing equation can be illustrated by the following equation written in the integral form:

$$\int_V \frac{\partial\rho\varphi}{\partial t}dV + \oint \rho\varphi\vec{v} \cdot d\vec{A} = \oint D_\varphi\nabla\varphi \cdot d\vec{A} + \int_V S_\varphi dV \tag{5}$$

In this equation the first term describes the rate of the change of mass accumulation in time in the given volume V. The second and third terms describe the mass transfer due to advection and diffusion, respectively. The last term is the source term, which describes the rate of the mass change in the volume V that results from the chemical reaction, phase mass transfer, or other processes. The discretization of the above equation on the given finite volume element leads to the following:

$$\frac{\partial \rho \varphi}{\partial t} V + \sum_{f}^{N_f} \rho_f \varphi_f \vec{v}_f \cdot \vec{A}_f = \sum_{f}^{N_f} D_\varphi \nabla \varphi_f \cdot \vec{A} + S_\varphi V \qquad (6)$$

The sums over the faces f substituted the surface integrals over the surfaces that enclose the volume. Equation (5) is valid for any finite volume element shape and that is assured by the use of the surface integrals. The discretized form (6) is valid only for the geometrical shape of the element for which the summation over the faces f can be done. Consequently, for Cartesian coordinate system, Eq. (6) can be used for the finite volumes that are constructed using planar faces in three dimensions, or straight edges in two dimensions.

The evolution of the interface due to the force and, consequently, the velocity fields can be simulated using several approaches in CFD. The front tracking methods (Unverdi and Tryggvason 1992), level set methods (Susman et al. 1994) and volume of fluid methods (Scardovelli and Zaleski 1999) are the most commonly used ones for interface modelling. For the purpose of interface capturing Eq. (7), which specifies the evolution of the volumetric amount of the phase contained by the cell, is introduced in addition to the above. This quantity is the volume fraction α of a given phase, and the sum of all phase volume fractions in a given volume must equal one.

$$\frac{\partial \alpha_q \rho_q}{\partial t} + \nabla \cdot \left(\alpha_q \rho_q \vec{v}_q \right) = S_{\alpha_q} + \sum_{p=1}^{n} \left(\dot{m}_{pq} - \dot{m}_{qp} \right) \qquad (7)$$

Equation (7) does not contain any diffusion term so its solution contains a discontinuity which is expressed as a sudden jump of computed quantity α. No numerical method for solving differential equations will give pure discontinuity in the solution. There is always a kind of numerical smoothing or sometimes numerical oscillations in the location where the discontinuity exists in the exact solution. This phenomenon is sometimes referred to as "numerical diffusion". Such a negative numerical phenomenon brings about the need to use higher order discretization schemes that are able to diminish artificial diffusion in the solution where it should not exist at all. Higher order discretization schemes should ensure that the numerical diffusion of the computed quantity is very small and the location of the interface can be reconstructed much more precisely. At first, the computational cells of the domain are initialized with values of every quantity, including also the phase volume fractions α. Because α may change in the range (0; 1) it is typical to assume that the space where α equals 0.5 locates the interface. This arbitrary assumption will give accurate results unless α changes its value from 0 to 1 along a very small distance, typically the size of one or two computational cells. The exact location of the interface is not known during calculation and is determined by an algorithm for geometrical reconstruction based on obtained results for every numerical timestep. There exist several algorithms that reconstruct the interface, having the distribution of α in the space. Noh and Woodward (1976), Hirt and Nichols (1981) and Jafari et al. (2007) presented several works on interface reconstruction. In practice, the SLIC (simple line interface calculation) and PLIC (Rider and Kothe 1998) (piecewise linear interface calculation) methods seem to be the most popular ones. The existence of spurious currents, negative numerical effect of interface location, is reported to be reduced by the use of PCIL (pressure calculation based on interface location) method (Jafari et al. 2007).

During interface reconstruction, the force source generated by the tension forces at the interface is a substantial quantity. Brackbill et al. (1992) proposed the *continuous surface force* method which is based on the assumption that the pressure jump across the interface is proportional to the curvature of the interface κ and the interfacial tension, according to the Young-Laplace equation:

$$p_2 - p_1 = \gamma \kappa \tag{8}$$

The planar nature of surface tension forces $\vec{F}_s$ must be expressed as the volumetric source term $\vec{F}$ in the Navier-Stokes Eq. (3). The continuous surface force method assumes that, for a control volume approaching zero, the volumetric force approaches the surface force:

$$\lim_{h \to 0} \int_{\Delta V} \vec{F} dV = \int_{\Delta A} \vec{F}_s dA \tag{9}$$

The above equation is accompanied by the assumption that the direction h and edges of the control volume are normal to the interface for the purpose of method derivation. Solving Eq. (9), applying additional mollifying interpolation function to smooth the changes in volume fraction α and locating the phases by phase density instead of α, one finally gets:

$$\vec{F} = \gamma \kappa \frac{\nabla \rho}{[\rho]} \frac{\rho}{\langle \rho \rangle} \tag{10}$$

The last term in Eq. (10) is the density correction term which is a weighting factor that does not change the total magnitude of the surface force but shifts the forces towards the region of higher density. This modification results in a more uniform acceleration field in the interface area. In this way, $\vec{F}$ defines the local force contribution applied to the point on the interface which appears in Eq. (3). The interface curvature κ is computed using formula (11):

$$\kappa = \nabla \cdot \tilde{n} \tag{11}$$

Assuming that the interface is normal to the gradient of α, the divergence of the interface normal unit vector $\tilde{n}$ is in fact the Laplacian (second order differential operator) of α:

$$\tilde{n} = \frac{\nabla\alpha}{|\nabla\alpha|} \tag{12}$$

The wall adhesion and, consequently, contact angle θ have a significant role in determining the fluid behaviour in the location of the fluid-fluid-wall contact line. It can be considered as a dynamic boundary condition which influences the curvature of the fluid-fluid interface near the wall. In the cells located at the wall, the interface curvature comes from the tension forces and additional wall adhesion forces. In the boundary cells, the latter is accounted for by the use of formula (13) for unit vector $\hat{n}$ normal to the interface at the wall contact point.

$$\hat{n} = \hat{n}_w \cos\theta_w + \hat{n}_t \sin\theta_w \tag{13}$$

where $\hat{n}_t$ is a normal vector to the contact line between the interface and the wall at the contact point and $\hat{n}_w$ is the unit wall normally directed to the wall.

3   CFD Modelling of a Multiphase Flow in MSR

The complexity of multiphase flow phenomena makes a multiphase flows modelling a challenge. There are sevearal approaches to this problem such as: Volume of Fluid (VOF) (Hirt and Nichols 1981), front-tracking method (Tryggvason et al. 2001), level set method (Sussman et al. 1994), particle-mesh method (Liu et al. 2005), lattice Boltzmann method and constrained interpolation profile (CIP) method (Yabe et al. 2001). The detailed review of the computational methods for multiphase flow CFD simulation is provided by Tang et al. (2004). Other techniques of interface reconstruction (Rider et al. 1998) and surface tension (Brackbill et al. 1992) modelling can be found in literature. Table 2 presents the comparison of MSR simulations with highlighted model, MSR type and simulation type.

4   Slug/Segmented Flow Generation and Mass Transfer in Microstructured Reactors

Mechanism of slugs generation goes as follows: one phase is flowing into reactor (Fig. 1a), cutting up flow of the second phase (Fig. 1b), which leads to an increase in pressure in blocked inlet-channel (Fig. 1c) and eventually to reverse the process (Fig. 1d). In the absence of mixing elements, mass transfer can be obtained only by

Table 2 Comparison of CFD simulations of liquid-liquid flow in microstructured reactors

Model	MSR type	Simulation type	Reaction	Publications
Volume of fluid, continuum surface force	Y-junction	Steady, 2D	Non-reacting system	Wang et al. (2013a, b)
Continuum surface force	Two adjacent fluid segments	Transient, 2D	Neutralization	Harries et al. (2003)
Piecewise linear interface calculation, continuum surface force	T-junction	Transient, 2D	Non-reacting system	Kashid et al. (2010)
Volume of fluid, continuum surface force	Y-junction	Transient, 2D	Non-reacting system	Kashid et al. (2007)
Volume of fluid	Y-junction	Transient, 2D	Non-reacting system	Wang et al. (2008)
Piecewise linear interface calculation, continuum surface force	T-junction	Transient, 2D	Non-reacting systems	Wang et al. (2013a, b)

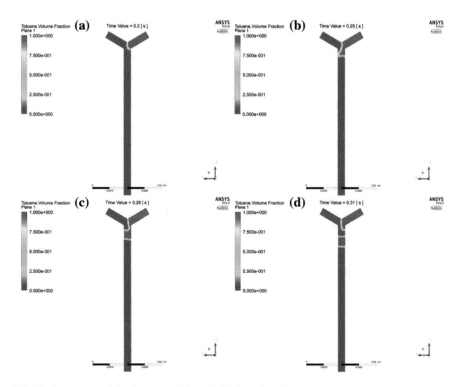

Fig. 1 Four steps of slug/segmented flow in Y-shape junction

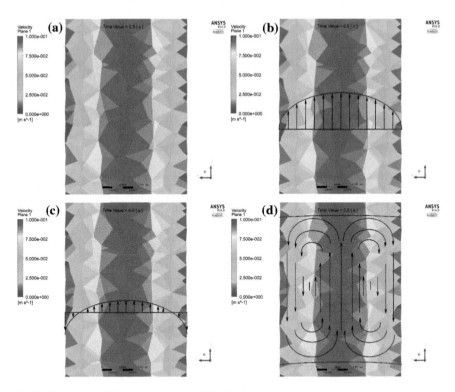

Fig. 2 Process of mixing and velocity field inside single slug

diffusion between two consecutive slugs through the interfacial surface (ends of slugs). In result, concentration gradient is axial and the mass transfer is slowed down. However, shear stress between stationary liquid near the channel wall and liquid in the slug axis generates internal circulation. The velocity gradient inside the slug is showed below (Fig. 2a). Shear stress slows down liquid near the walls, which results in big difference between liquid velocity close to the axis of the channel and the part of the liquid near the wall (Fig. 2b). Comparing liquid velocity inside the slug with velocity of whole slug, it can be observed that part of liquid far from axis flows from the front of slug to its back (Fig. 2c). This provides good mixing inside every slug (Fig. 2d) and creates a concentration gradient in radial direction. Ultimately, this internal circulation is a huge boost to the overall mass transfer.

5 Different Approaches to the Mass Transfer Simulations

Microstructured reactors are mainly used to enhance mass transfer in systems with two immiscible phases. If the reagents are initially dissolved in different phases, reaction can occur only near the interfacial surface. Lowering the concentration of one of the reagents near the interfacial surface results in creating concentration gradient between slug core and front (and also back) of the slug. Eventually whole reaction in slug flow can be divided into three parts: diffusion from the core of the phase to the interfacial surface, diffusion between phases and chemical reaction itself. Usually, it can be assumed that reaction and diffusion between phases are much faster than diffusion inside the slug.

Presentation of reagents concentrations in simulation volume as scalars have some advantages and disadvantages. Scalars are additional variables which can be regarded as simple mathematical information defined by user, in many cases independent of another parameters and models used in simulation. As a result the number of equations to balance is lower in comparison to full-defined components of the mixture and eventually whole model is simplified. On the other hand, every user-defined scalar is connected to one of the simulation threads (primary phase, secondary phase or mixture phase) and cannot be directly transported to the other thread. On top of this, "numerical diffusion" of every scalar works in way which results in averaging value of scalar in the entire thread volume which corresponds with single slug. It can be controlled by setting up values of the scalars diffusivity which corresponds to the dynamic viscosity. This results in three different possibilities of setting scalars.

Approach A:

One of the reagents can be assigned to the organic thread, which corresponds to its initial location, where the second one and the products of reaction are assigned to the water thread. In this approach reaction occurs only near the interfacial surface, where both thread are present. Diffusion of reagent inside organic slug and products inside water slug can be controlled only to certain point by the diffusivity values. At the end, diffusion inside independent slugs, diffusion between consecutive slugs and kinetic of reaction must be contained in effective reaction rate constant.

Approach B:

In this approach, a reagent that can diffuse between phases is assigned to the mixture thread, which results in unlimited numerical diffusion of this reagent between both phases. Consequently reaction occurs in whole water thread where second reagent can be found. This is similar to the nature of neutralization reaction, where acid is transferred from organic phase to water phase and then reacts with base, but because of numerical origins of diffusion in simulation, reaction rate calculated at the end of simulation can be less precise. Finally, like in approach A, both diffusion and kinetics of reaction must be contained in effective reaction rate constant.

Approach C:

There is a possibility to independently describe diffusion between phases and kinetics of reaction itself. In this approach one extra scalar variable is needed because reagent that can diffuse between phases will be assigned to the water thread and separately to the organic thread as two independent user-defined scalars (UDS). Diffusion of first reagent from water to organic through interfacial surface can be described as a transformation between reagent in water phase and reagent in organic phase due to the specific organization of data in the memory by the multiphase approach. To obtain full description, the area of interfacial surface must be calculated during simulation as well. After addition of this step, the whole chemical reaction is set up in water phase. Eventually, this approach results in full separate description of all three steps of a typical reaction in segmented flow: diffusion from the core of the phase to the interfacial surface is described by scalar diffusivity in phase, diffusion between phases is described as a transformation of reagent in one phase to the same reagent in the second phase and reaction itself described by its kinetics. It is also worth mentioning that this approach is more complicated than the others and from an experimental point of view, all data needed are more difficult to obtain.

References

Ahmed-Omer B, Barrow D, Wirth T (2006) Enhancement of reaction rates by segmented fluid flow in capillary scale reactors. Adv Synth Catal 348(9):1043–1048

Ahmed-Omer B, Barrow D, Wirth T (2007) Effect of segmented fluid flow, sonication and phase transfer catalysis on biphasic reactions in capillary microreactors. Chem Eng J 135(1):S280–S283

Antes J, Boskovic D, Krause H et al (2003) Analysis and improvement of strong exothermic nitration in microreactors. Chem Eng Res Des 81(7):760–765

Berčič G, Pintar A (1997) The role of gas bubbles and liquid slug lengths on mass transport in the Taylor flow through capillaries. Chem Eng Sci 52(21–22):3709–3719

Brackbill JU, Kothe DB, Zemach C (1992) A continuum method for modeling surface tension. J Comput Phys 100(2):335–354

Burns JR, Ramshaw C (1999) Development of a microreactor for chemical production. Chem Eng Res Des 77(3):206–211

Burns JR, Ramshaw C (2001) The intensification of rapid reactions in multiphase systems using slug flow in capillaries. Lab Chip 1(1):10–15

Burns JR, Ramshaw C (2002) A microreactor for the nitration of benzene and toluene. Chem Eng Commun 189(12):1611–1628

Chambers RD, Rolling D, Spink RCH et al (2001) Elemental fluorine part 13: gas-liquid thin film microreactors for selective direct fluorination. Lab Chip 1(2):132–137

Chambers RD, Holling D, Rees AJ et al (2003) Microreactors for oxidations using fluorine. J Fluorine Chem 119(1):81–82

de Mas N, Günter A, Schmidt MA et al (2003) Microfabricated multiphase reactors for the selective direct fluorination of aromatics. Ind Eng Chem Res 42(4):698–710

Dessimoz AL, Cavin L, Renken A et al (2008) Liquid-liquid two-phase flow patterns and mass transfer characteristics in rectangular glass microreactors. Chem Eng Sci 63(16):4035–4044

Doku GN, Haswell SJ, McCreedy T et al (2001) Electric field-induced mobilization of multiphase solution system based on the nitration of benzene in micro reactor. Analyst 126(1):14–20

Dummann G, Quittmann U, Gröschel L et al (2003) The capillary-microreactor: a new reactor concept for the intensification of heat and mass transfer in liquid-liquid reactions. Catal Today 79(1–4):433–439

Ehrich E, Linke D, Morgenshweis K et al (2002) Application of microstructured reactor technology for the photochemical chlorination of alkylaromatics. Chimia 56(11):647–653

Gibbs JW (1878) On the equilibrium of heterogeneous substances. Trans Conn Acad Art Sci 3:343–524

Gibbs JW (1906) Thermodynamics, vol 1. Longmans Green and Co., London New York and Bombay

Guggenheim EA (1940) The thermodynamics of interfaces in systems of several components. Trans Faraday Soc 35(3):397–412

Günter A, Khan SA, Thalmann M et al (2004) Transport and reaction in microscale segmented gas-liquid flow. Lab Chip 4(4):278–286

Gurtin ME, Voorhees PW (1996) The thermodynamics of evolving interfaces far from equilibrium. Acta Mater 44(1):235–247

Harries N, Burns JR, Barrow DA et al (2003) A numerical model for segmented flow in a microreactor. Int J Heat Mass Tran 46(17):3312–3322

Hirt CW, Nichols BD (1981) Volume of fluid (VOF) method for the dynamics of free boundaries. J Comput Phys 39(1):201–225

Irandoust S, Ertle S, Andersson B (1992) Gas-liquid mass transfer in Taylor flow through a capillary. Can J Chem Eng 70(1):115–119

Jafari A, Shirani E, Ashgriz N (2007) An improved three-dimensional model for interface pressure calculations in free-surface flows. Int J Comput Fluid D 21(2):87–97

Jähnish K, Baerns M, Hessel V et al (2000) Direct fluorination of toluene using elemental fluorine in gas/liquid microreactors. J Fluorine Chem 105(1):117–128

Jähnish K, Hessel V, Löwe H et al (2004) Chemistry in microstructured reactors. Angew Chem 43 (4):406–446

Kashid MN, Kiwi-Minsker L (2009) Microstructured reactors for multiphase reactions: State of the art. Ind Eng Chem Res 48(14):6465–6485

Kashid MN, Platte F, Agar DW et al (2007) Computational modelling of slug flow in a capillary microreactor. J Comput Appl Math 203(2):487–497

Kashid MN, Renken A, Kiwi-Minsker L (2010) CFD modelling of liquid-liquid multiphase microstructured reactor: slug flow generation. Chem Eng Res Des 88(3):362–368

Kashid MN, Renken A, Kiwi-Minsker L (2011) Gas-liquid and liquid-liquid mass transfer in microstructured reactors. Chem Eng Sci 66(17):3876–3897

Kiparissides C (1996) Polymerization reactor modeling: a review of recent developments and future directions. Chem Eng Sci 51(10):1637–1659

Liu J, Koshizuka S, Oka Y (2005) A hybrid particle-mesh method for viscous, incompressible, multiphase flows. J Comput Phys 202(1):65–93

Löwe H, Ehrfeld W (1991) State of the art in microreaction technology: concepts, manufacturing and applications. Electrochim Acta 44(21):3679–3689

Maruyama T, Sotowa JI, Kubota F et al (2003) Enzymatic degradation of p-chlorophenol in a two-phase flow microchannel system. Lab Chip 3(4):308–312

Müller A, Cominos V, Hessel V et al (2005) Fluidic bus system for chemical process engineering in the laboratory and for small-scale production. Chem Eng J 107(1–3):205–214

Nakashima T, Shimizu M, Kukizaki M (1991) Membrane emulsification by microporous glass. Key Eng Mat 61–62:513–516

Noh WF, Woodward PR (1976) Slic (simple line interface method). Lect Notes Phys Berlin/New York 59:330–340

Oversteegen SM, Barneveld PA, van Male J et al (1999) Thermodynamic derivation of mechanical expressions for interfacial parameters. Phys Chem Chem Phys 1(21):4987–4994

Rider WJ, Kothe DB (1998) Reconstructing volume tracking. J Comput Phys 141(2):112–152

Scardovelli R, Zaleski S (1999) Direct numerical simulation of free-surface and interfacial flow. Annu Rev Fluid Mech 31:567–603

Smirnova A, Mawatari K, Hibara A et al (2006) Micro-multiphase laminar flows for the extraction and detection of carbaryl derivative. Anal Chim Acta 558(1–2):69–74

Sobieszuk P, Pohorecki R, Cyganski P et al (2010) Marangoni effect in a falling film microreactor. Chem Eng J 164(1):10–15

Sussman M, Smereka P, Osher S (1994) A level set approach for computing solutions to incompressible two-phase floe. J Comput Phys 114(1):146–159

Tang H, Wrobel LC, Fan Z (2004) Tracking of immiscible interfaces in multiple-material mixing processes. Comp Mater Sci 29(1):103–118

Tryggvanson G, Bunner B, Esmaeeli A et al (2001) A front-tracking method for the computations of multiphase flow. J Comput Phys 169(2):708–759

Ueno M, Hisamoto H, Kitamori T et al (2003) Phase-transfer alkylation reactions using microreactors. Chem Commun 9(8):936–937

Unverdi SO, Tryggvason G (1992) Computations of multi-fluid flows. Physica D 60(1–4):70–83

van Baten JM, Krishna R (2004) CFD simulations of mass transfer from Taylor bubbles rising in circular capillaries. Chem Eng Sci 59(12):2535–2545

Wang X, Hirano H, Okamoto N (2008) Numerical investigation on the liquid-liquid two phase flow in a Y-shaped microchannel. Anziam J 48:963–976

Wang X, Hirano H, Xie G et al (2013a) VOF modeling and analysis of the segmented flow in Y-shaped microchannels for microreactor systems. Adv High Energy Phys 2013:1–6

Wang X, Hirano H, Xie GN (2013b) A PLIC-VOF based simulation of water-organic slug flow characteristics in a T-shaped microchannel. Adv Mech Eng 5:987428

Watts P, Wiles C (2007) Recent advantages in synthetic micro reaction technology. Chem Commun 5:443–467

Watts P, Wiles C (2012) Micro reactors, flow reactors and continuous flow synthesis. J Chem Res 36(4):181–193

Wiles C, Watts P (2011) Recent advantages in micro reaction technology. Chem Commun 47(23):6512–6535

Yabe T, Xiao F, Utsumi T (2001) The constrained interpolation profile method for multiphase analysis. J Comput Phys 169(2):556–593

Effect of Blade Shape on Unsteady Mixing of Gas-Liquid Systems

Sebastian Frankiewicz and Szymon Woziwodzki

1 Introduction

Mechanical mixing is one of most commonly used unit operations in the industry. It is conducted usually within turbulent flow regime in stirred vessels equipped with baffles. This mixing method in some processes is not recommended due to the presence of baffles. This applies to, for example, the pharmaceutical industry where particular attention is paid to cleanliness of apparatus (Yoshida et al. 2001a; Woziwodzki 2017). Turbulent mixing in stirred vessels without baffles causes many problems due to the primary circulation which results in lower mixing power and longer mixing time. For this reason, one of the major aspects in such systems is to improve the intensity and efficiency of mixing. Few methods can be used to achieve this, such as: eccentric positioning of the impeller and unsteady motion. In first solution, the impeller's position $E/R \approx 0.5$ generates higher power demand and improved intensity and efficiency of mixing (Karcz et al. 2005; Montante et al. 2006; Woziwodzki et al. 2010; Woziwodzki and Jędrzejczak 2011; Ng and Ng 2013). In the second solution, the unsteady motion of the impeller can be done in two ways: by reciprocating motion and by unsteady rotation of impeller. During reciprocating motion the impeller moves along vertical axis of the stirred vessel. These are usually disk impellers, disk impellers with flapping blades or impellers with complex dimensions depending on oscillation amplitude (Masiuk 1999, 2000, 2001; Komoda et al. 2000, 2001; Kamieński and Wójtowicz 2001, 2003; Masiuk and Rakoczy 2007; Masiuk et al. 2008; Wójtowicz 2012; Kordas et al. 2013).

In case of unsteady rotation of the impeller in a stirred vessel there is no formation of a central vortex but higher mixing turbulence with greater stress around the impeller is observed (Yoshida et al. 2008, 2009, 2010; Woziwodzki 2017).

S. Frankiewicz · S. Woziwodzki (✉)
Institute of Chemical Technology and Engineering,
Poznan University of Technology, Poznań, Poland
e-mail: szymon.woziwodzki@put.poznan.pl

© Springer International Publishing AG, part of Springer Nature 2018
M. Ochowiak et al. (eds.), *Practical Aspects of Chemical Engineering*,
Lecture Notes on Multidisciplinary Industrial Engineering,
https://doi.org/10.1007/978-3-319-73978-6_9

Unsteady rotation can be performed in many ways but given the structural limitations, usually the sinusoidal, triangular and rectangular time-course of impeller speed is used.

2   Basic Equations

To describe forces that occur in unsteady mixing, Morison equation is used (Morison 1953) determining torque T on the shaft of a stirred vessel (Woziwodzki 2017)

$$T = \frac{1}{16}D^5 C_1 \rho C_D |\omega|\omega + \frac{\pi D^5 \rho}{16}C_1 C_I \frac{d\omega}{dt} \tag{1}$$

where D is impeller diameter, ω is angular speed of impeller, C_D is drag coefficient and C_I inertia coefficient.

By knowing the impeller speed and so the angular speed, equations describing the given type of unsteady motion are obtained, such as, for triangular time-course of impeller speed (Woziwodzki 2017)

$$
\begin{aligned}
T = {} & \frac{16}{\pi^2}C_1 C_D N_{\max}^2 D^5 \rho \left| \sin(2\pi ft) - \frac{1}{9}\sin(6\pi ft) + \frac{1}{25}\sin(10\pi ft) \right| \\
& \times \left(\sin(2\pi ft) - \frac{1}{9}\sin(6\pi ft) + \frac{1}{25}\sin(10\pi ft) \right) \\
& + D^5 \rho C_1 C_I N_{\max}\frac{d\left(\sin(2\pi ft) - \frac{1}{9}\sin(6\pi ft) + \frac{1}{25}\sin(10\pi ft)\right)}{dt}
\end{aligned} \tag{2}
$$

where $N_{\max}$ is the maximum impeller speed and f is the oscillation frequency.

Finding of the torque allows to determine the mixing power variation over time (Woziwodzki 2017)

$$
\begin{aligned}
P = {} & \frac{32}{\pi}C_1 C_D N_{\max}^3 D^5 \rho \left| \sin(2\pi ft) - \frac{1}{9}\sin(6\pi ft) + \frac{1}{25}\sin(10\pi ft) \right| \\
& \times \left(\sin(2\pi ft) - \frac{1}{9}\sin(6\pi f)t + \frac{1}{25}\sin(10\pi ft) \right) \\
& + 2\pi D^5 \rho C_1 C_I N_{\max}^2 \frac{d\left(\sin(2\pi ft) - \frac{1}{9}\sin(6\pi f)t + \frac{1}{25}\sin(10\pi ft)\right)}{dt}
\end{aligned} \tag{3}
$$

An essential issue in unsteady mixing is to determine the drag force and inertia force domination ranges. Unsteady mixing studies with triangular time-course of impeller speed (Woziwodzki 2017) show that these ranges can be described with a Keulegan–Carpenter number

$$KC = \frac{N_{\max}}{f} \tag{4}$$

In respect of Keulegan–Carpenter numbers higher than $KC = 15$ the drag force is dominant and in respect of $KC < 4$ the inertia force prevails and in range of $KC\epsilon < 4; 15>$ both forces are important. This allows modifying the Morison equation accordingly and abandoning its elements that describe the force which is not prevailing.

3 Unsteady Mixing of Gas-Liquid Systems

One of the basic goals of mixing of gas-liquid systems is to ensure the appropriate development of interfacial area. However, this faces obstacles which are, among others, related to the decreased mixing power demand. This is due to the lower density of the system (compared to uniform system) resulting from presence of gas. Lower mixing power, in turn, causes the reduction of interfacial area and thus lower intensity of mass transfer in a stirred vessel. In addition, mixing of these type of two-phase systems has a tendency to form gas cavities (Kamieński 2004; Middleton and Smith 2004). These are formed due to the presence of low pressure areas behind the impeller blades and result in lower mixing power, such as, by about 65% for Rushton turbine (Middleton and Smith 2004). Low pressure areas are formed due to the liquid flowing around the blades. They can be reduced by modifying the shape of blades. These zones, i.e. behind the hollow blades, are smaller which contributes to the higher mixing power, such as, decrease in mixing power is about 30% for CD-6 (Smith turbine) and about 20% for BT-6 impeller (Fig. 1) (Bakker 2000). It can be concluded that the use of hollow (unsymmetrical elliptic) blades (BT-6) gives better results than cylindrical blades (CD-6).

Mixing of two-phase systems is accompanied by other issues such as the presence of areas characterized by lower homogeneity degree right behind the baffles, flooding of impeller, uneven gas dispersion for impellers with higher diameters at low impeller speeds or a rare turbulence near interphase areas in stirred vessels with larger impellers (Yoshida et al. 2001b).

Considering these effects, there is a need to solve these problems with a new mixing method. Oscillations can be successfully used in Oscillatory Baffled

Fig. 1 BT-6 impeller

Column (OBC) or Oscillatory Baffled Reactors (OBR). This allows for a significant increase of mass transfer coefficients. Ni and Gao (1996) point that a nearly 5-time increase is observed for two-phase water-air systems in OBCs. The use of oscillation in fermentation also contributes to about two-time increased mass transfer coefficient. For this reason, oscillation of impeller speed can also contribute to more intensive mass transfer.

In gas-liquid systems and unsteady mixing, usually the typical parameters related to the miscible liquid systems are used, such as the relative mixing power demand RPD (P_g/P_u) or the relative drag and inertia coefficients. Drag coefficient for gas-liquid system C_{dg} depends on the gas content and decreases with the increase of Reynolds number for $Re_{um} < 300$. Above $Re_{um} = 300$, C_{dg} coefficient is constant and independent of Reynolds number. For gas-liquid systems, the drag coefficient can be (just as mixing power) shown as a relative drag coefficient C_{dg}/C_{du} (RDC). RDC coefficient is dependent on impeller speed, impeller diameter, oscillation frequency as well as gas flow rate. This dependency can be shown with a general equation (Woziwodzki 2017):

$$RDC = C_1 Fl_{g,u}{}^{C_2} Fr_u{}^{C_3} \qquad (5)$$

where $Fl_{g,u}$ is unsteady gas flow number ($Fl_{g,u} = Q_g/fD^3$) and Fr_u is unsteady Froude number ($Fr_u = f^2D/g$).

In case of unsteady mixing (Fig. 2) the impeller speed changes constantly. This results in all regimes of gas-impeller interactions in a single oscillation cycle: from flooding to full dispersion. Therefore, the maximum mixing power P_{gmax} is important. Its determination requires determining time t after which P_g reaches its maximum value. For unsteady mixing with triangular time-course of impeller speed P_{gmax} can be determined with Eq. (6):

$$P_{gmax} = \frac{98.22}{\pi^2} C_1 C_{Dg} N_{max}^3 D^5 \rho \qquad (6)$$

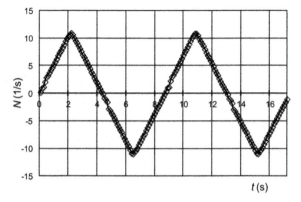

Fig. 2 Time-course of impeller speed for unsteady mixing (triangular wave)

The maximum mixing power number is determined with Eq. (7) (Woziwodzki 2017):

$$Ne_{g\max} = \frac{98.22}{\pi^2} C_1 C_{Dg} \tag{7}$$

and the ratio of mixing power $P_{g\max}$ and the average power is determined with Eq. (8) (Woziwodzki 2017):

$$\frac{P_{g\max}}{(P_g)_{av}} = 1.74\pi \tag{8}$$

Gas dispersion in a stirred vessel is affected by impeller type. Impellers with hollow, unsymmetrical blades (i.e. BT-6) are preferred to disperse larger amount of gas and work with higher gas load without flooding. However studies of unsteady mixing of two-phase gas-liquid systems in a stirred vessel with BT-6 are limited. There are also no data on the effect of the number of blades and their shape on gas dispersion. For this reason studies for air-water system for BT-6, BT-4 (Fig. 3) and BT-4E (Fig. 4) were conducted.

Unsteady mixing of miscible fluids is characterized by higher mixing power demand than steady mixing (Yoshida et al. 1999, 2001a; Woziwodzki 2011, 2017). Figures 5, 6 and 7 show the relation between power number *Ne* and Reynolds

Fig. 3 BT-4 impeller

Fig. 4 BT-4E impeller

Fig. 5 Relation between unsteady power number and Reynolds number for BT-4E

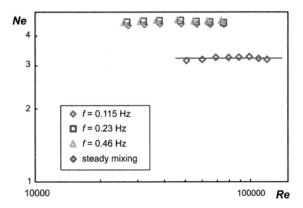

Fig. 6 Relation between unsteady power number and Reynolds number for BT-6; *CCD* counter-clockwise direction of rotation, *CD* clockwise direction of rotation

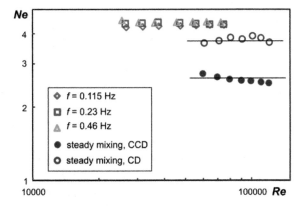

Fig. 7 Relation between unsteady power number and Reynolds number for BT-4

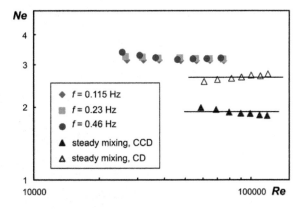

Table 1 Power number for unsteady mixing

Impeller	Power number
RT (Woziwodzki 2011)	7.14
ST (Woziwodzki 2011)	4.74
PBT (Woziwodzki 2011)	1.64
HE-3 (Woziwodzki 2017)	1.49
A315 (Woziwodzki 2017)	2.98
SC-3 (Woziwodzki 2017)	0.88

number Re for unsteady mixing. In all analyzed cases, oscillation frequency f does not affect the power number Ne. It was also observed for all impellers that the use of unsteady mixing caused increase in power demand in comparison to steady mixing. This increase was about 40% for BT-4E, about 16% for BT-6 and clockwise direction of impeller rotation (CD) and about 20% for BT-4 in relation to clockwise direction of rotation (CD). The increase in power demand is related to the need for changing the liquid circulation direction in a stirred vessel, which increases power demand, as well as related to higher drag force and larger disturbances are formed on both sides of blades. Exemplary power numbers for other impellers are shown in Table 1.

The highest turbulent power number was achieved for BT-4E impeller ($Ne = 4.54$), the lower power numbers were for BT-6 ($Ne = 4.37$) and BT-4 ($Ne = 3.19$) respectively.

The power number for BT-4E was higher about 42 and 4% in comparison to BT-4 and BT-6 impeller respectively.

The results obtained indicate that taking into account the unsteady mixing power demand, lower increase of power is observed, in relation to the steady mixing, for impellers with unsymmetrical hollow blades (BT-6, BT-4). These results imply that also for these impellers, the relative mixing power in gas-liquid systems should be lower. Figures 8, 9 and 10 show relation between RPD and gas flow number for the air-water system and BT-4E, BT-6 and BT-4 impellers.

Fig. 8 Relation between relative power demand (RPD) and gas flow number for BT-4E impeller; air-water system

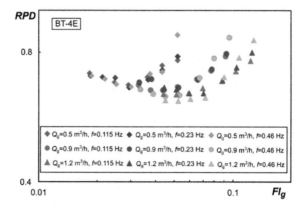

Fig. 9 Relation between relative power demand (RPD) and gas flow number for BT-6 impeller; air-water system

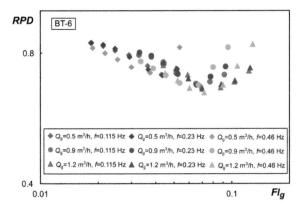

Fig. 10 Relation between relative power demand (RPD) and gas flow number for BT-4 impeller; air-water system

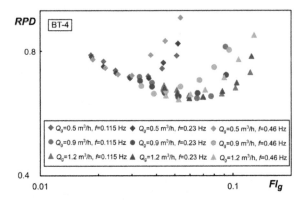

The analysis of unsteady mixing power $P_{g,u}$ in gas-liquid system shows that BT-4E impeller is characterized by much higher power demand than BT-4 for all tested gas flow rates, oscillation frequencies and impeller speeds. Comparing these results with BT-6, higher relative power compared to BT-4E at low speeds is observed. The impeller speed N range in which the power is higher for BT-4E is larger as the amount of supplied gas increases.

For impeller BT-4E (Fig. 8), an oscillation frequency affects RPD only slightly. With the increased impeller speed, the power initially drops and then starts to rise. This increase is lower in relation to BT-6 and BT-4 impellers. It is related to different types of gas cavities present behind the blades.

For BT-6 (Fig. 9), the effect of gas flow rate is smaller. The relative power demand decreases at low speeds and starts to rise rapidly after a certain impeller speed value is exceeded ($Fl_g < 0.07$) It is simply related to the reduction of gas cavities behind the blades. The data indicates that oscillation frequency affects the relative mixing power demand and it is notices for all gas flow rates. At low impeller speeds, the RPD is highest for highest oscillation frequencies and for higher impeller speeds, this dependence is smaller ($Fl_g < (0.04; 0.07)$).

In case of BT-4, as for other impellers, an initial decrease in relative power demand with the increase of impeller speed is observed and subsequent increase of power demand above a certain speed is observed, which is caused by changes of gas cavities. The oscillation frequency f also affects the RPD and is highest for highest oscillation rates but this dependence is smaller as the impeller speed increases. Highest relative power demand, at low impeller speeds, is noticed for BT-4E and at higher impeller speeds, for BT-6. This relation is valid for all gas flow rates. BT-6 impeller has the lowest power demand drop caused by supplied gas. BT-4E at higher impeller speeds for gas rate 0.5 m^3/h has the lowest RPD and in case of gas rates 0.9 and 1.2 m^3/h its relative power demand is close to that of BT-4.

The results obtained indicate that the blade shape affects the power demand for unsteady mixing. Comparison of relative power demand for BT-4 and BT-6 implies that, just as in case of steady mixing, impellers with more blades are preferred allowing for dispersion of more gas. Compared results for BT-4 and BT-4E imply that the use of an ellipsoidal blade allows obtaining higher mixing power than the ellipsoidal open blade. In case of relative power RPD, the results are similar. In case of all impellers, RPD values are higher than for Rushton turbine for which RPD was about 0.42 (Woziwodzki and Broniarz-Press 2014). Considering RPD, the recommended impeller is the one with no flat blades and BT-6 is preferred for higher speeds ($Fl_g < 0.07$) and BT-4E in case of lower speeds ($Fl_g < 0.07$).

Acknowledgements This work was supported by PUT research grant no. 03/32/DSPB/0702.

References

Bakker A (2000) A new gas dispersion impeller with vertically asymmetric blades. In: The online CFM book. http://www.bakker.org.cfm

Kamieński J (2004) Mieszanie układów wielofazowych. WNT, Warszawa

Kamieński J, Wójtowicz R (2001) Power input for a vibromixer. Inż Chem Proc 22:603–608

Kamieński J, Wójtowicz R (2003) Dispersion of liquid–liquid systems in a mixer with a reciprocating agitator. Chem Eng Process 42:1007–1017

Karcz J, Cudak M, Szoplik J (2005) Stirring of a liquid in a stirred tank with an eccentrically located impeller. Chem Eng Sci 60:2369–2380

Komoda Y, Inoue Y, Hirata Y (2000) Mixing performance by reciprocating disk in cylindrical vessel. J Chem Eng Jpn 33:879–885

Komoda Y, Inoue Y, Hirata Y (2001) Characteristics of turbulent flow induced by reciprocating disk in cylindrical vessel. J Chem Eng Jpn 34:929–935

Kordas M, Story G, Konopacki M et al (2013) Study of mixing time in a liquid vessel with rotating and reciprocating agitator. Ind Eng Chem Res 52:13818–13828

Masiuk S (1999) Power consumption measurements in a liquid vessel that is mixed using a vibratory agitator. Chem Eng J 75:161–165

Masiuk S (2000) Mixing time for a reciprocating plate agitator with flapping blades. Chem Eng J 79:23–30

Masiuk S (2001) Dissolution of solid body in a tubular reactor with reciprocating plate agitator. Chem Eng J 83:139–144

Masiuk S, Rakoczy R (2007) Power consumption, mixing time, heat and mass transfer measurements for liquid vessels that are mixed using reciprocating multiplates agitators. Chem Eng Process 46:89–98

Masiuk S, Rakoczy R, Kordas M (2008) Comparison density of maximal energy for mixing process using the same agitator in rotational and reciprocating movements. Chem Eng Process 47:1252–1260

Middleton JC, Smith JM (2004) Gas-liquid mixing in turbulent systems. In: Paul EL, Atiemo-Obeng V, Kresta SM (eds) Handbook of industrial mixing. Science and practice. Wiley, Hoboken, pp 585–635

Montante G, Bakker A, Paglianti A et al (2006) Effect of the shaft eccentricity on the hydrodynamics of unbaffled stirred tanks. Chem Eng Sci 61:2807–2814

Morison JR (1953) The force distribution exerted by surface waves on piles. University of California, Institute of Engineering Research, Berkeley

Ng KC, Ng EYK (2013) Laminar mixing performances of baffling, shaft eccentricity and unsteady mixing in a cylindrical vessel. Chem Eng Sci 104:960–974

Ni X, Gao S (1996) Scale-up correlation for mass transfer coefficients in pulsed baffled reactors. Chem Eng J Biochem Eng J 63:157–166

Wójtowicz R (2012) Wizualizacja procesu wytwarzania emulsji w mieszalniku z mieszadłem wykonującym ruch posuwisto-zwrotny. Inż Apar Chem 51:402–403

Woziwodzki S (2011) Unsteady mixing characteristics in a vessel with forward-reverse rotating impeller. Chem Eng Technol 34:767–774

Woziwodzki S (2017) Mieszanie nieustalone - analiza i wybrane zastosowania. Wydawnictwo Politechniki Poznańskiej, Poznań

Woziwodzki S, Broniarz-Press L (2014) Power characteristics of unsteadily rotating Rushton turbine in aerated vessel. Tech Trans 2–Ch:155–164

Woziwodzki S, Jędrzejczak Ł (2011) Effect of eccentricity on laminar mixing in vessel stirred by double turbine impellers. Chem Eng Res Des 89:2268–2278

Woziwodzki S, Broniarz-Press L, Ochowiak M (2010) Effect of eccentricity on transitional mixing in vessel equipped with turbine impellers. Chem Eng Res Des 88:1607–1614

Yoshida M, Yamagiwa K, Ohkawa A et al (1999) Torque of drive shaft with unsteadily rotating impellers in an unbaffled aerated agitation vessel. Mater Technol 17:19–31

Yoshida M, Ito A, Yamagiwa K et al (2001a) Power characteristics of unsteadily forward-reverse rotating impellers in an unbaffled aerated agitated vessel. J Chem Technol Biotechnol 76: 383–392

Yoshida M, Yamagiwa K, Ito A et al (2001b) Flow and mass transfer in aerated viscous Newtonian liquids in an unbaffled agitated vessel having alternating forward-reverse rotating impellers. J Chem Technol Biotechnol 76:1185–1193

Yoshida M, Hiura T, Yamagiwa K et al (2008) Liquid flow in impeller region of an unbaffled agitated vessel with an angularly oscillating impeller. Can J Chem Eng 86:160–167. https://doi.org/10.1002/cjce.20028

Yoshida M, Nagai Y, Yamagiwa K et al (2009) Turbulent and laminar mixings in an unbaffled agitated vessel with an unsteadily angularly oscillating impeller. Ind Eng Chem Res 48: 1665–1672

Yoshida M, Wakura Y, Yamagiwa K et al (2010) Liquid flow circulating within an unbaffled vessel agitated with an unsteady forward-reverse rotating impeller. J Chem Technol Biotechnol 85:1017–1022

Supercritical Fluids in Green Technologies

Marek Henczka, Małgorzata Djas and Jan Krzysztoforski

1 Introduction

The principles of green chemistry introduced nearly 20 years ago (Anastas and Warner 1998) and the principles of green engineering (Anastas and Zimmermann 2003) have greatly increased the awareness of chemical engineers towards developing industrial processes that are safe and environmentally friendly. The implementation of green technologies minimizes the use of materials that are hazardous to human health and environment, decrease energy and water consumption and maximizes process efficiency. Development of green technologies implies using the principles of green chemistry and engineering, from process inception to process application in a commercial scale. Solvent substitution remains an important method for making industrial processes more environmentally friendly and for reducing the amounts of hazardous wastes. Supercritical fluids, ionic liquids, aqueous solutions, immobilized solvents, solvent-free conditions, low-toxicity organic solvents, and fluorous solvents were identified as alternatives to using volatile, aromatic, chlorinated and chlorofluorocarbon solvents (Anastas 2002).

Supercritical fluids are nowadays involved in numerous industrial processes and have a potential wide field of new applications in pharmaceutical, food and textile industries. The supercritical state of a substance is achieved when the temperature and the pressure is raised over its critical values, t_c and p_c. The value of the pressure, p, and temperature, T, divided by its corresponding critical value is called the reduced pressure, p_r, and temperature, T_r.

M. Henczka (✉) · J. Krzysztoforski
Faculty of Chemical and Process Engineering, Warsaw University
of Technology, Warsaw, Poland
e-mail: marek.henczka@pw.edu.pl

M. Djas
Department of Chemical Technologies, Institute of Electronic
Materials Technology, Warsaw, Poland

© Springer International Publishing AG, part of Springer Nature 2018
M. Ochowiak et al. (eds.), *Practical Aspects of Chemical Engineering*,
Lecture Notes on Multidisciplinary Industrial Engineering,
https://doi.org/10.1007/978-3-319-73978-6_10

As shown in Fig. 1, under supercritical conditions (i.e. when $p_r > 1$ and $T_r > 1$) substance exists in one phase of properties between those of a gas and a liquid. By changing the pressure and the temperature of the fluid, the properties can be "tuned" to be more liquid- or gas-like. In addition, there is no surface tension in a supercritical fluid, as there is no liquid to gas phase boundary.

One of the most important properties of supercritical fluids is their ability to act as solvents. Solubility in a supercritical fluid tends to increase with the density of the fluid (at constant temperature). Since the density increases with pressure, solubility tends to increase with pressure. Another major benefit refers to the possibility of adjusting the properties of SCFs, such as diffusivity, viscosity or density, by simply varying the operating pressure and/or temperature. Moreover, SCFs have excellent heat transfer properties, and have been studied as environmentally benign heat transfer fluids. Most of industrial applications make use of carbon dioxide (CO_2) as the supercritical fluid due to its moderate critical temperature of 31.1 °C and pressure 7.4 MPa. Carbon dioxide is non-carcinogenic, nontoxic, non-mutagenic, non-flammable and thermodynamically stable. It has a zero surface tension, which provides a good wetting of the surface and facilitates a better penetration into porous structures. Supercritical CO_2 is easy separable by releasing pressure so the final products do not contain any residuals. Therefore $scCO_2$ processing often results in obtaining materials with superior properties with regards to those obtained using conventional solvents at ambient temperature. Although carbon dioxide is "greenhouse" gas, it can be obtained from other industrial processes and recycled, so the use of $scCO_2$ in high pressure technologies is the way to recover industrial emissions prior to their release to the environment. This chapter presents some of the SCF technologies being in use today from the perspective of their environmental and energetic advantages in comparison to the classical processes. This overview represents an attempt of highlighting the advantages of high pressure technologies and of contributing to their recognition as green and sustainable alternatives to current technological approaches.

Recent research on the application of SCFs showed that they could be used as new reaction media for chemical and biochemical reactions, for synthesis of new materials and new catalyst supports such as aerogels, for special separation

Fig. 1 Temperature-pressure diagram for a single component system: C-critical point, T-triple point

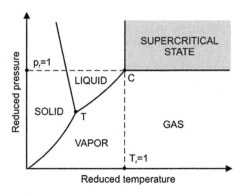

techniques such as chromatography using SCFs and extraction processes, and for particle formation and product formulation. There are several new applications developed to the industrial scale, such as extraction of solids and liquids, dry cleaning, high pressure sterilization, jet cutting, thin-film deposition for microelectronics, separations of value added products from fermentation broths in biotechnology fields and as solvents in a broad range of syntheses involving transition metals (Knez et al. 2014). There is a great variety of potential applications of SCFs in the industrial processing of fats, oils and their derivatives. SCFs are considered alternative refrigerants for automotive air conditioners and working fluids in power cycles. There is also an increasing interest in chemical reactions involving SCFs, especially supercritical water, for the treatment of wastes and by-products and which may generate value added products such as energy carrier compounds (bio-oils and permanent gases such as hydrogen and methane). In the industrial processes related to energy production, liquid rocket, diesel and gas turbine engines operate at supercritical conditions for the injected fuel (Ahn et al. 2015).

2   Reactive Extraction of Carboxylic Acids

Carboxylic acids are used on a large scale in the food, chemical and pharmaceutical industries. Applications of carboxylic acids include the production of detergents, pharmaceuticals, plastics, dyes, textile, perfumes, and animal feed (Soccol et al. 2006; Liu et al. 2012). Currently, other advanced applications of carboxylic acids are being developed, e.g. biopolymer production, drug delivery and tissue engineering (Dhillon et al. 2011; Song and Lee 2006). Most carboxylic acids are produced on an industrial scale by chemical synthesis. However, the high cost of petrochemical products, depletion of world petroleum reserves and requirements for the development of environmentally friendly technologies have led to dynamic research of bio-based production of carboxylic acids by fermentation. The biological production cost of carboxylic acids is dominated by feedstock cost and downstream processing. Even more than half of the cost of producing carboxylic acids by fermentation is associated with downstream processing (Straathof 2011). The industrial method of carboxylic acid separation is precipitation using calcium hydroxide, which is environmentally unfriendly, as it requires large amounts of sulphuric acid and generates solid waste in the form of calcium sulphate (Kurzrock and Weuster-Botz 2010; Lopen-Garzon and Straathof 2014). Recently, reactive extraction process has received increasing attention as an alternative and effective method of carboxylic acid separation from fermentation broth (Kurzrock and Weuster-Botz 2010; Wasewar 2012; Li et al. 2016). In general, reactive extraction of carboxylic acids is commonly performed using organic solvents. Recent trends in extraction processes are mainly focused on finding methods that minimize the use of organic solvents. Therefore, the elimination of organic solvents from industrial processes and replacing them with supercritical fluids, in particular supercritical carbon dioxide ($scCO_2$), is intensively investigated.

2.1 Mechanism of Reactive Extraction

The scheme of reactive extraction of the carboxylic acid (H_3A) from an aqueous solution using complexing reactant (G) is shown in Fig. 2.

The reactive extraction process of the carboxylic acid involves two immiscible phases: a stationary phase (aqueous phase) that includes the carboxylic acid and a transport phase (organic solvent or supercritical fluid) containing a complexing reactant. The key step of the reactive extraction process is formation of a complex consisting of carboxylic acid and complexing reactant $(H_3A)_a(G)_b$, which unlike the carboxylic acids is soluble in the transport phase. Complexing reactant forms a complex with a carboxylic acid of structure (a, b), where a is the number of molecules of carboxylic acid, and b is the number of molecules of the complexing reactant (Tamada et al. 1990). The most often applied reactants in reactive extraction of carboxylic acids processes as complexing agents are long-chain tertiary amines having more than six carbon atoms in chain (e.g. tri-n-octylamine, tri-n-butylamine).

2.2 Reactive Extraction Efficiency

The influence of operating parameters on the course of reactive extraction of carboxylic acids with different molecular structures from aqueous solution using supercritical CO_2 as the solvent was experimentally investigated (Henczka and Djas 2016; Djas and Henczka 2016). The final efficiency of the separation processes performed in batch and semi-continuous modes using supercritical CO_2 was compared with that obtained using 1-octanol. It was concluded that the efficiency of separation processes of some carboxylic acids from aqueous solution using

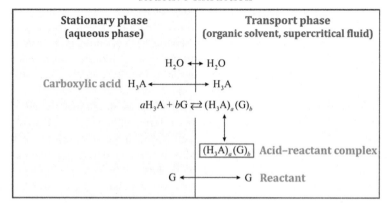

Fig. 2 The scheme of reactive extraction process of carboxylic acid from aqueous solution

supercritical CO_2 is comparable to that obtained in separation processes performed with the use of 1-octanol. Moreover, as shown in Fig. 3, reactive extraction using supercritical CO_2 and tri-n-octyloamine allows to improve the efficiency of carboxylic acid separation as compared to the efficiency of physical extraction.

It was observed that the efficiency of physical and reactive extraction using supercritical CO_2 performed in batch mode is low, due to the limited solubility of TOA in supercritical CO_2, and depends on the partition equilibrium. To increase the efficiency of the extraction using supercritical CO_2 it is recommended to perform the process in semi-continuous mode. Applying the flow of CO_2 through the system, the extraction equilibrium is displaced as a fresh solvent and the complexing reactant is continuously delivered to the aqueous solution of carboxylic acid so the solute is removed from the reaction mixture. The course of extraction processes of carboxylic acids with the flow of supercritical CO_2 saturated and unsaturated with TOA were considered. The comparison of process efficiencies of carboxylic acid separation using supercritical CO_2 performed in batch mode and in semi-continuous mode is shown in Fig. 4.

The molecular structure of carboxylic acids significantly influences the efficiency of supercritical extraction. Monocarboxylic acids are separated using the physical extraction process, but the application of the reactive extraction process is necessary for separation of di- and tricarboxylic acids. It was found that the reactive extraction of acetic acid and propionic acid using TOA as a reactant allows for a significant increase in the efficiency of acid recovery as compared to physical extraction. Performing the process in semi-continuous mode results in a higher efficiency of

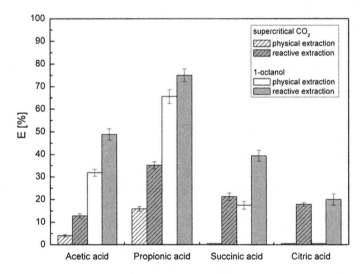

Fig. 3 Efficiency of carboxylic acid extraction using supercritical CO_2 and 1-octanol—effect of solvent (16 MPa, 308 K, 60 min, batch mode)

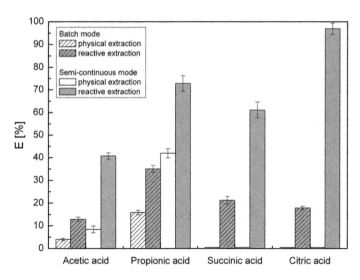

Fig. 4 Efficiency of carboxylic acid extraction using supercritical CO_2—effect of process mode (16 MPa, 308 K, 60 min)

physical and reactive extraction of carboxylic acids as compared to the batch mode. The highest efficiency of physical extraction equal to 42% was achieved for propionic acid, while the highest reactive extraction efficiency equal approximately to 97% was achieved for citric acid. It can be concluded that the efficiency of supercritical extraction performed in semi-continuous mode is similar or higher than that obtained in process performed using 1-octanol. To avoid the negative effect of low solubility of carboxylic acids and TOA in $scCO_2$ it is recommended to perform the reactive extraction process in the semi-continuous mode. Both pressure and temperature have a significant influence on the separation efficiency of carboxylic acids due to the effect on the density and solvating power of $scCO_2$. Performing the process at high pressure is favourable for reactive extraction of carboxylic acids as the solubility of TOA in $scCO_2$ increases along with an increase in pressure, which causes an increase in the efficiency of reactive extraction. The feasibility of reactive extraction using $scCO_2$ on an industrial scale requires the economical assessment of the process. It was shown that the cost of supercritical reactive extraction slightly increases with the pressure increase (Djas and Henczka 2016). However, regardless of the applied process pressure, this cost is significantly smaller than that of the process performed using organic solvent.

Reactive extraction using $scCO_2$ is an emerging and promising technology for the recovery of carboxylic acids from aqueous solutions. The application of this method allows for the elimination of environmentally unfriendly organic solvents from industrial technologies, which is in agreement with the strategy of sustainable development and with Green Chemistry principles.

3 Supercritical Fluids in Membrane Technologies

Supercritical fluids (SCFs), especially supercritical carbon dioxide ($scCO_2$), have found numerous applications in the field of membrane production, maintenance and modification technologies. In many of them, SCFs have replaced organic solvents used in standard techniques, leading to higher process safety, reduction of waste material and lower process costs due to the possibility of solvent regeneration. Moreover, due to the unique features of SCFs, including the high tunability of processes involving SCFs, novel membranes with superior properties can be obtained.

3.1 Membrane Production

Supercritical fluids are used in many membrane production techniques, e.g. for the manufacturing of porous membranes by phase separation. SCFs can be used for induction of the phase separation, or as cleaning fluids for removal of the primary solvent present in the pores of the porous matrix after phase separation.

The application of supercritical fluids as anti-solvents in membrane production was studied by many researchers. Li et al. (2007), performed formation of polyvinyl butyral (PVB) membranes by contacting an organic solution of PVB with $scCO_2$. The organic solvent gradually diffuses to the SCF phase and the solution becomes supersaturated, leading finally to phase separation. Using this method, the solvent can be completely removed from the porous structure and no additional cleaning step is needed. Using a similar technique, Baldino et al. (2016) produced antimicrobial cellulose acetate membranes containing potassium sorbate. Xinli et al. (2009) obtained porous PVA membranes using $scCO_2$ as porogenic agent, by contacting a thin PVA film with $scCO_2$. In all the presented studies, the morphology and properties of the resulting membranes strongly depend on the conditions, in which the polymer solutions were contacted with $scCO_2$. Moreover, in the work of Baldino et al. various release rates of potassium sorbate could be obtained depending on the process conditions. These examples show the great potential of SCFs, as by changing the process conditions, the characteristics of the produced membranes can be precisely adjusted to specific needs and applications.

After the phase separation during membrane production via Temperature Induced Phase Separation (TIPS), a liquid contaminant is present in the porous membrane and it has to be removed (Berghmans et al. 1996). In the standard solution, large amounts of organic solvents at elevated temperatures are used, which increases the process cost and contributes to lower process safety due to fire hazard. Zhang et al. (2007) investigated the application of $scCO_2$ in the cleaning step of the TIPS technique. They found that the morphology of the resulting membrane can be tailored by adjustment of supercritical fluid extraction conditions. Moreover, the membrane produced using $scCO_2$ shows less shrinkage and better water permeability. Krzysztoforski (2016) performed experimental and numerical investigation of the transport phenomena during porous membrane cleaning using $scCO_2$ and compared the novel solution with the standard technique utilizing liquid isopropyl alcohol as cleaning fluid. The application of $scCO_2$, when compared to liquid isopropyl alcohol, shows significant advantages in terms of process safety and environmental load and it can be used as replacement in the membrane cleaning process. However, the disadvantage of using pure $scCO_2$ as solvent in the membrane cleaning process is a substantially lower cleaning rate than in the standard process utilising liquid organic solvent due to limited solubility of the contaminants in $scCO_2$. Krzysztoforski et al. (2013) and Michałek et al. (2015) investigated the effect of addition of organic co-solvents to $scCO_2$ in the membrane cleaning process in order to increase the process rate. Minor amounts of methyl, ethyl or isopropyl alcohol as co-solvents induce increase solvating power of $scCO_2$ and, therefore, result in reduced cleaning time.

The described technologies for membrane cleaning during their production can be also applied for regeneration of used membranes in order to restore their initial performance after realisation of a membrane separation process.

3.2 Membrane Modification

Membranes can be subjected to physical and/or chemical modification after their production in order to alter their separation properties such as the tendency to fouling, antibacterial properties, or permeability. In these modification technologies, supercritical fluids can be applied as reaction media and carriers of the active substances, which have to be grafted onto the membrane surface. Wang et al. (2004), modified porous polypropylene membranes by grafting acrylic acid in $scCO_2$, resulting in an uniformly grafted, pH-sensitive membrane. The method enables precise adjustment of the grafting degree using process parameters such as monomer concentration, temperature, pressure, and soaking time.

However, the interaction of CO_2 molecules with the polymeric matrix of the membrane, which occurs during a modification process, can also change key properties of the membrane. In the worst case, the membrane's performance deteriorates and the membrane becomes useless. Therefore, one has to assess whether treatment of a membrane with $scCO_2$ is safe with regards to its future application (Tarabasz et al. 2016).

3.3 Membrane Separation Processes Involving Supercritical Fluids

Another group of processes involving the interaction of SCFs and membranes are membrane separation processes performed in the presence of supercritical fluids. In these processes, the advantages of membrane separation can be combined with the unique features of supercritical fluids, leading to higher competitiveness of supercritical fluid technologies. Lohaus et al. (2015) studied drying of water-loaded $scCO_2$ using a dense polymeric membrane. With this process, $scCO_2$ can be regenerated and reused in supercritical fluid extraction (SFE) without the need of its decompression. This significantly reduces the energy cost of supercritical solvent regeneration. Membranes can also be used as contactors in SFE processes, such as SFE of acetone (Miramini et al. 2013). The increase of the contact area of the fluids thanks to the membrane contractor enables the increase of the process rate.

4 Manufacturing of Porous Materials for Tissue Engineering

Numerous methods have been developed to generate highly porous biodegradable polymer or its composites for tissue engineering, which include phase separation, emulsion freeze-drying, solvent casting/salt leaching, fibre forming, and 3D printing. These conventional methods, however, generally involve the use of organic solvents in the fabrication process. Therefore, further purification and drying steps are often needed. The residues of organic solvent remained in the matrix may promote bioincompatibility and inactivity of the growth factors. Moreover, some of these methods are often employed at high temperatures, which may degrade the incorporated thermosensitive components, such as pharmaceutical drugs and bioactive compounds. For several decades many techniques of preparation of porous structures using supercritical fluids have been developed to obtain polymeric materials such as microparticles, fibres or foams (Wu and Meng 2004; Tomasko et al. 2003). Furthermore, the application of supercritical fluids is preferred especially in the processing of polymers dedicated for biomedical applications (Liao et al. 2012). Polymer foaming using supercritical fluids is an essential process for bone scaffold fabrication because it does not require the use of potentially harmful organic solvents and further purification and drying steps to obtain the final product.

Carbon dioxide used as a plasticizer and a foaming agent to create 3D scaffolds is an attractive choice to overcome limitations of conventional methods (Yoganathan et al. 2010). Although $scCO_2$ is a weak solvent for polymers except for some fluoropolymers and silicones, it has substantial solubility in amorphous and semicrystalline polymers. The absorption of CO_2 in polymers results in the decrease of glass transition temperature, viscosity, interfacial tension, and increase of polymer chain flexibility because of an increase in the free volume fraction. This helps to avoid the use of high temperature and decrease the saturation time compared with the conventional methods. In order to produce polymer foams, gas bubbles must nucleate and grow within the molten or plasticized polymer. In this process, firstly amorphous or semicrystalline polymer is saturated with CO_2 at a certain temperature and pressure. The amount of absorbed CO_2 depends on temperature, pressure, and the intermolecular interactions between CO_2 and a polymer. Once the system reaches equilibrium, the phase separation can be induced by either reducing pressure (pressure quench method) or increasing temperature (temperature soak method). The sudden reduction in pressure or increase in temperature leads to the generation of nuclei due to the supersaturation. Finally, these nuclei grow to form the porous structure. By changing the processing parameters such as pressure, temperature, and pressure drop rate the foam morphology and structure can be controlled.

Unfortunately, for biomedical applications, the disadvantage of the CO_2 foaming process is that the foams are often lack of interconnectivity between pores, which lowers viability of seeded cells and hence results in a non-uniform distribution of seeded cells throughout the matrix. To improve pore interconnectivity, a combination technique of CO_2 gas foaming with particulate leaching (Aydin et al. 2009) or particle seeding (Collins et al. 2010) is utilized. These techniques allow creating porous matrix with well-controlled porosity and pore morphology. Lately, there are many publications on preparation of composite materials using SCFs based on polymers enriched with various substances (Diaz-Gomez et al. 2016; Nofar 2016). The carbon nanomaterial-coated substrates showed high biocompatibility and enhanced gene transfection efficiency. Chung et al. (2013) presented results of investigation of graphene/chitosan hybrid films prepared, which also showed promising applicability to tissue engineering to repair and improve tissue functions. Interestingly, the neurite sprouting and outgrowth were also promoted on graphene surface compared to conventional tissue culture plate made of polystyrene.

5 Conclusions

The main advantages of supercritical fluids applications refer to their specific thermodynamic and heat transfer properties, the possibility of avoiding the use of organic solvents, obtaining products with high purity, and the lower energy consumption. In recent years SCFs were proposed for different applications in the energy field. Supercritical steam plant technology is today the option of choice for most new coal-fired power stations. These technically-advanced plants offer greater efficiency than older sub-critical designs and, most importantly, lower emissions. Therefore, it can be concluded that advances in the area of high pressure technologies have opened up new pathways for products obtained with "green" and environmentally friendly methods.

Acknowledgements Research financed by the National Science Centre, Poland, Project No. 2014/15/N/ST8/01516.

References

Ahn Y, Jun Baea S, Minseok K et al (2015) Review of supercritical CO_2 power cycle technology and current status of research and development. Nucl Eng Technol 47:647–661

Anastas PT (2002) Green chemistry as applied to solvents. Clean Solvents 819:1–9

Anastas PT, Warner JC (1998) green chemistry: theory and practice. Oxford University Press, New York

Anastas PT, Zimmermann JB (2003) Design through the 12 principles of green engineering. Environ Sci Technol 37:94a–101a

Aydin HM, El Haj AJ, Pişkin E et al (2009) Improving pore interconnectivity in polymeric scaffolds for tissue engineering. J Tissue Eng Regener Med 3(6):470–476

Baldino L, Cardea S, Reverchon E (2016) Production of antimicrobial membranes loaded with potassium sorbate using a supercritical phase separation process. Innov Food Sci Emerg 34:77–85

Berghmans S, Berghmans H, Meijer HEH (1996) Spinning of hollow porous fibres via the TIPS mechanism. J Membr Sci 116(2):171–189

Chung C, Kim YK, Shin D et al (2013) Biomedical applications of graphene and graphene oxide. Acc Chem Res 46(10):2211–2224

Collins NJ, Bridson RH, Leeke GA et al (2010) Particle seeding enhances interconnectivity in polymeric scaffolds foamed using supercritical CO_2. Acta Biomater 6(3):1055–1060

Dhillon GS, Brar SK, Verma M et al (2011) Recent advances in citric acid bio-production and recovery. Food Bioprocess Technol 4:505–529

Diaz-Gomez L, Concheiro A, Alvarez-Lorenzo C et al (2016) Growth factors delivery from hybrid PCL-starch scaffolds processed using supercritical fluid technology. Carbohydr Polym 142:282–292

Djas M, Henczka M (2016) Reactive extraction of citric acid using supercritical carbon dioxide. J Supercrit Fluids 117:59–63

Henczka M, Djas M (2016) Reactive extraction of acetic acid and propionic acid using supercritical carbon dioxide. J Supercit Fluids 110:154–160

Knez Z, Markocic E, Leitgeb M et al (2014) Industrial applications of supercritical fluids: a review. Energy 77:235–243

Krzysztoforski J (2016) Transport phenomena in the process of porous membrane cleaning using supercritical fluids. PhD Dissertation. Warsaw University of Technology, Poland

Krzysztoforski J, Krasiński A, Henczka M et al (2013) Enhancement of supercritical fluid extraction in membrane cleaning process by addition of organic solvents. Chem Process Eng 34(3):403–414

Kurzrock T, Weuster-Botz D (2010) Recovery of succinic acid from fermentation broth. Biotech Lett 32:331–339

Li Z, Tang H, Liu X et al (2007) Preparation and characterization of microporous poly(vinyl butyral) membranes by supercritical CO_2-induced phase separation. J Membr Sci 312(1–2):115–124

Li Q-Z, Jiang X-L, Feng X-J et al (2016) Recovery processes of organic acids from fermentation broths in the biomass-based industry. J Microbiol Biotechnol 26(1):1–8

Liao X, Zhang H, He T (2012) Preparation of porous biodegradable polymer and its nanocomposites by supercritical CO_2 foaming for tissue engineering. J Nanomater 2012:1–12

Liu L, Zhu Y, Li JH et al (2012) Microbial production of propionic acid from propionibacteria: current state, challenges and perspectives. Crit Rev Biotechnol 32:374–381

Lohaus T, Scholz M, Koziara BT et al (2015) Drying of supercritical carbon dioxide with membrane processes. J Supercrit Fluid 98:137–146

López-Garzón CS, Straathof AJJ (2014) Recovery of carboxylic acids produced by fermentation. Biotechnol Adv 32:873–904

Michałek K, Krzysztoforski J, Henczka M et al (2015) Cleaning of microfiltration membranes from industrial contaminants using "greener" alternatives in a continuous mode. J Supercrit Fluid 102:115–122

Miramini SA, Razavi SMR, Ghadiri M et al (2013) CFD simulation of acetone separation from an aqueous solution using supercritical fluid in a hollow-fiber membrane contactor. Chem Eng Process 72:130–136

Nofar M (2016) Effects of nano-/micro-sized additives and the corresponding induced crystallinity on the extrusion foaming behavior of PLA using supercritical CO_2. Mat Des 101:24–34

Soccol CR, Vandenberghe LPS, Rodrigues C et al (2006) New perspectives for citric acid production and application. Food Technol Biotechnol 44:141–149

Song H, Lee SY (2006) Production of succinic acid by bacterial fermentation. Enzyme Microbial Technol 39:352–361

Straathof AJJ (2011) The proportion of downstream costs in fermentative production processes in book: comprehensive biotechnology. Elsevier, pp 811–814

Tamada JA, Kertes AS, King CJ (1990) Extraction of carboxylic acids with amine extractants. 1. Equilibria and law of mass action modelling. Ind Eng Chem Res 29:1319–1326

Tarabasz K, Krzysztoforski J, Szwast M et al (2016) Investigation of the effect of treatment with supercritical carbon dioxide on structure and properties of polypropylene microfiltration membranes. Mater Lett 163:54–57

Tomasko LD, Liu H, Li D et al (2003) A review of CO_2 applications in the processing of polymers. Ind Eng Chem Res 42:6431–6456

Wang Y, Liu Z, Han B et al (2004) pH sensitive polypropylene porous membrane prepared by grafting acrylic acid in supercritical carbon dioxide. Polym 45(3):855–860

Wasewar KL (2012) Reactive extraction: an intensifying approach for carboxylic acid separation. Int J Chem Eng Appl 3:249–255

Wu D, Meng Q (2004) A study of bubble inflations in polymer and its applications. Phys Lett A 327:61–66

Xinli Z, Xiaoling H, Ping G et al (2009) Preparation and pore structure of porous membrane by supercritical fluid. J Supercrit Fluid 49(1):111–116

Yoganathan RB, Mammucari R, Foster NR (2010) Dense gas processing of polymers. Polym Rev 50(2):144–177

Zhang CF, Zhu BK, Ji GL et al (2007) Supercritical carbon dioxide extraction in membrane formation by thermally induced phase separation. J Appl Polym Sci 103(3):1632–1639

The Application of CFD Methods for Modeling of a Three-Phase Fixed-Bed Reactor

Daniel Janecki, Grażyna Bartelmus and Andrzej Burghardt

1 Introduction

Processes based upon multiphase reactions predominate in the majority of chemical technologies and form the basis for manufacture of a large variety of intermediate and consumer end-products.

The heart of installations in which the aforementioned processes are carried out is the multiphase reactor in which the transformation of matter takes place, resulting in the production of new qualitative products.

Efficiency and selectivity of the process carried out in the reactor determine the load of all separation units and apparatuses in which unit operations, preceding or following the transformation of matter in reactor, are performed. So, lack of thorough understanding of the phenomena occurring in multiphase reactors can lead to disasters in scale-up or design.

The subject of the analysis contained in the present study is modeling of processes carried out in the reactor operating at gas-liquid co-current down-flow through the stationary bed of catalyst particles. Reactors of this type, called "trickle bed reactors" (TBRs), form a very important group of apparatuses used in quite a few branches of chemical industry as well as in the dynamically developing field of biotechnology and environmental engineering. The main advantage of these reactors is the absence of the catalyst separation, which facilitates a continuous-mode operation and possibility of using high flow rates of both phases without causing the onset of flooding in the column—a phenomenon which considerably limits the operating range of the counter-current flow. Therefore, these apparatuses are used

D. Janecki (✉)
Department of Process Engineering, University of Opole, Opole, Poland
e-mail: zecjan@uni.opole.pl

G. Bartelmus · A. Burghardt
Institute of Chemical Engineering, Polish Academy of Sciences, Gliwice, Poland

© Springer International Publishing AG, part of Springer Nature 2018
M. Ochowiak et al. (eds.), *Practical Aspects of Chemical Engineering*,
Lecture Notes on Multidisciplinary Industrial Engineering,
https://doi.org/10.1007/978-3-319-73978-6_11

extensively in those branches of industry which process high streams of substrates (mainly in petrochemical and petroleum industries). The main group of processes performed in TBRs are the hydrotreating processes of various fractions of crude oil (hydrodesulfurization, hydrodenitrification, removal of heavy metals (nickel, vanadium), reactions of hydrocracking and hydrogenation of aromatic compounds and olefins) (Duduković et al. 2002; Shah 1979).

As the range of applications of TBRs is immense, it is no surprise that for a dozen or so years intensive research has been conducted in order to improve our understanding of the phenomena occurring in the trickle—bed reactors. Special attention has been devoted to the hydrodynamics of the TBR (Szady and Sundaresan 1991; Attou et al. 1999; Gunjal et al. 2003, 2005). The experimental results clearly indicated that the computational fluid dynamics (CFD) has to be introduced into the quantitative description of the process. CFD is a tool to solve mathematical models of fluid dynamics limiting the number of experiments essential to determine the influence of individual operational parameters on the efficiency of the process. The momentum balance enables the estimation of the dynamic parameters of the fluid flow, i.e. the velocity profile of the two phases, the hold-up profile and the pressure gradient in the reactor. These are strictly connected with the phenomena of mass and energy transport between the phases, and thus indirectly influence the yield of the chemical process (Burghardt 2014).

2   Conservation Equation of Trickle-Bed Reactors

From the mathematical point of view the multiphase flow can be considered as a region divided into one-phase sub-regions with moving and continuously changing interfaces which is the dynamical structure of the separated flow systems (to which TBR belongs) (Ishii and Hibiki 2011).

The attempts at developing a multiphase flow model as a result of an appropriate compilation of the local, instantaneous equations which constitute the mass and momentum balances for each phase did not lead to satisfactory results.

Therefore, in the last four decades the investigations, both theoretical and experimental, focusing on the development of a general and exact quantitative description of the multiphase flow have been considerably intensified (Soo 1990; Enwald et al. 1996; Lyczkowski 2010; Ishii and Hibiki 2011).

This research led to the development of a macroscopic model of multiphase flow by means of applying appropriate averaging methods with respect to local, instantaneous balance equations of mass and momentum. The majority of averaging methods applied treat the multiphase flow as a mixture of quasi-continuous fluids. This approach, called Computational Fluid Dynamics (CFD), yields averaged values of the state variables, flow parameters and physicochemical properties. It requires however, additional relationships in order to close the system of the averaged differential equations analyzed. These relationships are mostly empirical and demand additional experimental data to be verified.

They define the two types of closure laws: constitutive and transfer laws. The constitutive laws relate the fluxes and sources in the bulk of the fluid phases to their driving forces composed of state variables. The transfer laws are empirical equations obtained mostly on the basis of experiments and define the interactions occurring between phases at the interface. The selection of the appropriate closure laws is crucial for accurate modeling of the reactor process.

However, a considerable majority of the publications cited above (if not all) present investigations which concern exclusively two-phase flows and, particularly, the dispersed flows like bubbling gas in liquid, liquid droplets in gas or solid particles in gas or liquid. Similarly, the general classification of multiphase flows proposed by Ishii and Hibiki (2011) also includes only two-phase flows (gas—liquid, solid particles—liquid or gas), assembling them in three groups depending on the geometry of interfaces.

However, in the chemical and petrochemical industries there is a very broad range of processes carried out in three-phase fixed-bed reactors (Satterfield 1975; Bartelmus et al. 1998; Burghardt et al. 1999).

In the following part of the study the macroscopic averaged balances of mass and momentum which describe quantitatively the processes occurring in the trickle-bed reactor will be presented and discussed.

2.1 Mass and Momentum Balances

The Eulerian multi-fluid model is used to simulate fluid flow in the bed (Jiang et al. 1999, 2001, 2002a, b; Gunjal et al. 2003, 2005; Janecki et al. 2008). This model treats every phase as a continuous fluid of various velocities, volume fractions and physicochemical properties. The volume averaged equations of mass and momentum balances expressed by means of the volume fractions occupied by each phase and their local velocities take the following form:

- continuity equation:

$$\frac{\partial}{\partial t}(\alpha_k \rho_k) + \nabla \cdot (\alpha_k \rho_k \vec{u}_k) = 0; \quad k = L, g \tag{1}$$

- momentum balance equation:

$$\frac{\partial}{\partial t}(\alpha_k, \rho_k \vec{u}_k) + \nabla \cdot (\alpha_k \rho_k \vec{u}_k \vec{u}_k) = -\alpha_k \nabla P + \nabla \cdot \alpha_k \bar{\bar{\tau}}_k$$
$$+ \alpha_k \rho_k \vec{g} + \sum_{j=1}^{n} \vec{R}_{jk}; \quad k = g, L; \quad j = g, L, s \tag{2}$$

where ε_k, ρ_k, u_k represent volume fraction, density (kg m^{-3}) and velocity (m s^{-1}) of phase k, P is pressure (Pa), $\bar{\bar{\tau}}_k$ stress tensor (Nm^{-2}), $\vec{g}$ acceleration due to gravity (m s^{-2}) and $\vec{R}_{jk}$ drag interaction between phases (kg m^{-2} s^{-2}); subscripts g, L, s denote gas, liquid and solid, respectively.

Equation (1) determines the net rate of increase of mass per unit volume (first term) as a result of the net rate of mass addition per unit volume by convection (second term).

The first term on the left-hand side of Eq. (2) determines the net rate of the momentum increase per unit volume of the phase k and the second one the rate of momentum addition by convection per unit volume. The right-hand side presents the body forces exerted on the fluid phase k. ∇P is a vector force of the scalar P acting perpendicular to the exposed surface, $\nabla \cdot \alpha_k \bar{\bar{\tau}}_k$ (a vector) is the divergence of the stress tensor whose components, τ_{ij}, determine the force in the j direction on a unit area perpendicular to the i direction. The third term defines the body forces per unit volume (e.g. gravity) and the last term represents the interphase molecular momentum exchange which determines the interphase friction force and has to be estimated experimentally.

It is the subject of the following conditions:

$$\bar{R}_{jk} = -\bar{R}_{kj} \text{ and } \bar{R}_{kk} = \bar{R}_{jj} = 0 \tag{3}$$

and is usually defined as (transfer law):

$$\bar{R}_{jk} = -F_{\substack{jk \\ j \neq k}} \left(\overline{u}_j - \overline{u}_k \right) \quad k = g, L \quad j = g, L, s \tag{4}$$

The detailed formulae of the exchange coefficients F_{jk} (kg m^{-3} s^{-1}) will be given in the next section, which presents the parametric sensitivity studies of the CFD model taking also into account other empirical parameters necessary in the solution procedure of the CFD equations. The viscous stress tensor $\bar{\bar{\tau}}$ is defined by the stress-strain relation (constitutive closure law):

$$\bar{\bar{\tau}}_k = \mu_k \left(\nabla \overline{u}_k + \nabla \overline{u}_k^T \right) + (\chi_k - \tfrac{2}{3}\mu_k) \nabla \cdot \vec{u}_k I \tag{5}$$

where μ_k is the shear and χ_k the dilatation viscosity (Pas).

For incompressible fluids (liquids) $\nabla \cdot \vec{u}_k = 0$ and the second term of Eq. (5) is discarded. For ideal monoatomic gases χ_k is identically equal to zero and in many other computations it is usually neglected because of the lack of reliable correlations.

A two-dimensional (2D) axisymmetric domain based on the cylindrical coordinate system is usually applied for the solution of the balance Eqs. (1) and (2). Axisymmetric boundary conditions of the variables are assumed: along the axis of the column symmetry, at the walls—no slip conditions of the velocities of the both phases. A flat velocity profile of the liquid at the inlet into the column is assured by using in experiments multipoint distributor which evenly distributes the liquid over the surface of the packing. Similarly, a flat velocity profile of gas is assumed at the inlet into the column.

The mean values of the gas (u_g) and liquid (u_L) velocities supplied to the column and the mean value of the volume fraction of the liquid phase are used as initial conditions in the column. It has to be noted that the pressure appearing in the momentum balance equations for the two phases is usually treated as identical, though there is a slight difference caused by capillary pressure (P_c):

$$P_C = P_g - P_L = 2\sigma\left(\frac{1}{d_1} + \frac{1}{d_2}\right) \tag{6}$$

where d_1 and d_2 are the main diameters of the curvature of the surface(m) and σ is surface tension (Nm^{-1}). Attou and Ferschneider (2000) obtained following expression for the capillary pressure by means of estimating the geometric formulae for d_1 and d_2 of spherical particles and introducing an empirical factor to take into account the high pressure operations:

$$P_g - P_L = 2\sigma\left(\frac{1-\varepsilon}{1-\alpha_g}\right)^{0.333}\left(\frac{5.416}{d_p}\right)F\left(\frac{\rho_g}{\rho_L}\right) \tag{7}$$

where ε is bed porosity, d_p particle diameter (m) and

$$F\left(\frac{\rho_g}{\rho_L}\right) = 1 + 88.1\frac{\rho_g}{\rho_L} \quad \text{for} \quad \frac{\rho_g}{\rho_L} < 0.025 \tag{8}$$

The order of magnitude analysis indicates that the capillary pressure forces are rather small in comparison to the magnitude of the interphase drag forces and can be reasonably neglected in simulations of pressure drop and liquid holdup, though they play a significant role in estimating the limit of the onset of pulsing flow (Grosser et al. 1988). Jiang et al. (2001), proposed a relation to calculate the capillary pressure by introducing an empirical parameter—the wetting efficiency of the bed (f):

$$P_g - P_L = (1-f)P_C \tag{9}$$

For a pre-wetted bed, f is set equal to one implying a zero capillary pressure and for non-wetted bed f is set to zero. For $0 \leq r \leq R$ and $z = H$ (outlet from the column):

$$P_g = P_L = P^0 \quad \text{and} \quad \frac{\partial P}{\partial z} = 0 \tag{10}$$

where R is the radius of the column (m), r—radial coordinate (m), z—axial coordinate (m), H—bed height (m) and P^0 is the pressure of the environment (Pa).

Fast development of the Computational Fluid Dynamics Model (CFD), which is a quantitative formulation of the momentum balance, caused a more frequent application of the CFD model to simulate the dynamic phenomena in the TBR. In this respect the studies of Jiang et al. (1999, 2001, 2002a, b), are fundamental since they have formulated a quantitative description of a co-current flow of gas and liquid through a fixed bed and its implementation into the CFDLIB program. Gunjal et al. (2003, 2005), analyzed the influence of initial wetting of the bed on the hydrodynamic parameters of the reactor (mean pressure drop and liquid hold-up) using the Eulerian three-phase model of fluid flow. They applied three sets of experimental data taken from the literature in order to verify their computational results by the CFD code. However, a fairly good agreement between the experimental data and computational results could be reached only by applying different pairs of Ergun constants for each set of experiments. Similarly, the influence of a one-point distributor of liquid on the liquid volume fraction distribution in the bed was analyzed in the study by Boyer et al. (2005). Janecki et al. (2008) employed the three-phase Eulerian model to simulate the hydrodynamic parameters of a TBR operating at periodically changing the feeding of the bed with liquid. The computational model formulated and used in the study shows a reasonable compatibility with the experimental data illustrating properly the tendency of the changes in hydrodynamic quantities in the bed as a function of operating parameters.

Lopes and Quinta-Ferreira (2008, 2009, 2010), in their simulations of pressure drop and liquid holdup in TBR used a model which is a combination of the discrete particle scale model with macroscopic friction terms as boundary conditions at the surfaces of the spherical particle. The implementation of this model into the CFD code demands a very complicated grid generation and large computation power and is therefore restricted to small dimensions of reactors, thus unrealistic for practical applications. Atta et al. (2007a), developed a two-phase Eulerian CFD model based on the porous media concept to simulate gas-liquid flow through packed beds. The friction terms have been defined based on relative permeability model by Saez and Carbonell (1985). An improved hydrodynamic model for wetting efficiency, pressure drop and liquid holdup in the trickle-bed reactor has been presented by Lappalainen et al. (2009), elaborated based on a wide range of experimental data. Hamidipour et al. (2013), studying the liquid, gas and gas/liquid alternating cyclic operation in the trickle-bed reactor used the standard volume averaged conservation equations of mass and momentum and the friction terms by Attou et al. (1999). They compared the one-dimensional, two-dimensional and three dimensional models obtaining nearly identical results for all the geometries.

2.2 Parametric Sensitivity of the CFD Model

Although the CFD model is based on fundamental principles, some empirical relations must be implemented into the momentum balance in order to ensure a proper description of the dynamics of very complex three-phase system in an intricate geometrical structure (Janecki et al. 2008). These relations determine the interactions between the phases (drag forces), the capillary pressure, wetting efficiency of the pellet's surface as well as the axial and radial profiles of the porosity in the bed and are usually taken from the literature.

As it has been mentioned before, the base hydrodynamic parameters are pressure drop (ΔP) and liquid holdup (ε_L). In the literature on the subject, a large number of studies have been presented in which, for various experimental systems (Charpentier and Favier 1975; Midoux et al. 1976; Specchia and Baldi 1977; Mills and Dudukovic 1981; Sai and Varma 1987; Bartelmus and Janecki 2003), specific range of operating conditions (Szlemp et al. 2001; Burghardt et al. 2002, 2005), atmospheric and elevated pressure (Larachi et al. 1991; Wammes et al. 1991; Al-Dahhan et al. 1997), the said parameters have been determined experimentally. The correlations which have been established on the basis of experimental data are summarized in the articles of Soroha and Nigam (1996) and Al-Dahhan et al. (1997).

2.2.1 The Influence of the Form of Equations Defining Drag Forces

As a result of the studies cited above three one-dimensional two-phase models have been developed which enable the prediction of the pressure gradient and liquid saturation in TBR and provide formulae for coefficients which determine the interphase momentum exchange. These include the relative permeability model of Saez and Carbonell (1985), the single slit model of Holub et al. (1992), and the two-fluid phase interaction model of Attou et al. (1999). The form of the relationships which define the interaction forces between phases is based on the Ergun equations.

The relative permeability model as well as the slit model neglects the interphase force between the gas and liquid phases, thus assuming zero drag force at the gas and liquid interface. However, the experimental studies have shown (Wammes et al. 1991) that the gas flow has considerable influence on the hydrodynamics of the trickle-bed reactor, especially at high operating pressures. Accordingly, the interactions between the gas and liquid phases are not negligible with regard to the other momentum transfer mechanisms. So, only the two-fluid phase interaction model of Attou et al. (1999), provides the formulae for the coefficients of interphase momentum exchange which determine all the interaction forces in gas-liquid-solid particles system.

The main goal of the study by Janecki et al. (2016a), was to compare the experimental and computational results obtained for different forms of equations defining drag forces between the phases based on specially performed experimental results. In this study the mean relative error as well as standard deviation of determined experimentally and computed values of pressure drop and average liquid holdup have been used as criterion for the validation of the model. The choice of appropriate equations defining the drag forces is a very significant element of uncertainty in the modeling of processes in a TBR. Therefore, their selection is crucial for accurate modeling of TBR processes. Thus, numerous authors, in order to elaborate the best approximation of experimentally obtained hydrodynamic parameters transformed and compiled the relationships of friction factors adapted from the three models cited above, but mainly from the model of Attou et al. (1999).

The relations tested by Janecki et al. (2016a), are listed in Table 1. The following constants were introduced into the relations listed in Table 1.

$$A_g = E_1 \mu_g \frac{(1 - \alpha_g)^2}{\alpha_g d_p^2}; \quad A_L = E_1 \mu_L \frac{\alpha_S^2}{\alpha_L^2 d_p^2} \tag{11}$$

Table 1 Interaction forces equations tested in calculations (Janecki et al. 2016a)

Case	Exchange coefficients F_{jk}	References
I	$F_{gL} = A_g \left(\frac{\alpha_S}{1-\alpha_g}\right)^{2/3} + B_g \left(\frac{\alpha_S}{1-\alpha_g}\right)^{1/3} (u_g - u_L)$ $F_{gS} = A_g \left(\frac{\alpha_S}{1-\alpha_g}\right)^{2/3} + B_g \left(\frac{\alpha_S}{1-\alpha_g}\right)^{1/3} u_g \quad F_{LS} = (1 - \alpha_S)(A_L + B_L u_L)$	Attou et al. (1999) applied by Hamidipour et al. (2013) and Janecki et al. (2014)
II	$F_{gL} = A_g \left(\frac{\alpha_S}{1-\alpha_g}\right)^{2/3} + B_g \left(\frac{\alpha_S}{1-\alpha_g}\right)^{1/3} (u_g - u_L)$ $F_{gS} = A_g \left(\frac{\alpha_S}{1-\alpha_g}\right)^{2/3} + B_g \left(\frac{\alpha_S}{1-\alpha_g}\right)^{1/3} u_g \quad F_{LS} = \alpha_L (A_L + B_L u_L)$	Gunjal et al. (2005), Gunjal and Ranade (2007), Lappalainen et al. (2009), Kuzeljevic and Dudukovic (2012)
III	$F_{gL} = \frac{\alpha_g}{(1-\alpha_S)} \left[A_g \left(\frac{\alpha_S}{1-\alpha_g}\right)^{2/3} + B_g \left(\frac{\alpha_S}{1-\alpha_g}\right)^{1/3} (u_g - u_L) \right]$ $F_{gS} = \frac{\alpha_g}{(1-\alpha_S)} \left[A_g \left(\frac{\alpha_S}{1-\alpha_g}\right)^{2/3} + B_g \left(\frac{\alpha_S}{1-\alpha_g}\right)^{1/3} u_g \right]$ $F_{LS} = \alpha_L (A_L + B_L u_L)$	Jiang et al. (2002a, b)
IV	$F_{gL} = \frac{\alpha_g}{(1-\alpha_S)} \left[A_g \left(\frac{\alpha_S}{1-\alpha_g}\right)^{2/3} + B_g \left(\frac{\alpha_S}{1-\alpha_g}\right)^{1/3} (u_g - u_L) \right]$ $F_{gS} = \frac{\alpha_S}{1-\alpha_g} \left[A_g \left(\frac{\alpha_S}{1-\alpha_g}\right) + B_g u_g \right] \quad F_{LS} = \alpha_L (A_L + B_L u_L)$	Jiang et al. (2001, 2002a, b)
V	$F_{gL} = \frac{\alpha_g}{(1-\alpha_S)} \left[A_g \left(\frac{\alpha_S}{1-\alpha_g}\right)^{2/3} + B_g \left(\frac{\alpha_S}{1-\alpha_g}\right)^{1/3} (u_g - u_L) \right]$ $F_{gS} = \frac{\alpha_S (1-\alpha_S)^{1.8}}{(1-\alpha_g) \alpha_g^{1.8}} \left[A_g \left(\frac{\alpha_S}{1-\alpha_g}\right) + B_g u_g \right] F_{LS} = \alpha_L \left(\frac{\varepsilon - \alpha_L^0}{\alpha_L - \alpha_L^0}\right)^{2.43} (A_L + B_L u_L)$	Jiang et al. 2002a, b)
VI	$F_{gS} = \frac{\alpha_S (1-\alpha_S)^{1.8}}{(1-\alpha_g) \alpha_g^{1.8}} \left[A_g \left(\frac{\alpha_S}{1-\alpha_g}\right) + B_g u_g \right]$ $F_{LS} = \frac{\alpha_L^4}{(1-\alpha_S)^3} \left(\frac{\varepsilon - \alpha_L^0}{\alpha_L - \alpha_L^0}\right)^{2.43} (A_L + B_L u_L)$	Atta et al. (2007a, b, 2010a, b)

$$B_g = E_2 \rho_g \frac{1 - \alpha_g}{d_p}; \quad B_L = E_2 \rho_L \frac{\alpha_S}{\alpha_L d_p} \quad (12)$$

where E_1 and E_2 are Ergun constants, d_p is particle diameter (m) and ε_L^0 (Table 1) is static liquid holdup.

Taking into account the results of simulations presented by Janecki et al. (2014), the values of Ergun constants equal to $E_1 = 180$ and $E_2 = 1.8$ as well as constant bed porosity [D/d_p used in the experiments exceeds the limiting value $D/d_p = 18$, where D is column diameter (m)] have been applied in the computations.

Experimental data-base obtained as a result of measurements carried out for varying flow rates of both phases and for the systems with various physicochemical properties (water and solutions of glycerol) was the frame of reference for calculations. The mean relative error (e_Y) as well as the standard deviation (σ_{st}) of the averaged liquid holdup (ε_L) and the pressure drop (ΔP) with respect to the experimental values have been determined for each set and listed in Table 2.

Additionally, special diagrams (parity plots) have been prepared comparing the computed results with experimental data (Fig. 1) which illustrate clearly the discrepancies between calculated and experimental values for each set of friction factors.

Analyzing the errors of the hydrodynamic parameters listed in Table 2, one should distinguish the set of friction factors of Case I which exhibits the lowest errors among the sets analyzed and so approximates the experimental values with the best agreement. Striking and quite inexplicable is the large error of the pressure drop in Case V. In the modified Case V equation proposed by Atta et al. (2007a, b, 2010a, b) (Case VI) has been applied for calculation F_{LS} values causing a considerable decrease of the pressure drop error.

Simulations performed by Janecki et al. (2016a), indicate clearly that the classical equations of Attou et al. (1999), (Case I), defining the friction factors F_{jk} approximate the experimental values of the hydrodynamic parameters with the best agreement and were recommended to apply in the momentum balances of TBRs.

Table 2 Mean relative error (e_Y) and standard deviation (σ_{st}) of the gas pressure drop and liquid holdup values obtained experimentally and calculated from CFD model applying in simulations the values of F_{jk} from Table 1 (Janecki et al. 2016a)

Cases	ΔP (Pa)		ε_L	
	e_Y (%)	σ_{st} (%)	e_Y (%)	σ_{st} (%)
Case I	16.61	15.93	6.76	2.67
Case II	28.65	6.12	24.8	4.26
Case III	49.57	4.18	19.4	4.53
Case IV	43.21	4.75	20.22	4.37
Case V	119.91	27.17	15.2	3.43
Case VI	42.71	5.54	12.05	2.18
Modified Case V	30.18	8.66	18.9	2.5

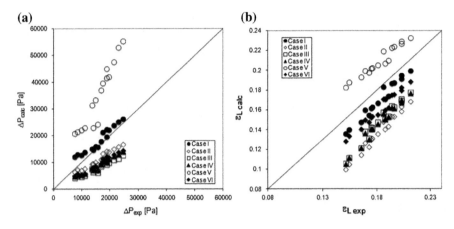

Fig. 1 Parity plot for the prediction of the pressure drop **a** and total liquid holdup **b** using in CFD model the values of F_{jk} from Table 1 (Janecki et al. 2016a)

2.2.2 The Influence of the Porosity Profiles

The studies published as well as our own experience concerning the application of the CFD model to the TBR allow us to conclude that the porosity and its distribution in the bed are the key parameters in determining the flow distribution within the bed. Therefore, the two very important relations which have to be implemented into the CFD model are the axial and radial porosity of the bed. These relations can influence considerably the computed values of the average holdup and the pressure drop especially in the laboratory reactors where the ratio of the column diameter to the particle diameter is small $(D/d_p < 20)$.

In the study by Janecki et al. (2014), six correlations for the radial variation of the axially averaged radial porosity have been chosen (Table 3). They comprise 3 representatives of the oscillatory correlations and 3 of the exponential correlations (Fig. 2).

Applying the axially averaged radial porosity and assuming a standard deviation of the Gaussian probability density function, the axial porosity profile has been developed for each radial section in the column.

The computations determining the averaged holdup in the bed and the pressure drop have been performed for four experimental systems (water and solutions of glycerol: 30, 35 and 45 wt%) applying for each of them six radial porosity profiles (Table 3) and three sets of Ergun constants ($E_1 = 180$, $E_2 = 1.8$—experimentally determined; $E_1 = 150$, $E_2 = 1.8$—suggested by Macdonald et al. (1979); $E_1 = 235$, $E_2 = 1.59$—determined from relations estimated by Iliuta et al. (1998)) in the equations defining the interphase momentum exchange.

In order to introduce a quantitative criterion for the evaluation and comparison of the computed results the mean relative error of the averaged holdup and the pressure drop with respect to the experimental values have been determined for each set.

Table 3 Porosity correlations (reprinted from Janecki et al. 2014, with permission from Elsevier)

Authors	Relationship
Mueller (1992)	$\varepsilon(r) = \varepsilon + (1 - \varepsilon)J_0(ar^*)e^{-br}$ where: $a = 8.243 - \dfrac{12.98}{\left(D/d_p - 3.156\right)}$ for $2.61 \leq D/d_p \leq 13$ $a = 7.383 - \dfrac{2.932}{\left(D/d_p - 9.864\right)}$ for $13 \leq D/d_p.$
Martin (1978)	$\varepsilon(x) = \varepsilon_{min} + (1 - \varepsilon_{min})x^2$ for $-1 \leq x \leq 0$ $\varepsilon(x) = \varepsilon_b + (\varepsilon_{min} - \varepsilon)\exp\left(\frac{-x}{4}\right)\cos\left(\frac{\pi x}{0.876}\right)$ for $x \geq 0$ where: $x = 2\frac{R-r}{d_p} - 1;$ $\varepsilon_{min} = 0.2$
de Klerk (2003)	$\varepsilon(r) = 2.14z^2 - 2.53z + 1$ for $z \leq 0.637$ $\varepsilon(r) = \varepsilon + 0.29\exp(-0.6z)[\cos(2.3\pi(z - 0.16))]$ $+ 0.15\exp(-0.9z)$ for $z \geq 0.637$ where: $z = \frac{R-r}{d_p}$
Sun et al. (2000)	$\varepsilon(r) = 1 - (1 - \varepsilon)\left[1 - \exp\left(-2\left(\frac{R-r}{d_p}\right)^2\right)\right]$
Hunt and Tien (1990)	$\varepsilon(r) = \varepsilon\left[1 + \left(\frac{1-\varepsilon}{\varepsilon}\exp\left(-N\left(\frac{R-r}{d_p}\right)\right)\right)\right](N = 2, N = 6)$

where J_0 is zeroth order Bessel function, D and d_p are column and particle diameters (m)

As the values of the errors of the averaged holdup are comprised for all the analyzed systems within the range of several per cents (2–4% for E_1 – 235.53 and $E_2 = 1.59$), which can be treated as a good agreement with experiments, the authors (Janecki et al. 2014) decided to choose the errors of the pressure drop as the criterion for selecting of the optimum system (Fig. 3). The values of the errors of the pressure drop determined experimentally and computed by implementing the Martin (1978) correlation as well as the Hunt and Tien (1990) correlation ($N = 6$) into the CFD model were the lowest among the systems analyzed independently of the set of Ergun constants applied. Therefore, these radial porosity correlations in connection with the relations defining the interphase momentum exchange by Attou et al. (1999), and using the Ergun constants $E_1 = 235.53$ and $E_2 = 1.59$ (Iliuta et al. 1998) can be recommended for modelling of the hydrodynamics in TBRs with the low value of the ratio D/d_p, i.e. especially in laboratory reactors by means of CFD.

The analysis of Fig. 4 presenting the gas velocity profiles in the cross-section of the bed enables the conclusion that the radial porosity profile correlations investigated (except the Martin 1978; Hunt and Tien 1990 ($N = 6$) correlation) overpredict the porosity in the wall region influencing considerably the distribution of the fluid flow in the cross-section of the bed which in consequence leads to the decrease of the computed pressure drop with respect to the experimental values. Therefore, the use of the porosity profile in the wall region of the bed in the model of the

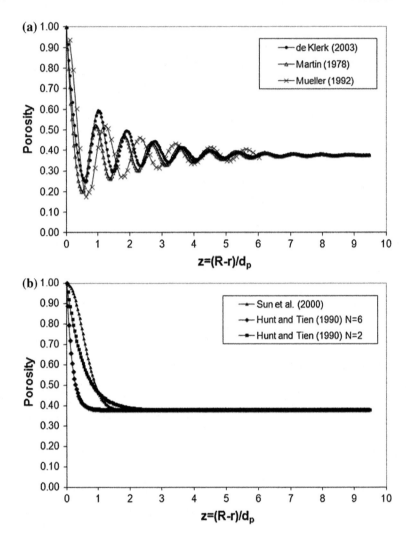

Fig. 2 Radial variation in porosity of oscillatory (**a**) and exponential (**b**) correlations (reprinted from Janecki et al. 2014, with permission from Elsevier)

hydrodynamics for the trickle-bed reactor is crucial for the estimation of correct values of the pressure drop especially for low values of the ratio D/d_p (laboratory reactors).

The computations performed by Janecki et al. (2014), for varying ratios of D/d_p showed that for values of $D/d_p > 25$ the error applying in computations constant bed porosity is lower than 10%. According to Herskowitz and Smith (1978) the wall effects can be ignored for $D/d_p > 18$.

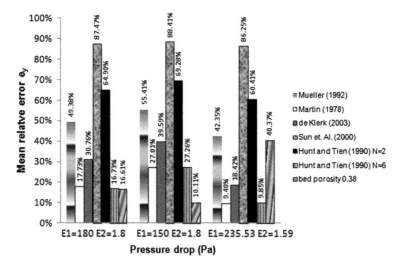

Fig. 3 Mean relative error (e_y) of the gas pressure drop values obtained experimentally and calculated from CFD model

3 Mathematical Model of a Reaction Process in the TBR

The CFD model regarding a chemical reaction process in TBR consists of the continuity equations (1) and the momentum balances of each phase (2) which must be supplemented by addition of mass balances of components for each phase of the fluid mixture.

Each fluid phase is treated as a continuous phase of various velocities, volume fractions, compositions and physicochemical properties. The solid catalyst phase is also considered as a continuous pseudohomogeneous phase connected with the fluid phase by means of transport phenomena.

The macroscopic averaged mass balances equations of component i in the phase k is:

$$\frac{\partial}{\partial t}\left(\alpha_k \rho_k Y_k^i\right) + \nabla \cdot \left(\alpha_k \rho_k Y_k^i \overline{u_k}\right) = -\nabla \cdot \left(\alpha_k \overline{J}_k^i\right) + \alpha_k \rho_k S_k^i + \sum_{\substack{j \\ j \neq k}}^{\vee} J_{jk}^i$$

$$k = 1, 2, 3 \text{ phases}, \quad i = 1, 2, \ldots, n_k \text{ components of phases k}$$

(13)

where Y_k^i is the mass fraction of component i in the phase k and α_k, ρ_k and $\overline{u}_k$ the volume fraction, density (kg m^{-3}) and velocity (m s^{-1}) of phase k.

The first term of the left-hand side of Eq. (13) presents the net rate of accumulation of component i in phase k per unit volume and the second term the net rate of addition of the mass of component i per unit volume by convection within phase

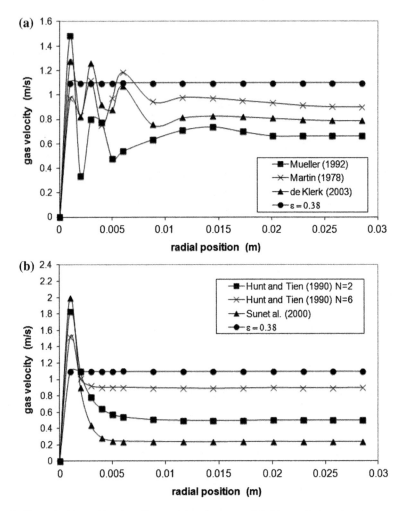

Fig. 4 Gas velocity profile of oscillatory (**a**) and exponential (**b**) correlations for nitrogen-water system for $w_g = 0.25$ m/s, $w_L = 0.0045$ m/s depending on radial position in the column (reprinted from Janecki et al. 2014, with permission from Elsevier)

k. $\vec{J}_k^i$ is the diffusional mass flux of component i in phase k (kg m^{-2} s^{-1}) defined by the relationship:

$$\vec{J}_k^i = -\rho_k D_k^i \nabla Y_k^i \tag{14}$$

where D_k^i is kinematic diffusion coefficient (m^2 s^{-1}) and S_k^i presents the source term of component i in phase k (kg$i$ kg^{-1} s^{-1}) which expresses the production of component i as a result of chemical reactions taking place in the phase k. Similarly, $\overset{\vee}{J}_{jk}^i$

defines the mass flux of component i from phase j into phase k (kg m^{-2} s^{-1}) by means of empirical equation which uses mass transfer coefficients:

$$\overset{\vee}{J^i_{jk}} = K^i_{jk}\left(\frac{Y^i_j}{N^i_{jk}} - Y^i_k\right) \tag{15}$$

where N^i_{jk} is the slope of the equilibrium curve of component i between phases j and k and K^i_{jk} is overall mass transfer coefficient (kg m^{-2} s^{-1}).

Janecki et al. (2016b), investigated the process of catalytic wet air oxidation (CWAO) for phenol in a trickle-bed reactor using active carbon as a catalyst (operating parameters: $P = 1.85$ MPa, $T = 393$; 413; 433 K). The reaction system comprised two main reagents: oxygen in the gas and liquid phase and phenol in the liquid phase. The detailed mass balances of the two reagents take the following form:

- oxygen balance in the gas phase:

$$\frac{\partial}{\partial t}\left(\alpha_g \rho_g Y^{O_2}_g\right) + \nabla \cdot \left(\alpha_g \rho_g Y^{O_2}_g \bar{u}_g\right) = \nabla \cdot \left(\alpha_g \rho_g D^{O_2}_g \nabla Y^{O_2}_g\right)$$
$$- (K_L a_{gL})^{O_2}\left(\frac{Y^{O_2}_g}{N^{O_2}_{gL}} - Y^{O_2}_L\right) - (k_{gS} a_{gS})^{O_2} \rho_g \left(Y^{O_2}_g - Y^{O_2}_{gS}\right) \tag{16}$$

$$\frac{1}{K^{O_2}_L} = \frac{1}{k^{O_2}_g \rho_g N^{O_2}_{gL}} + \frac{1}{k^{O_2}_L \rho_L} \tag{17}$$

- oxygen balance in the liquid phase:

$$\frac{\partial}{\partial t}\left(\alpha_L \rho_L Y^{O_2}_L\right) + \nabla \cdot \left(\alpha_L \rho_L Y^{O_2}_L \bar{u}_L\right) = \nabla \cdot \left(\alpha_L \rho_L D^{O_2}_L \nabla Y^{O_2}_L\right)$$
$$+ (K_L a_{gL})^{O_2}\left(\frac{Y^{O_2}_g}{N^{O_2}_{gL}} - Y^{O_2}_L\right) - (k_{LS} a_{LS})^{O_2} \rho_L \left(Y^{O_2}_L - Y^{O_2}_{LS}\right) \tag{18}$$

phenol balance in the liquid phase:

$$\frac{\partial}{\partial t}\left(\alpha_L \rho_L Y^{Ph}_L\right) + \nabla \cdot \left(\alpha_L \rho_L Y^{Ph}_L \bar{u}_L\right) = \nabla \cdot \left(\alpha_L \rho_L D^{Ph}_L \nabla Y^{Ph}_L\right)$$
$$- (k_{LS} a_{LS})^{Ph} \rho_L \left(Y^{Ph}_L - Y^{Ph}_{LS}\right) - (1-f) M^{Ph} \eta \rho_p \left[-r^{Ph}\left(\frac{C^{O_2}_{gS}}{H^{O_2}_{gL}}, C^{Ph}_{LS}\right)\right] \tag{19}$$

where f is the wetting efficiency of the catalyst pellet, η the effectiveness factor, $H^{O_2}_{gL}$ Henry's constant, k^i_{jk} mass transfer coefficient (m s^{-1}), a_{jk} specific surface area (m^{-1}), K_L overall mass transfer coefficient (kg m^{-2} s^{-1}) and M molar mass (kg k mol^{-1}). The boundary conditions at the inlet into the reactor are: ($z = 0$; $0 \leq r \leq R$):

$$Y_g^{O_2} = Y_g^{0,O_2}; Y_L^{O_2} = Y_L^{0,O_2}; Y_L^{Ph} = Y_L^{0,Ph} \tag{20}$$

and the fluxes of components perpendicular to the cylindrical walls of the reactor are equal to zero.

In Eqs. (16–20) superscript i denotes i component, O_2 oxygen, Ph phenol.

Values of velocities and volume fraction which appear in the mass balances of components were determined as a result of solving the equations of continuity (1) and the momentum balances (2) presented in Sect. (2).

The experimental results have been simulated by two models: the two-dimensional, axisymmetric Eulerian multi-fluid model (CFD) and, for comparison, by the plug flow model. From the parity plot (Fig. 5), which compares directly the experimental and predicted results, as well as from the estimated mean errors and standard deviations (11.6 and 5.2% for the CFD model and 16.8 and 12.2% for the plug flow model) it is evident that the CFD model approximates better the experimental data than the plug flow model.

The plug flow model underpredicts the computed results with respect to the experimental ones, especially in the range of low liquid velocities which lead to low values of wetting efficiency of the bed. The porosity profile in the wall region influences in the presented case considerably the distribution of the fluid flow in the cross section of the bed ($D/d_p = 13.3 < 25$) so that the wall region occupies a significant part of the whole section of the bed and influences considerably the computed values of phenol concentrations.

The case study confirms that the use of porosity profiles in modelling TBR is crucial for estimating the correct values of process variables especially for laboratory-scale reactors.

Fig. 5 Parity plot comparing the values of outlet phenol concentrations obtained experimentally and predicted by the two models (Janecki et al. 2016b)

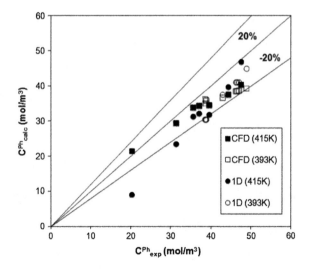

References

Al-Dahhan MH, Larachi F, Duduković MP et al (1997) High pressure trickle-bed reactors: a review. Ind Eng Chem Res 36:3292–3314

Atta A, Shantanu R, Nigam KDP (2007a) Prediction of pressure drop and liquid holdup in trickle bed reactor using relative permeability concept in CFD. Chem Eng Sci 62:5870–5879

Atta A, Shantanu R, Nigam KDP (2007b) Investigation of liquid maldistribution in trickle bed reactor using porous media concept in CFD. Chem Eng Sci 62:7033–7044

Atta A, Hamidipour M, Roy S et al (2010a) Propagation of slow/fast-mode solitary liquid waves in trickle beds via electrical capacitance tomography and computational fluid dynamics. Chem Eng Sci 65:1044–1150

Atta A, Roy S, Nigam KDP (2010b) A two-phase Eulerian approach using relative permeability concept for modeling of hydrodynamics in trickle-bed reactors at elevated pressure. Chem Eng Res Des 88:369–378

Attou A, Ferschneider G (2000) A two-fluid hydrodynamic model for the transition between trickle and pulse flow in cocurrent gas-liquid packed-bed reactor. Chem Eng Sci 55:491–511

Attou A, Boyer C, Ferschneider G (1999) Modeling of the hydrodynamics of the cocurrent gas-liquid trickle flow through a trickle-bed reactor. Chem Eng Sci 54:785–802

Bartelmus G, Janecki D (2003) Hydrodynamics of a cocurrent downflow of gas and foaming liquid through the packed bed. Part II. Liquid holdup and gas pressure drop. Chem Eng Process 42:993–1005

Bartelmus G, Gancarczyk A, Stasiak M (1998) Hydrodynamics of cocurrent fixed-bed three-phase reactors, the effect of physicochemical properties of the liquid on pulse velocity. Chem Eng Proc 37:331–341

Boyer C, Koudil A, Chen P et al (2005) Study of liquid spreading from a point source in a trickle bed via gamma-ray tomography and CFD simulation. Chem Eng Sci 60:6279–6288

Burghardt A (2014) Eulerian three-phase flow model applied to trickle-bed reactors. Chem Process Eng 35(1):75–96

Burghardt A, Bartelmus G, Gancarczyk A (1999) Hydrodynamics of pulsing flow in three-phase catalytic reactors. Chem Eng Proc 38:441–426

Burghardt A, Bartelmus G, Janecki D et al (2002) Hydrodynamics of a three-phase fixed-bed reactor operating in the pulsing flow regime at an elevated pressure. Chem Eng Sci 57:4855–4863

Burghardt A, Bartelmus G, Szlemp A et al (2005) Analiza matematycznych kryteriów zmiany reżimów hydrodynamicznych w reaktorach trójfazowych. Chem Process Eng 26:259–279

Charpentier JC, Favier M (1975) Some liquid holdup experimental data in trickle-bed reactors for foaming and nonfoaming hydrocarbons. AJChE J 21:1213–1218

de Klerk A (2003) Voidage variation in packed beds at small column to particle diameter ratio. AIChE J 49:2022–2029

Duduković MP, Larachi F, Mills PL (2002) Multiphase catalytic reactors: a perspective on concurrent knowledge and future trends. Catal Rev 44:123–146

Enwald H, Peirano E, Almsted AE (1996) Eulerian two-phase flow theory applied to fluidization. Int J Multiphase Flow 22:21–66

Grosser K, Carbonell RG, Sandaresan S (1988) Onset of pulsing flow in two-phase cocurrent down flow through a packed bed. AIChE J 34:1850–1860

Gunjal PR, Ranade VV (2007) Modeling of laboratory and commercial scale hydro-processing reactors using CFD. Chem Eng Sci 62:5512–5526

Gunjal PR, Ranade VV, Chaudhari RV (2003) Liquid distribution and RTD in trickle bed reactors: experiments and CFD simulations. Can J Chem Eng 81:821–830

Gunjal PR, Kashid MN, Ranade VV et al (2005) Hydrodynamics of trickle-bed reactors: experiments and CFD modeling. Ind Eng Chem Res 44:6278–6294

Hamidipour M, Chen J, Larachi F (2013) CFD study and experimental validation of trickle bed hydrodynamics under gas, liquid and gas/liquid alternating cyclic operations. Chem Eng Sci 89:158–170

Herskowitz M, Smith JM (1978) Liquid distribution in trickle-bed reactors. AIChE J 24:439–450

Holub RA, Duduković PA, Ramachandran PA (1992) A phenomenological model for pressure drop, liquid holdup and flow regime transition in gas-liquid trickle flow. Chem Eng Sci 47:2343–2348

Hunt ML, Tien CL (1990) Non-Darcian flow, heat and mass transfer in catalytic packed-bed reactors. Chem Eng Sci 45:55–63

Iliuta I, Larachi F, Grandjean BPA (1998) Pressure drop and liquid holdup in trickle flow reactors: improved Ergun constants and slip correlations for the slit model. Ind Eng Chem Res 37:4542–4550

Ishii M, Hibiki T (2011) Thermo-fluid dynamics of two-phase flow, 2nd edn. Springer, New York

Janecki D, Burghardt A, Bartelmus G (2008) Computational simulations of the hydrodynamic parameters of a trickle-bed reactor operating at periodically changing feeding the bed with liquid. Chem Proc Eng 29:583–596

Janecki D, Burghardt A, Bartelmus G (2014) Influence of the porosity profile and sets of Ergun constants on the main hydrodynamic parameters in the trickle-bed reactors. Chem Eng J 237:176–188

Janecki D, Burghardt A, Bartelmus G (2016a) Parametric sensitivity of CFD model concerning the hydrodynamics of trickle-bed reactor (TBR). Chem Proc Eng 37:97–107

Janecki D, Szczotka A, Burghardt A et al (2016b) Modelling of wet-air oxidation of phenol in a trickle-bed reactor using active carbon as a catalyst. J Chem Technol Biotechnol 91:59–607

Jiang Y, Khadikar MR, Al-Dahhan MH et al (1999) CFD modeling of multiphase flow distribution in catalytic packed bed reactors: Scale down issues. Catal Today 66:209–218

Jiang Y, Khadilkar MR, Al-Dahhan MH et al (2001) CFD modelling of multiphase flow distribution in catalytic packed bed reactors: scale down issues. Catal Today 66:209–218

Jiang Y, Khadilkar MR, Al-Dahhan MH et al (2002a) CFD of multiphase flow in packed-bed reactors: I. K-fluid modelling issues. AIChE J 48:701–715

Jiang Y, Khadilkar MR, Al-Dahhan MH et al (2002b) CFD of multiphase flow in packed-bed reactors: II. Results and applications. AIChE J 48:716–730

Kuzeljevic ZV, Dudukovic MP (2012) Computational modeling of trickle bed reactors. Ind Eng Chem Res 51:1663–1671

Lappalainen K, Manninen M, Alopaeus V (2009) CFD modeling of radial spreading of flow in trickle-bed reactors due to mechanical and capillary dispersion. Chem Eng Sci 64:207–218

Larachi F, Laurent A, Midoux N et al (1991) Experimental study of a trickle-bed reactor operating at high pressure: two phase pressure drop and liquid saturation. Chem Eng Sci 46:1233–1246

Lopes RJG, Quinta-Ferreira RM (2008) Three dimensional numerical simulation of pressure drop and liquid holdup for high-pressure trickle-bed reactor. Chem Eng J 145:112–120

Lopes RJG, Quinta-Ferreira RM (2009) CFD modeling of multiphase flow distribution in trickle beds. Chem Eng J 147:342–355

Lopes RJG, Quinta-Ferreira RM (2010) Hydrodynamic simulation of pulsing-flow regime in high-pressure trickle-bed reactor. Ind Eng Chem Res 49:1105–1112

Lyczkowski RW (2010) The history of multiphase computational fluid dynamics. Ind Eng Chem Res 49:5029–5036

MacDonald LF, El-Sayed MS, Mow K et al (1979) Flow through porous media—the Ergun equation revisited. Ind Eng Chem Fundam 18:199–208

Martin H (1978) Low Peclet number particle to fluid heat and mass transfer. Chem Eng Sci 33:913–919

Midoux N, Favier M, Charpentier JC (1976) Flow pattern, pressure loss and liquid holdup data in gas-liquid downflow packed beds with foaming and nonfoaming hydrocarbons. J Chem Eng Jpn 9:350–356

Mills PL, Dudukovic MP (1981) Evaluation of liquid-solid contacting in trickle-bed reactors by tracer methods. AIChE J 27:893–904

Mueller G (1992) Radial void fraction distribution in randomly packed fixed beds of uniformly sized spheres in cylindrical containers. Powder Technol 72:269–275

Saez AE, Carbonell RG (1985) Hydrodynamic parameters for gas-liquid cocurrent flow in packed beds. AIChE J 31:52–62

Sai PST, Varma YBG (1987) Pressure drop in gas-liquid downflow through packed beds. AIChE J 33:2027–2036

Saroha AK, Nigam KDP (1996) Trickle bed reactors. Rev Chem Eng 12:207–347

Satterfield CN (1975) Trickle-bed reactors. AIChE J 21:209–228

Shah YT (1979) Gas-liquid-solid reactor design. Mc Graw Hill, New York

Soo SL (1990) Multiphase fluid dynamics. Science Press, Gower Technical, New York

Specchia V, Baldi G (1977) Pressure drop and liquid hold-up for two phase concurrent flow in packed beds. Chem Eng Sci 32:515–532

Sun CG, Yin FH, Afacan AKT et al (2000) Modelling and simulation of flow maldistribution in random packed columns with gas-liquid countercurrent flow. Trans IChemE 78A:378–388

Szady MJ, Sundaresan S (1991) Effect of boundaries on trickle bed hydrodynamics. AIChE J 37:1237–1241

Szlemp A, Bartelmus G, Janecki D (2001) Hydrodynamics of a co-current three-phase solid-bed reactor for foaming systems. Chem Eng Sci 56:1111–1116

Wammes WJA, Middelkamp J, Huisman WJ et al (1991) Hydrodynamics in a concurrent gas-liquid trickle bed at elevated pressures. AIChE J 37:1849–1861

The Use of Spray Drying in the Production of Inorganic-Organic Hybrid Materials with Defined Porous Structure

Teofil Jesionowski, Beata Michalska, Marcin Wysokowski
and Łukasz Klapiszewski

1 Spray Drying in Materials Science

In the spray drying process, a raw material in the form of a liquid or suspension is
converted into a substance in powder form. This technique has found applications
in many branches of science, and at present it is a commonly used process in the
manufacture of food, pharmaceuticals (Vehring 2008), ceramics and chemicals
(Jesionowski and Krysztafkiewicz 1996; Trandafir et al. 2010; Pilarska et al. 2011;
Zellmer et al. 2015). Of particular interest is the use of spray drying processes in
contemporary materials chemistry and engineering. The technique serves to meet
the growing demand for powdered materials with controlled morphology and
porous structure (Kowalski 2007). It is therefore of key. This fact provides moti-
vation for further research into the impact of the process parameters that control the
formation and drying of the particles. Importance to identify and of the process and
the mechanisms responsible for controlling the shape, structure and size of the final
particles of the dried products (Vehring et al. 2007). In view of the wide range of
variables having an impact on the results of spray drying, and the interdependences
between them, the modeling of the process is a significant.

2 Fundamentals of the Spray Drying Process

The first person to patent a spray drying process was Samuel Percy in 1872, who
titled his work *Improvements in Drying and Concentrating Liquid Substances by
Atomizing* (Percy 1872). The industrial use of the spray drying technique began in

T. Jesionowski (✉) · B. Michalska · M. Wysokowski · Ł. Klapiszewski
Institute of Chemical Technology and Engineering,
Poznan University of Technology, Poznań, Poland
e-mail: teofil.jesionowski@put.poznan.pl

© Springer International Publishing AG, part of Springer Nature 2018
M. Ochowiak et al. (eds.), *Practical Aspects of Chemical Engineering*,
Lecture Notes on Multidisciplinary Industrial Engineering,
https://doi.org/10.1007/978-3-319-73978-6_12

1920, when it was applied to milk and detergents. Since then it has been used in a wide variety of types of food production (Keshani et al. 2015).

The spray drying process consists of three main stages:

- breaking up of the liquid source material into a spray (*atomization*)
- contact with the hot drying medium
- evaporation of solvent (moisture).

The process uses convection spray dryers (Vicente et al. 2013). By selecting an appropriate dryer design and altering the process settings, it is possible to exert effective control over the properties of the product (Warych 2004). The material undergoing the drying process may be supplied in the form of a solution, emulsion, suspension or paste (Li et al. 2010). The breaking up of the material into small droplets, which are then well mixed with the drying gas, is a key point in the process (Lintingre et al. 2016) which has a significant effect on the properties of the dried product, in view of the behavior of the droplets during the drying process. At the spray-forming stage, suitable atomizers are selected for the applicable drying conditions, leading to economically desired products. Contact between the atomizer and the air is defined by the position of the atomizer relative to the drying air inlet (Keshani et al. 2015). The process of mixing the liquid product with the stream of drying air serves to define the method of spray drying (Rosenberg et al. 1990). Spray-forming devices include rotary (disk) atomizers, hydraulic or pneumatic nozzles, and ultrasonic atomizers (Bittner and Kissel 1999; Cal and Sollohub 2010; Kemp et al. 2016). Rotary atomizers can produce sprays with droplet diameters in the range 10–250 µm. The stream of atomized liquid in laboratory dryers is measured in liters per hour, but in industrial dryers may reach values of over 200,000 kg/h. The size of the droplets formed is affected by the rotational speed, the mass flow rate, the concentration of dispersed particles, the viscosity and the surface tension. In hydraulic nozzles the liquid flows under appropriate pressure through a hole approximately a millimeter in diameter, where it is broken up into small droplets. Higher pressures lead to droplets with smaller diameters; the minimum achievable diameter is several tens of micrometers. A single industrial nozzle can provide a flow capacity up to approximately 750 kg/h. For large streams, combined nozzles are used. In the case of hydraulic nozzles it is not possible to atomize suspensions, which may block the holes. Pneumatic nozzles are used with dryers of small capacities, where the spray is formed using compressed air at pressures up to several megapascals. The ratio of the air flow rate to that of the atomized liquid determines the size of the final particles, the minimum droplet diameter being measured in micrometers. Ultrasonic atomizers are used for small streams of liquid (Bittner and Kissel 1999). The cloud of droplets is fed into the dryer chamber towards the hot drying gas, where in a short time the two are mixed (Warych 2004). During evaporation there must be no disturbance of the motion of the droplets in the dryer or contact with the wall of the apparatus, and the degree of drying must be adequate (Walton 2002). For this reason the gas inlet must be suitably designed—it is usually a channel in the form of an Archimedean spiral

(Doolittle 1931). The evaporating particles have the form of droplets until the moisture content becomes sufficiently low that the droplets are dispersed by the dried surface (Mezhericher et al. 2010). To achieve complete drying of the droplets without the material settling on the walls of the apparatus, the system consisting of the drying chamber, atomizing device and gas distribution device must be suitably constructed. Spray dryers may use parallel currents, opposite currents or mixed current. The dried powder is obtained in a cyclone, filter bag or electrostatic dust collector (Keshani et al. 2015). The solid particles formed, with micrometric sizes, are similar to each other in terms of size, shape, porosity, density and chemical composition (Walker et al. 1999). The spent gases, following dust removal, may be returned to circulation or released into the atmosphere. Diaphragm heating of the gases (air) takes place in heat exchangers with jacketed pipes, although in low-capacity dryers electrical heating may be used. The flow of the drying medium through the dryer is enabled by blowers or radial fans, which clean the drying air of aerosol particles and other impurities (Warych 2004). The difference in the temperature of the product at the entry to and the exit from the dryer as well as the mass flow rate of the material affect the rate of drying, which is directly proportional to these values. The initial temperature is dependent on the thermal resistance of the product, which must not undergo thermal degradation, while the exit temperature is determined based on the humidity of the final product. Based on the temperature difference it is possible to determine the required gas flow for a given rate. The time for which the material remains in the dryer is estimated experimentally for a given particle size and given properties of the drying process (Chew et al. 2015; Keshani et al. 2015). The spray drying process is a complex phenomenon involving multiphase flow—a gas phase (drying air), a liquid phase (droplets) and a solid phase (particles). Furthermore each phase is not a pure substance, but a mixture of several components (Mezhericher et al. 2010). Evaporation of droplets of the solution during drying is accompanied by heat and mass transport. The driving force of the process is the difference between the partial pressure of the solvent vapor and its partial pressure in the gas phase. The rate of evaporation is determined by the equilibrium between the stream of evaporation enthalpy and the stream of energy being transported to the surface of the droplets. The latter may be supplied from the gas phase and from the thermal capacity of the droplets as a result of cooling. The rate of evaporation determines the rate of vanishing of the droplet surface area. The decrease in surface area causes diffusional motion of the dissolved substances from the surface to the interior of the droplets. If there is no internal convection, the separation of the chemical components in the evaporating droplet is described by the equation of nonlinear diffusion found in Fick's Second Law. It is assumed here that there is no interaction between the dissolved substances, and that diffusion coefficients are constant.

Given radial symmetry, applying the normalized radial coordinate $R = (r/r_s)$ the equation may be written as:

$$\frac{\partial c_i}{\partial t} = \frac{D_i}{r_s^2}\left(\frac{\partial^2 c_i}{\partial R^2} + \frac{2\partial c_i}{R\partial R}\right) + \frac{R\partial c_i\partial r_s}{r_s\partial R\partial t} \qquad (1)$$

where: c_i is the concentration of dissolved substance i, D_i is the diffusion coefficient of dissolved substance i in the liquid phase, r_s is the droplet radius, t is the evaporation time of the droplet.

Equation (1) has an analytical solution in the equilibrium state:

$$r_s\frac{\partial r_s}{\partial t} = const. \qquad (2)$$

In the equilibrium state, the evaporation of a water droplet of diameter d is proportional to the surface area of the droplet. The evaporation rate κ is defined as:

$$\partial^2(t) = d_0^2 - \kappa t \qquad (3)$$

The evaporation rate constant satisfies condition (2), because

$$\frac{\partial d^2}{\partial t} = -\kappa = 8r_s\frac{\partial r_s}{\partial t} \qquad (4)$$

Assuming constant evaporation, a solution to (1) may be given in the following form (Leong 1987):

$$c_i = c_{c,i}\exp\left(-\frac{r_s\partial r_s}{2D_i\partial t}R^2\right) \qquad (5)$$

Equation (5) expresses the concentration c as a function of the concentration in the center of the droplet, c_c. It is thus more useful to determine the concentration as a function of the mean concentration in the droplet, c_m. It is then possible to use in the calculations the initial conditions (diffusion coefficient, droplet radius, evaporation time, concentration of dissolved substance) at any moment at which they occur in the droplet (as long as the rate of evaporation is constant) using the mass balance in combination with Eq. (3). Hence Eq. (5) is integrated with the volume of a sphere, so as to introduce the mean concentration c_m:

$$c_i = c_{m,i}\frac{\exp\left(\frac{Pe_iR^2}{2}\right)}{3\int_0^1 R^2\exp\left(\frac{Pe_iR^2}{2}\right)dR} \qquad (6)$$

The dimensionless Péclet's number Pe is introduced to simplify the equation:

$$Pe_i = -\frac{r_s \partial r_s}{D_i \partial t} = \frac{\kappa}{8D_i} \tag{7}$$

Péclet's number describes the interaction between the evaporation rate of the liquid and the diffusion rate in the droplet of the nanoparticles formed as a result of atomization (Belotti et al. 2015). Hence, for $R = 1$ the surface concentration may be found (Vehring et al. 2007):

$$c_{s,i} = \frac{c_{m,i}}{3\beta_i \exp\left(\frac{Pe_i}{2}\right)} \tag{8}$$

Here β is given by:

$$\beta_i = \int_0^1 R^2 \exp\left(\frac{Pe_i}{2} R^2\right) dR \tag{9}$$

β is a function which must be numerically integrated with the number Pe to obtain an exact solution (Vehring et al. 2007).

In such an analytical evaporation model it is assumed that for most of the time the temperature of the gas, and thus also the temperature of the droplets, is constant. No account is taken of the fact that at the start of the non-stationary evaporation process the droplet temperature is subject to rapid change, or that the evaporation rate changes immediately prior to the solidification of the particles. Usually the diffusion coefficients depend on the concentration of the dissolved substances and vary significantly in the course of drying, particularly when the concentration of a dissolved substance reaches a state of saturation or when the viscosity of the liquid phase increases prior to solidification. For these reasons the above simplification makes it impossible to apply the analytical model during the solidification of the particles (Vehring et al. 2007).

Usually, when Péclet's number is large (meaning that the drying time is short and the rate of drying high) the nanoparticles have less time to diffuse into the interior and may accumulate on the droplet surface, increasing the likelihood of their combination. Moreover the capillary pressure generated during the evaporation of water is the main driving force bringing the nanoparticles close to each other, which again favors their combination. When Péclet's number is low, however, the nanoparticles have more time to diffuse from the surface to the core of the droplets, causing them to be distributed more homogeneously throughout the formed powder. The degree of combination of the nanoparticles is smaller, and hence they are more dispersible (Zhang et al. 2014). During the spray drying process an increase in pressure causes a decrease in the radius of the formed droplets, and therefore a decrease in the value of Péclet's number. This means that the nanoparticles have a shorter time of diffusion to the interior of the droplet, a smaller concentration and a more even distribution within the spray-dried particles. Moreover, with further reduction in the size of the droplets, smaller microparticles are formed, containing fewer nanoparticles. For this reason the probability of particle aggregation decreases and the dispersibility of the dried powders improves (Jaskulski et al. 2015).

A major advantage of the spray drying technique is the ability to produce powders with defined granularities and moisture content, irrespective of the capacity of the dryer. The process offers continuity, ease of operation, automatic control, low costs, speed, and the ability to work with materials both sensitive and resistant to high temperatures. An excellent example of the usefulness of the spray drying technique for preparing advanced powder materials is its use in the synthesis of nanoscale metal-organic frameworks and their assembly into hollow super-structures (Carne-Sanchez et al. 2013).

3 Spray Drying of Silica-Based Materials

Silica is a multifunctional material widely used in many fields (Bula et al. 2007; Jesionowski et al. 2009; Osińska et al. 2009; Zalewska et al. 2010) owing to its good physical and chemical properties, which can lead to the formation of products with desired morphology (Iskandar et al. 2001a). The development of production methods for morphologically defined silica nanostructures with controlled porous structure parameters is the subject of ongoing research. For a process to have the potential for practical application it must be economical as well as offering a high degree of control over the morphology of the products. Spray drying remains on of the solutions proposed to eliminate some defects of earlier methods of producing nanostructured particles (Cho 2016).

It has been proved that spray drying enables accurate control of the morphology and porous structure of the resulting particles, and can be used to obtain hollowed spherical structures of biocompatible silica (Cheow et al. 2010). The morphology of the product has been found to be strongly influenced by three key parameters: the concentration of the dried substance, the solution pH, and the ratio of the flow intensity of the atomizing gas to the rate of dosing of the sample. In general, a low concentration of the dried substance leads to a relatively homogeneous particle size distribution, particularly at low pH. In such conditions particles with small diameter are formed. Spray drying with a higher pressure of the dosed substance can lead to a product with larger particles, although they then exhibit high polydispersity. Moreover the parameters of the products depend significantly on the atomizing nozzle used and the temperature on entering the dryer (Katta and Gauvin 1975).

It has also been found that adding various stabilizers and pore-forming agents, such as SDS and polystyrene latex, provides further possibilities of control of the aforementioned parameters. Iskandar et al. (2001a, b, 2002), presented in their initial research a process of spray drying of macroporous particles using a colloidal mixture of silica with polystyrene latex as a pore-forming agent (Iskandar et al. 2001b, 2002). They showed that drying at a temperature sufficient to cause

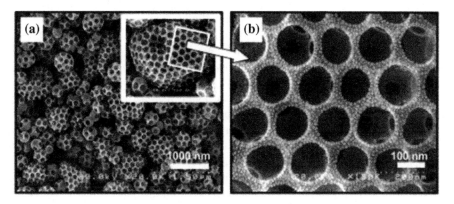

Fig. 1 SEM images of silica powders prepared using 5 nm silica particle size and 178 nm PSL particle size: **a** low magnification, **b** high magnification of surface particles. The close-packed hexagonal pores in the produced powders can be clearly observed, indicating that a self-organization of PSL particles had taken place during the drying process in the reactor tube. Image reproduced from (Iskandar et al. 2002) by permission of the American Chemical Society, Copyright 2002

decomposition of the polymer led to homogeneous spherical silica particles with an ordered mesoporous structure, as shown in Fig. 1.

The use of the spray drying technique has increased the potential for application of silica materials as fillers for chromatography columns (Dziadas et al. 2011), but above all has led to more intense work on their use as carriers for drugs and biologically active substances.

4   Spray Drying of Natural Polymers: Chitin and Lignin

Natural bioproducts and processes are becoming increasingly important for the production of new materials. Modern nanomaterial technologies have attracted a great deal of interest due to the high level of functionality of their products, related to their large and active surfaces. Most biopolymers, such as cellulose, lignin, chitin and silk, serve as supports for life and form hierarchical structures in plant and animal tissues. In terms of size, they range from simple particles and complex crystalline fibers at a nanometric level to composites on micro- and macrometric scales (Wysokowski et al. 2015; Zhu et al. 2016; Klapiszewski and Jesionowski 2017). Since nanostructures have an extremely high surface-to-volume ratio and form a highly porous mesh, their properties are different from those of micro-sized structures (Ifuku and Saimoto 2012). Therefore, in recent years, interest in using biopolymer nanoparticles—chitin (Ifuku and Saimoto 2012), cellulose, lignin (Lievonen et al. 2016)—in more advanced applications has increased rapidly. For example, lignin-based nanoparticles may find potential uses in functional surface coatings (Richter et al. 2016), nanoglue (El Mansouri and Salvadó 2006), drug delivery (Figueiredo et al. 2017), and microfluidic devices (Lievonen et al. 2016). Despite the obvious need for well-defined aqueous lignin nanoparticle dispersions, only a few scientific articles on lignin-based nanoparticles have been published. Lievonen et al. (2016) present a straightforward method for the preparation of lignin nanoparticles from waste lignin that yields a spherical particle morphology (Fig. 2a) and colloidal stability over time. Spherical lignin nanoparticles were obtained by dissolving softwood kraft lignin in tetrahydrofuran (THF) and subsequently introducing water into the system through dialysis. Water acts as a non-solvent reducing the lignin's degrees of freedom, causing the segregation of hydrophobic regions to compartments within the forming nanoparticles.

Similarly, chitin nanofibers have recently gained a great deal of attention and are now considered to have great potential in modern biomaterials science (Anitha et al. 2014; Ding et al. 2014; Mutsenko et al. 2017; Oh et al. 2016) and other sophisticated technologies such as the creation of flexible OLED-devices (Jin et al. 2016). To meet demand for chitin nanofibers, several methods have been developed for their preparation, including TEMPO-mediated oxidation (Fan et al. 2008a), acid hydrolysis at 90 °C (Fig. 2b) (Zeng et al. 2012; Villanueva et al. 2015) and a sonochemical fibrillation technique (Fan et al. 2008b; Nata et al. 2012). However, the nanofibers obtained by these methods are substantially different from the native chitin nanofibers in terms of crystallinity, chemical structure, and/or homogeneity. Therefore, the development of methods that can be considered "non-invasive" with respect to chitin's crystalline and chemical structure is still a topic with far-reaching impact (Dotto et al. 2015).

Nanofibrils as well as nanoparticles in aqueous suspension are sensitive to the presence of organic solvents, dyes and metal ions or electrolytes in general, which,

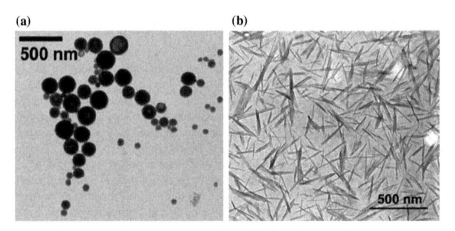

Fig. 2 TEM images of **a** lignin nanoparticles prepared from softwood, kraft lignin dissolved in THF, image reproduced from Lievonen et al. (2016) by permission of the Royal Society of Chemistry, Copyright 2016, **b** dilute suspension of chitin whiskers prepared by acid hydrolysis, reproduced from Gopalan Nair and Dufresne (2003) by permission of the American Chemical Society, Copyright 2002

by disturbing the blanket of water molecules surrounding them, cause precipitation from the suspensions. Moreover, heat treatments lead to aggregation—sometimes irreversible—of the nanostructures. Thus, the procedure of isolating and drying nanofibrils and nanoparticles from colloidal suspensions is a challenging task. For instance, evaporation produces rigid films of ordered nanofibrils (Ifuku and Saimoto 2012), whereas lyophilization produces highly dehydrated scaffolds or aerogels (Heath et al. 2013), prone to difficult redispersion.

A method of spray drying of chitin nanofibrils has recently been patented (Morganti and Muzzarelli 2013). This technique enables the isolation of nanofibers with a very high degree of crystallinity not obtainable by previously known methods. The nanofibrils obtained by spray drying contain water of crystallization (5–10%). This ensures their immediate and homogeneous dispersion in water. As a result, they may be stored at room temperature for an indefinite time in the form of dry powder, in conditions of limited humidity. The dried chitin nanoparticles retain all of the basic properties of native chitin, and may be used as carriers for controlled drug release. Similarly, spray drying has been used to isolate lignin nanoparticles from stable suspensions of that biopolymer. It should be noted that the spray drying method, apart from being economically advantageous, leads to lignin materials with increased surface areas (Dimitri 1974).

5 Production of Silica-Chitin-Based and Silica-Lignin-Based Composites

Intensive and successful work has been done in recent years on the synthesis of functional inorganic-organic hybrid systems based on polymers of natural origin, including chitin (Retuert et al. 1997; Alonso and Belamie 2010; Wysokowski et al.

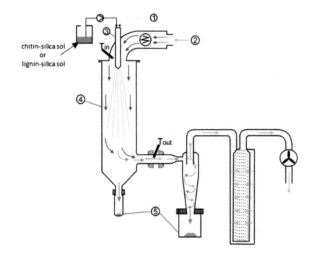

Fig. 3 Scheme of spray drying of chitin-silica-based and lignin-silica-based hybrid materials; T_{in}—inlet temperature; T_{out}—outlet temperature. 1 spray-gas in; 2 drying-gas in; 3 nozzle; 4 drying chamber; 5 product vessels

2015; Smolyakov et al. 2017) and lignin (Klapiszewski et al. 2015, 2017). By combining multiple components, materials are produced with a wide range of applications in various branches of science and industry. A particular challenge in the synthesis of such hybrids is the suitable design of their morphology on nano- and micrometric scales, in addition to control of the porous structure parameters of the resulting powders. The engineering of powdered materials is taking on particular significance in the synthesis of materials used as drug carriers. The development of techniques for the synthesis of functional inorganic-organic nano- and microcapsules makes possible the creation of innovative drug delivery systems (Muzzarelli et al. 2014; Figueiredo et al. 2017). Spray drying is one of the most important techniques used in the controlled preparation of innovative hybrid materials (Carne-Sanchez et al. 2013). Figure 3 gives a schematic representation of a method of preparing the chitin-silica and lignin-silica composites using spray drying.

It has been shown using scanning electron microscopy (SEM) that a chitin-silica composite obtained by spray drying has a homogeneous spherical morphology and a relatively homogeneous particle distribution (see Fig. 4c). The monodisperse

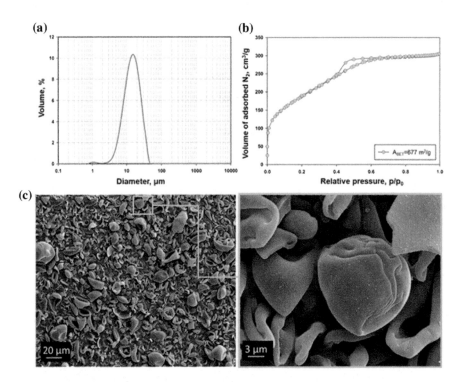

Fig. 4 **a** Particle size distribution of chitin-SiO$_2$ composite by volume contribution, determined using a Mastersizer 2000; **b** nitrogen adsorption-desorption isotherm and graph showing the dependence of volume on pore diameter; **c** SEM images showing the microstructure of the chitin-SiO$_2$ product

nature of the product is confirmed by laser diffraction (Fig. 4a). It should also be noted that the product takes the form of a hollow spherical capsule. The unique morphological structure of the product is reflected in its specific surface area, which in this case is as high as 677 m²/g (Fig. 4b).

The porous structure parameters and the initial results of testing indicate that the hybrid systems obtained may be attractive for applications in sustained drug release.

Researchers have also produced lignin-silica composites. As can be seen from the particle size distribution graph, the material obtained has a homogeneous structure with a mean particle size of 20 μm (see Fig. 5a). This is confirmed by SEM images (Fig. 5c). These can be used to draw conclusions concerning the morphology of the product, which consists of almost spherical capsules with a clearly hollow interior. The effectiveness of the synthesis of the lignin-silica composite using a spray drying process has also been confirmed (see Fig. 5b). The spectrum for the composite contains signals generated by chemical groups from both silica and lignin, which confirms that the hybrid material was correctly obtained (Klapiszewski et al. 2015, 2017).

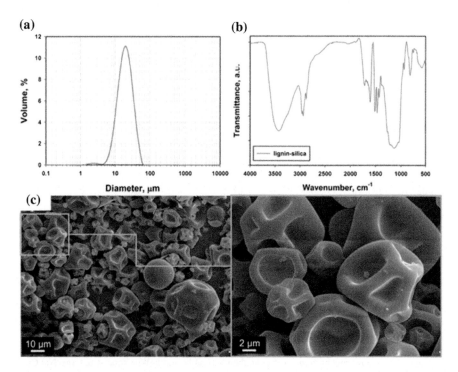

Fig. 5 **a** Particle size distribution of lignin-SiO₂ composite by volume contribution, determined using a Mastersizer 2000, **b** FTIR spectrum, **c** SEM images showing the microstructure of the lignin-SiO₂ product

In view of their specific morphological and microstructural properties, the composites will find numerous practical applications in many areas, in both science and industry.

6 Summary and a Look into the Future

Spray drying is a process for transforming liquid source materials into powders, which on being dissolved in water largely recover the qualities of the original substance. The process is coming to be used more and more often in modern chemical technology and engineering, particularly in the production of advanced functional composites and biocomposites with defined structures and morphology. A very significant role in this work is played by polymers of natural origin, including chitin and lignin, which can be combined with silica to form products with improved functional properties. Further research in this area will undoubtedly confirm the suitability of such systems for being used in the processes of sustained drug release. There would therefore appear to be good future prospects for the application of spray drying in advanced technologies for the production of inorganic-organic hybrids and biocomposites.

Acknowledgements This work was partially supported by PUT research grant no. 03/32/DSPB/0706.

References

Alonso B, Belamie E (2010) Chitin-silica nanocomposites by self-assembly. Angew Chemie Int Ed 49:8201–8204

Anitha A, Sowmya S, Kumar P et al (2014) Chitin and chitosan in selected biomedical applications. Prog Polym Sci 39:1644–1667

Belotti S, Rossi A, Colombo P et al (2015) Spray-dried amikacin sulphate powder for inhalation in cystic fibrosis patients: the role of ethanol in particle formation. Eur J Pharm Biopharm 93:165–172

Bittner B, Kissel T (1999) Ultrasonic atomization for spray drying: a versatile technique for the preparation of protein loaded biodegradable microspheres. J Microencapsul 16:325–341

Bula K, Jesionowski T, Krysztafkiewicz A et al (2007) The effect of filler surface modification and processing conditions on distribution behaviour of silica nanofillers in polyesters. Colloid Polym Sci 285:1267–1273

Cal K, Sollohub K (2010) Spray drying technique. I: hardware and process parameters. J Pharm Sci 99:575–586

Carne-Sanchez A, Imaz I, Cano-Sarabia M et al (2013) A spray-drying strategy for synthesis of nanoscale metal-organic frameworks and their assembly into hollow superstructures. Nat Chem 5:203–211

Cheow WS, Li S, Hadinoto K (2010) Spray drying formulation of hollow spherical aggregates of silica nanoparticles by experimental design. Chem Eng Res Des 88:673–685

Chew JH, Woo MW, Chen XD et al (2015) Mapping the shrinkage behavior of skim milk droplets during convective drying. Dry Technol 33:1101–1113

Cho Y-S (2016) Fabrication of hollow or macroporous silica particles by spray drying of colloidal dispersion. J Dispers Sci Technol 37:23–33

Dimitri MS (1974) US Patent 3,808,192, 30 Apr 1974

Ding F, Deng H, Du Y et al (2014) Emerging chitin and chitosan nanofibrous materials for biomedical applications. Nanoscale 6:9477–9493

Doolittle AK (1931) US Patent 1,973,051, 9 Jul 1931

Dotto GL, Cunha JM, Calgaro CO, Tanabe EH, Bertuol DA (2015) Surface modification of chitin using ultrasound-assisted and supercritical CO_2 technologies for cobalt adsorption. J Hazard Mater 15:29–36

Dziadas M, Nowacka M, Jesionowski T et al (2011) Comparison of silica gel modified with three different functional groups with C-18 and styrene–divinylbenzene adsorbents for the analysis of selected volatile flavor compounds. Anal Chim Acta 699:66–72

El Mansouri NE, Salvadó J (2006) Structural characterization of technical lignins for the production of adhesives: application to lignosulfonate, kraft, soda-anthraquinone, organosolv and ethanol process lignins. Ind Crops Prod 24:8–16

Fan Y, Saito T, Isogai A (2008a) Chitin nanocrystals prepared by TEMPO-mediated oxidation of α-chitin. Biomacromol 9:192–198

Fan Y, Saito T, Isogai A (2008b) Preparation of chitin nanofibers from squid pen β-chitin by simple mechanical treatment under acid conditions. Biomacromol 9:1919–1923

Figueiredo P, Lintinen K, Kiriazis A et al (2017) In vitro evaluation of biodegradable lignin-based nanoparticles for drug delivery and enhanced antiproliferation effect in cancer cells. Biomaterials 121:97–108

Gopalan Nair K, Dufresne A (2003) Crab shell chitin whisker reinforced natural rubber nanocomposites. 1. Processing and swelling behavior. Biomacromolecules 4:657–665

Heath L, Zhu L, Thielemans W (2013) Chitin nanowhisker aerogels. Chemsuschem 6:537–544

Ifuku S, Saimoto H (2012) Chitin nanofibers: preparations, modifications, and applications. Nanoscale 4:3308–3318

Iskandar F, Lenggoro IW, Xia B et al (2001a) Functional nanostructured silica powders derived from colloidal suspensions by sol spraying. J Nanoparticle Res 3:263–270

Iskandar F, Mikrajuddin, Okuyama K (2001b) In situ production of spherical silica particles containing self-organized mesopores. Nano Lett 1:231–234

Iskandar F, Mikrajuddin, Okuyama K (2002) Controllability of pore size and porosity on self-organized porous silica particles. Nano Lett 2:389–392

Jaskulski M, Wawrzyniak P, Zbiciński I (2015) CFD model of particle agglomeration in spray drying. Dry Technol 33:1971–1980

Jesionowski T, Krysztafkiewicz A (1996) Production of a highly dispersed sodium-aluminium silicate to be used as a white pigment or as a polymer filler. Pigment Resin Technol 25:4–14

Jesionowski T, Krysztafkiewicz A, Żurawska J et al (2009) Novel precipitated silicas—an active filler of synthetic rubber. J Mater Sci 44:759–769

Jin J, Lee D, Im HG et al (2016) Green electronics: Chitin nanofiber transparent paper for flexible green electronics. Adv Mater 28:5169–5175

Katta S, Gauvin WH (1975) Some fundamental aspects of spray drying. AIChE J 21:143–152

Kemp IC, Hartwig T, Herdman R et al (2016) Spray drying with a two-fluid nozzle to produce fine particles: Atomization, scale-up, and modeling. Dry Technol 34:1243–1252

Keshani S, Daud WRW, Nourouzi MM et al (2015) Spray drying: An overview on wall deposition, process and modeling. J Food Eng 146:152–162

Klapiszewski Ł, Jesionowski T (2017) Novel lignin-based materials as products for various applications. In: Thakur VK, Thakur MK, Kessler MR (eds) Handbook of composites from renewable materials, polymeric composites. Wiley, New Jersey, pp 519–554

Klapiszewski Ł, Bartczak P, Wysokowski M et al (2015) Silica conjugated with kraft lignin and its use as a novel "green" sorbent for hazardous metal ions removal. Chem Eng J 260:684–693

Klapiszewski Ł, Siwińska-Stefańska K, Kołodyńska D (2017) Preparation and characterization of novel TiO$_2$/lignin and TiO$_2$-SiO$_2$/lignin hybrids and their use as functional biosorbents for Pb (II). Chem Eng J 314:169–181

Kowalski SJ (2007) Drying of porous materials. Springer, New York

Leong KH (1987) Morphological control of particles generated from the evaporation of solution droplets: experiment. J Aerosol Sci 18:525–552

Li X, Anton N, Arpagaus C et al (2010) Nanoparticles by spray drying using innovative new technology: the büchi nano spray dryer B-90. J Control Release 147:304–310

Lievonen M, Valle-Delgado JJ, Mattinen M-L et al (2016) A simple process for lignin nanoparticle preparation. Green Chem 18:1416–1422

Lintingre E, Lequeux F, Talini L et al (2016) Control of particle morphology in the spray drying of colloidal suspensions. Soft Matter 12:7435–7444

Mezhericher M, Levy A, Borde I (2010) Spray drying modelling based on advanced droplet drying kinetics. Chem Eng Process 49:1205–1213

Morganti P, Muzzarelli C (2013) US Patent 8,552,164 B2, 8 Oct 2013

Mutsenko VV, Gryshkov O, Lauterboeck L et al (2017) Novel chitin scaffolds derived from marine sponge *Ianthella basta* for tissue engineering approaches based on human mesenchymal stromal cells: biocompatibility and cryopreservation. Int J Biol Macromol 1–11

Muzzarelli R, Mehtedi M, Mattioli-Belmonte M (2014) Emerging biomedical applications of nano-chitins and nano-chitosans obtained via advanced eco-friendly technologies from marine resources. Mar Drugs 12:5468–5502

Nata IF, Wang SSS, Wu TM et al (2012) β-Chitin nanofibrils for self-sustaining hydrogels preparation via hydrothermal treatment. Carbohydr Polym 90:1509–1514

Oh DX, Cha YJ, Nguyen H-L et al (2016) Chiral nematic self-assembly of minimally surface damaged chitin nanofibrils and its load bearing functions. Sci Rep 6:23245–23250

Osińska M, Walkowiak M, Zalewska A et al (2009) Study of the role of ceramic filler in composite gel electrolytes based on microporous polymer membranes. J Membr Sci 326:582–588

Percy SR (1872) US Patent 125,406, 9 Apr 1872

Pilarska A, Markiewicz E, Ciesielczyk F et al (2011) The influence of spray drying on the dispersive and physicochemical properties of magnesium oxide. Dry Technol 29:1210–1218

Retuert J, Nuñez A, Martínez F et al (1997) Synthesis of polymeric organic-inorganic hybrid materials. Partially deacetylated chitin-silica hybrid. Macromol Rapid Commun 18:163–167

Richter AP, Bharti B, Armstrong HB et al (2016) Synthesis and characterization of biodegradable lignin nanoparticles with tunable surface properties. Langmuir 32:6468–6477

Rosenberg M, Kopelman IJ, Talmon YJ (1990) Factors affecting retention in spray-drying microencapsulation of volatile materials. J Agric Food Chem 38:1288–1294

Smolyakov G, Pruvost S, Cardoso L et al (2017) PeakForce QNM AFM study of chitin-silica hybrid films. Carbohydr Polym 166:139–145

Trandafir DL, Turcu RVF, Simon S (2010) Structural study of spray dried silica-germanate nanoparticles. Mater Sci Eng B 172:68–71

Vehring R (2008) Pharmaceutical particle engineering via spray drying. Pharm Res 25:999–1022

Vehring R, Foss WR, Lechuga-Ballesteros D (2007) Particle formation in spray drying. J Aerosol Sci 38:728–746

Vicente J, Pinto J, Menezes J et al (2013) Fundamental analysis of particle formation in spray drying. Powder Technol 247:1–7

Villanueva ME, Salinas A, Díaz LE et al (2015) Chitin nanowhiskers as alternative antimicrobial controlled release carriers. New J Chem 39:614–620

Walker WJ, Reed JS, Verma SK (1999) Influence of slurry parameters on the characteristics of spray-dried granules. J Am Ceram Soc 82:1711–1719

Walton DE (2002) Spray-dried particle morphologies. Dev Chem Eng Miner Process 10:323–348

Warych J (2004) Aparatura chemiczna i procesowa. OWPW, Warszawa

Wysokowski M, Petrenko I, Stelling A et al (2015) Poriferan chitin as a versatile template for extreme biomimetics. Polymers 7:235–265

Zalewska A, Walkowiak M, Niedzicki L et al (2010) Study of the interfacial stability of PVdF/
    HFP gel electrolytes with sub-micro- and nano-sized surface-modified silicas. Electrochim
    Acta 55:1308–1313
Zellmer S, Garnweitner G, Breinlinger T et al (2015) Hierarchical structure formation of
    nanoparticulate spray-dried composite aggregates. ACS Nano 9:10749–10757
Zeng JB, He YS, Li SL et al (2012) Chitin whiskers: an overview. Biomacromol 13:1–11
Zhang X, Guan J, Ni R et al (2014) Preparation and solidification of redispersible nanosuspen-
    sions. J Pharm Sci 103:2166–2176
Zhu H, Luo W, Ciesielski PN et al (2016) Wood-derived materials for green electronics, biological
    devices, and energy applications. Chem Rev 116:9305–9374

Applications and Properties of Physical Gels Obtained on the Basis of Cellulose Derivatives

Patrycja Komorowska and Jacek Różański

1 Cellulose and Its Derivatives

Cellulose is a linear natural polymer with high molecular weight. It consists of D-anhydroglucopyranose unit connected by $\beta(1 \rightarrow 4)$-glycosidic bonds. Cellulose is a renewable, biodegradable and bio-compatible material. This polysaccharide is characterised by high cristallinity and strong inter- and intramolecular hydrogen bonds (Pingping and Ruigang 2015). Cellulose is not water-soluble and it does not solve in the majority of organic solvents, which limits the range of its applications. Each link in the cellulose chain has three hydroxyl groups in positions: 2, 3, 6. It is possible to introduce other functional groups in these places, owing to which cellulose derivatives are formulated characterised by higher solubility and, hence, wider application range for the cellulose itself (Jaworska and Vogt 2013).

Table 1 presents the listing of cellulose derivatives most frequently applied in practice. They include cellulose ethers formed as a result of alcohol group etherification (Jaworska and Vogt 2013). Hydroxyl groups are entirely or partially replaced by ether substituents. The degree of substitution (DS) is defined as an average number of etherified hydroxyl groups in a glucose unit. DS for the cellulose derivatives influences their solubility (Pingping and Ruigang 2015). From among the cellulose ethers specified by Table 1, only NaCMC is a polyelectrolyte. Cellulose and its derivatives may be additionally joined with natural and/or synthetic polymers in order to obtain the materials bearing specific properties (Sannino et al. 2009).

P. Komorowska · J. Różański (✉)
Institute of Chemical Technology and Engineering, Poznan University
of Technology, Poznań, Poland
e-mail: jacek.rozanski@put.poznan.pl

© Springer International Publishing AG, part of Springer Nature 2018
M. Ochowiak et al. (eds.), *Practical Aspects of Chemical Engineering*,
Lecture Notes on Multidisciplinary Industrial Engineering,
https://doi.org/10.1007/978-3-319-73978-6_13

Table 1 Exemplary cellulose derivatives

Type	Functional group	Solubility	Publication
Sodium carboxymethylcellulose (CMC)	$-CH_2COONa$	Water $DS = 0.4-1.3$	Ramli et al. (2015), Abdulmumin et al. (2016) and Benslimane et al. (2016)
Methylcellulose (MC)	$-CH_3$	Hot water $DS = 1.4(1.7)-2$, organic solvents $DS = 2-3$	Li et al. (2001), Kuang et al. (2006) and Fahad et al. (2017)
Hydroxypropylmethylcellulose (HPMC)	$-CH_3$, $-CH_2CH(OH)CH_3$	Cold water, some organic solvents	Ford (2014), Wuestenberg (2014) and Wang et al. (2016)
Hydroxyethylcellulose (HEC)	$-CH_2CH_2OH$	Water	Shifeng et al. (2010) and Dashtimoghadam et al. (2016)
Hydroxypropylcellulose (HPC)	$-CH_2CH_2CH_2OH$	Water, some organic solvents	Sarode et al. (2013) and Zhang et al. (2016)
Ethylcellulose (EC)	$-CH_2CH_3$	Organic solvents	Heng et al. (2005) and Murtaza (2012)

2 Physical Gels on the Basis of Cellulose Derivatives

Gels in which covalent bonds have been formed between the structural elements are referred to as "chemical" ones. However, if the interactions between the structural elements have a non-covalent character are called "physical" ones (hydrogen bonds, electrostatic, hydrophobic and van der Waals interactions) (Gulrez et al. 2011). Chemical and physical gels can be achieved based on the cellulose derivatives. Chemical gels are relatively well known and described in the literature, however the reasons behind the formation of physical gels may have verified characters and in numerous instances have not been clarified entirely. Clark and Ross-Murphy (1987) additionally applied the division of physical gels into strong (genuine) and weak ones. The differentiation into these two kinds of gelled systems is based on the operational definition of a gel from mechanical spectroscopy (more information on the subject will be provided in Sect. 3).

Gels which are created due to weak physical bonds are reversible and may be easily transformed into sol by changing the environmental conditions. The conditions which contribute to the transformation sol-gel include: temperature, pH, ionic strength, electric and magnetic field (Barros et al. 2015; Pingping and Ruigang 2015).

Physical gels may be obtained by means of various techniques: by ordinary mixing of ingredients, freezing and defrosting, heating and cooling, casting-solvent evaporation, pH lowering (Hoffman 2012; Barros et al. 2014; Pingping and Ruigang 2015). In case of cellulose derivatives, physical cross-linking occurs most frequently due to (Nishinari 2000):

- hydrophobic interactions (MC, HEC, HPMC, NaCMC), in order to formulate the gel it is indispensable to increase the temperatures (Balaghi et al. 2014; Karlsson et al. 2015; Benslimane et al. 2016)
- hydrogen bonds, gels are created in within the ambient temperature or lower ("cold mode") (Fahad et al. 2013; Dashtimoghadam et al. 2016).

Table 2 presents exemplary cellulose derivatives, in solutions of which physical gels are formed due to hydrophobic interactions between the polymer chains. The mechanism of hydrophobic association in aqueous solutions of MC is relatively

Table 2 Physical gels formulated due to hydrophobic interaction

Type	Addition	Solvent	Method	Publication
MC	–	H_2O	Heating	Kobayashi et al. (1991), Li et al. (2001), Joshi et al. (2008), Lam et al. (2007), Schupper et al. (2008), Desbrieres et al. (2000), Bayer and Knarr (2012) and Edelby et al. (2014)
MC	NaN$_3$	H_2O	Heating	Funami et al. (2007)
MC	–	D_2O	Heating	Miura (2014)
MC	Glycerine	H_2O	Heating	Kuang et al. (2006)
MC	SDS	H_2O	Heating	Li et al. (2012)
MC	Cetyltrimethylammonium bromide (CTAB)	H_2O	Heating	Li et al. (2007)
MC	Phosphate buffer saline (PBS)	H_2O	Heating	Zheng et al. (2004)
MC	K$_2$HPO$_4$ or KH$_2$PO$_4$	H_2O	Heating	Alamprese and Mariotti (2015)
MC	NaCl	H_2O	Heating	Joshi and Lam (2005)
MC	CS and NaCl/Na$_3$PO$_4$/NaHCO$_3$/ glycerophosphate	H_2O	Heating	Tang et al. (2010)
MC HPMC	–	H_2O	Heating	Bodvik et al. (2010) and Balaghi et al. (2014)
HPMC	–	H_2O	Heating	Hussain et al. (2002), Silva et al. (2008) and Veríssimo et al. (2010)
HPMC	CS, glycerine, acetic acid	H_2O	Heating	Wang et al. (2016)
HPMC	SDS	H_2O	Heating	Silva et al. (2011)
HPMC	Hydroxypropyl starch (HPS)	H_2O	Heating	Zhang et al. (2015)

(continued)

Table 2 (continued)

Type	Addition	Solvent	Method	Publication
HPMC	Salts (NaCl, KCl, NaBr, NaI, Na_2HPO_4, K_2HPO_4, Na_2SO_4, Na_3PO_4) surfactants (SDS, sodium n-decyl sulfate (SDeS) and Triton X-100, sodium n-hexadecyl sulfate (SHS)	H_2O	Heating	Joshi (2011)
HPMC	Regenerated silk fibroin (RSF)	H_2O	Heating	Luo et al. (2016)
HEC	CS, glycerophosphate	H_2O	Heating	Dashtimoghadam et al. (2016)
CMC	Bentonite	H_2O	Heating	Benslimane et al. (2016)
CMC HPMC	–	H_2O	Heating	Alvarez-Lorenzo et al. (2001)
CMC	Pectin	H_2O	Heating	Bekkour et al. (2014)

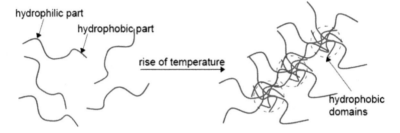

Fig. 1 Physical cross-linking of chains under the influence of temperature. Adapted from Gyles et al. (2017)

best known. The process of MC dissolution in water remains exothermic in character, hence in order to obtain a homogenous solution it must be cooled down. With low temperatures, hydrogen bonds are probably formed between MC hydrophobic groups and water particles. As a result of heating, the hydrogen bonds disintegrate (Tang et al. 2010), which results in dehydration of MC chains and their subsequent hydrophobic association. With properly high polymer concentration, hydrophobic interactions lead to the formulation of a three-dimensional network (Joshi et al. 2008; Pingping and Ruigang 2015). The mechanism of hydrophobic association has been presented in schematic form in Fig. 1 (Gyles et al. 2017).

Forming physical gels due to hydrophobic association has also been observed during the heating of other cellulose derivatives [HPMC and NaCMC (Veríssimo et al. 2010; Benslimane et al. 2016)], whereby the gelation temperature is dependent on the type of substituents. For instance, the gelation temperature for HMPC is higher that the gelation temperature for MC, which is related to the presence of polar hydroxyproyl blocks. Their occurrence causes weaker hydrophobic association between the hydrophobic blocks. HPMC solutions, contrary to the once of MC, return entirely to their original state after cooling. Probably, maintaining MC in high temperature leads to irreversible changes of its structure (Balaghi et al. 2014).

Gelation temperature of aqueous solutions of cellulose derivatives depends also on the polymer concentration, the degree of its substitution, the positions of the substituents or molecular weight (Nishinari 2000; Hussain et al. 2002; Funami et al. 2007; Silva et al. 2008; Bayer and Knarr 2012; Edelby et al. 2014). It may be modified by adding salt, surfactants or other additives (Zheng et al. 2004; Kuang et al. 2006; Wang et al. 2006; Su et al. 2008; Tang et al. 2010; Li et al. 2012; Bekkour et al. 2014).

The increase in the concentration of MC, HPMC and CMS brings about a noticeable decrease in the gelation temperature (Edelby et al. 2014). By adding some salts to aqueous solutions of nonionic cellulose derivatives, their gelation temperature may be regulated (Zheng et al. 2004; Wang et al. 2006; Su et al. 2008; Tang et al. 2010). Phosphate ions (Alamprese and Mariotti 2015) as well as chloride ions (Joshi and Lam 2005) lowers the temperature for the transfer sol-gel in MC solution. Solvation of salt ions is a competitive process to MC chain hydration by water particles, which leads to the decrease in polymer solubility. Consequently, the solution containing the salts has more of hydrophobic aggregates of MC, which results in the formulation of gel in lower temperature (Alamprese and Mariotti 2015). The majority of salts added to aqueous solutions of HPMC leads to salting-out effects, consequently to which the lowering of the gelation temperature occurs. The salts NaI and NaSCN shall serve as exceptions here as they damp the gelation effect (Joshi 2011). Lowering of gelation temperature for MC and HPMC may be achieved by adding glycerin (Kuang et al. 2006). Glycerol has a great affinity for water, which, on a par with salts, leads to dehydration of the hydrophobic chain segments (Wang et al. 2016).

Increasing the gelation temperature for the solutions of MC, HPMC as well as other cellulose derivatives may be achieved by adding the surfactants (Joshi 2011; Silva et al. 2011; Li et al. 2012). The association of charged amphiphiles to cellulose ethers leads to the deactivation of polymer's hydrophobic regions and enhances their solubility due to the presence of charges along the chains. This brings about inhibition of hydrophobic association among polymer segments, which is demonstrated by increased gelation temperature and may completely inhibit gel formation with high surfactant concentrations (Silva et al. 2011).

Another reason behind physical cross-linking of cellulose derivatives is the formation of intramolecular hydrogen bonds. Table 3 shows exemplary systems containing the cellulose and its derivatives, in which hydrogen bonds play a key role during the formation of physical gels. MC formulates the physical gels of this type in non-aqueous solvents (mainly propylene glycol and butylene glycol) (Fahad et al. 2017).

Heating and subsequent cooling off of MC solutions in glycols shall lead to the formulation of a higher number of hydrogen bonds. It is likely that the increase in

Table 3 Exemplary gels on the basis of cellulose and its derivatives cross-linked due to hydrogen bonds

Type	Addition	Solvent	Method	Publication
MC	–	GP GB	Heating—cooling	Fahad et al. (2013)
MC	–	GP GB Ethylene glycol (GE)	Heating—cooling	Fahad et al. (2017)
MC	CS	H_2O	Mixing	Tang et al. (2010)
MC	Glycerine	H_2O	Mixing	Kuang et al. (2006)
MC	PVA, NaCl	H_2O	Freezing—thawing	Bain et al. (2012)
HEC	CS	H_2O	Mixing	Dashtimoghadam et al. (2016)
HPMC	HA	H_2O	Mixing	Hoare et al. (2010, 2014)
HPMC	Hydroxypropyl starch (HPS)	H_2O	Mixing	Zhang et al. (2015)
EC	PGD	H_2O	Mixing	Heng et al. (2005) and Bruno et al. (2012)
CMC	PVA	H_2O	Freezing—thawing	Congming and Yongkang (2007)
CMC	HCl	H_2O	Mixing	Gulrez et al. (2011)
Carboxylated methyl cellulose (CLMC)	PVA	H_2O	Freezing—thawing	Congming et al. (2012)
RC	PVA	NaOH/urea/H_2O	Freezing—thawing	Ren et al. (2014)

the temperature leads to breaking the intramolecular hydrogen bonds within unmodified cellulose units. Subsequently, during the solution cooling interchain hydrogen bonds are formed between OH groups of glycol and ether oxygen in anhydroglucopyranose unit (Fahad et al. 2017).

Hydrogen bonds play a crucial role in the mixtures of cellulose derivatives and other polymers for examples CS, PVA and HA (Kuang et al. 2006; Tang et al. 2010; Dashtimoghadam et al. 2016). Formulation of physical gels in the mixtures of polymers frequently results from the occurrence of inter- and intramolecular hydrogen bonds, together with the hydrophobic interactions.

3   Application of Rheology in the Research Over Physical Gels

Rheological properties of gels are most frequently determined in oscillatory flow within the linear viscoelastic limit [small-amplitude oscillatory shear (SAOS)]. During the measurements of this sort, the gel is subject to very small deformations, owing to which its structure is not destroyed. Additionally, the time involved in the measurements is short relative to the characteristic times of the gelation. Oscillatory measurement outcomes allow us to acquire information on the properties of polymer network microstructure (Lopes da Silva and Rao 2007). Theoretical basis for oscillatory measurements may be encountered in numerous reference sources (Lopes da Silva and Rao 2007; Osswald and Rudolph 2014). In general, the sinusoidally varying shear stress τ within time t can be represented as:

$$\tau(t) = \gamma_0 [G' \sin(\omega t) + G'' \cos(\omega t)] \tag{1}$$

where γ_0 is the strain amplitude, ω is the angular frequency, G' is the storage modulus and G'' is the loss modulus. The storage modulus expresses storage of elastic energy, while the loss modulus is a measure of the viscous dissipation of the energy. The ratio G''/G' is called the loss tangent δ, where δ is the phase angle taking on the values within the range from $0°$ to $90°$. Generally, $\tan \delta < 1$ indicates a dominance of elastic fluid properties, and $\tan \delta > 1$ shows a dominance of viscous fluid characteristics. Two basic types of experiments in oscillatory flow are performed: strain sweep test, during which the dependence of modules G' and G'' is registered in the function of the deformation amplitude, and frequency sweep test in which G' and G'' are determined as a function of frequency. The strain sweep test is used predominantly to determine the range for strain amplitude or stress amplitude, in which the linear viscoelastic limit occurs.

Frequency sweep experiment has been more widely applied in rheological research over gels. Depending on the type of mechanical spectrum within the frequency range between 0.001 and 100 Hz Clark and Ross-Murphy (1987) suggested dividing the polymer solutions into four groups: strong gel (or true gel), weak-gel, entangled polymer solution and non-entangled polymer solution. In this approach, the term gel is understood as a substance for which plateau module of G' and $G' \gg G''$ ($\tan \delta < 0.1$) occurs within the entire range of frequencies (in practice from 0.001 to 100 Hz). According to Nishinari (2009) such a definition of gel shall be regarded as an operational definition, as it cannot be ruled out that with the frequencies lower than 0.001 Hz a given substance shall demonstrate a liquid-like behaviour. This remark refers predominantly to physical gels.

Clark and Ross-Murphy (1987) introduced also the term "weak-gel", as a substance for which G' module remains slightly bigger than G'' module ($\tan \delta$ is greater than 0.1) and both modules increase slightly with the increase in the frequency (Mandala et al. 2004; Razmkhah et al. 2017). As an example of weak-gel aqueous solutions of xanthan gum are provided most frequently. As weak-gel do not remain true gels, Picout and Ross-Murphy (2003) suggest that "structured fluids" shall serve as a more appropriate name for this type of substances.

Rheological research have been applied to the determination of a gel point. In line with the classical definition given by Paul J. Flory, the gel point is defined as a condition at which a system manifests an infinite steady-shear viscosity and a zero equilibrium modulus (Flory 1953). On the basis of the criteria set forth in the aforementioned definition, the gel point is not determined for physical gels. As it was noted by Schwittay et al. (1995) near the GP the time to reach steady shear flow might be similar or longer than the lifetime of the physical junction, therefore the zero shear viscosity might remain finite at the GP because the material can still completely relax (Schwittay et al. 1995).

The literature on the subject suggests several procedures and criteria to determine gel point for physical gels. The one which is very frequently applied includes small amplitude oscillatory shear technique to trace the gelation process. The gel point is defined as a temperature, concentration or time at which modules G' and G'' intersect (Tung and Dynes 1982) or at which a rapid increase of G' module starts (Fig. 2).

The said methods are frequently applied to determine the gelation temperature of MC, CMC, EC, HPMC solutions (Kuang et al. 2006; Tang et al. 2010; Bain et al. 2012; Balaghi et al. 2014; Edelby et al. 2014; Alamprese and Mariotti 2015; Davidovich-Pinhas et al. 2015; Karlsson et al. 2015; Wang et al. 2016). The values

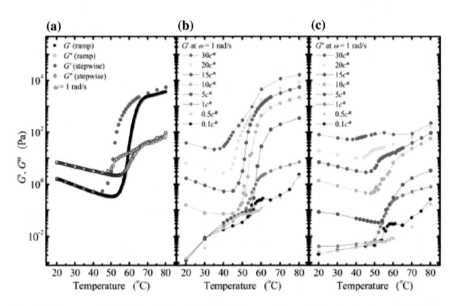

Fig. 2 Storage modulus (G') and loss modulus (G'') versus temperature for the sample MC. The gel point is the point of crossover of the shear storage (G') and loss (G'') modulus: **a** 2.1 wt% (15c^*, where c^* is the critical overlap concentrations) for two heating methods (ramp at a rate of 1 °C/min and stepwise at an effective rate of $\sim$2 °C/h in the vicinity of the gel point), **b** G', **c** G'' for a range of concentrations (Arvidson et al. 2013)

of G' and G'' modules are registered with the temperature increasing at a steady speed. The measurement is performed with the pre-set value of frequency and strain amplitude (or shear stress). Strain (or shear stress) amplitude shall keep within the linear viscoelastic limit along the entire range of temperature range. This way to determine the gelation temperature is very simple and relatively quick, however it does not consider the influence of the frequency over the value of temperature at which the modules G' and G'' intersect. The experimental research shows that the gelation temperature measured in this manner apparently depends on the frequency (Bruno et al. 2012).

The method which allows to determine the gel point which value is independent of the oscillation frequency was suggested by Winter and Chambon (1986). The authors based their considerations on the observation that at the gel point, G' and G'' exhibit a power law behaviour over an extended frequency range, whose mathematical expression is given below:

$$G'(\omega) = G''(\omega) \propto \omega^n \tag{2}$$

The value of exponent n keeps within a wide spectrum from 0.19 to 0.9 (Scalan and Winter 1991). This generates a frequency independent value of $\tan \delta$ that can be expressed as follows:

$$\tan \delta = \frac{G''}{G'} = \tan\left(\frac{n\pi}{2}\right) \tag{3}$$

This criterion was initially developed for chemical networks but, subsequently, was extended to physical gels (Nijenhuis and Winter 1989; Cuvelier et al. 1990). Many results have shown that this method is reliable and valid for the determination of the gel point such as the gel concentration and the gel temperature (Lue and Zhang 2008). In order to determine the gel point, the values of $\tan \delta$ must be determined in the function of temperatures or concentrations at various oscillation frequencies. Next, the diagram for the dependences of $\delta = f(T)$ for various values of ω must be produced. Temperature or concentration at which the curves intersect is referred to as the gel point (Fig. 3).

In practice, it has turned out that the method suggested by Winter and Chambon (1986) has got limited application. For many biopolymers gelation begins from a sol state characterized by a very low viscosity that is often below the resolution of the rheometer. Additionally, the lowest detected value of G' may already be greater than G'' (Gosal et al. 2004).

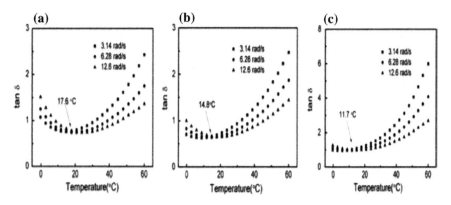

Fig. 3 Determination of point by means of Winter and Chambon criterion for solutions of cellulose/[BMIM]Cl prepared from different dissolving process at various angular frequencies: **a** solution A (cellulose was dispersed in [BMIM]Cl by glass rod in a bottle), **b** solution B (cellulose was dispersed in [BMIM]Cl and stirred by an electric paddle stirrer), **c** solution C (cellulose and [BMIM]Cl were poured into the kneader system) (Xia et al. 2014)

4 Examples of Physical Gel Applications

Physical gels which contain cellulose derivatives are widely applied in chemical, food processing, cosmetics and pharmaceutical industries (Barros et al. 2014). Such properties like non-toxicity, biocompatibility and biodegradability constitute the major reasons for their common usage. Table 4 shows examples application of physical gel.

Table 4 Exemplary application of physical gels based on cellulose derivatives

Type	Addition	Application	Publication
MC	Water	Viscous ultrasonic coupling medium for transdermal sonophoresis for insulin and vasopressin, building a tissue-engineering scaffold in injured brain, production tablets for the release of drugs	Joshi et al. (2008)
MC	GP	Support materials in 3D printing	Fahad et al. (2013, 2017)
MC	CS	Injectable product	Tang et al. (2010)
MC	PVA, NaCl, water	Drug delivery	Bain et al. (2012)
MC	PVA, water	Porous composites with high performance mortars for laying tiles and coating walls and ceilings, adhesives, drug delivery	Bain et al. (2012) and Nasatto et al. (2015)

(continued)

Table 4 (continued)

Type	Addition	Application	Publication
MC	PBS, water	Novel cell sheet harvest systems	Chen et al. (2006)
HPMC	Water	Hydrogel as a drug-delivery device, thickeners, binders, film formers, water-retention agents, food ingredient, engineering scaffolds for tissue growth	Baumgartner et al. (2002), Gulrez et al. (2011), Veríssimo et al. (2010), Siepmann and Peppas (2012), Balaghi et al. (2014) and Joshi (2011)
HPMC	CS	In textile applications, as active species (antiperspirants, moisturisers, scents)	Barros et al. (2014)
HPMC	HPS, water	Production of capsules	Zhang et al. (2015)
HPMC	HA, water	Drug delivery, matrices for the delivery of anti-adhesion drugs loaded in solution	Hoare et al. (2010, 2014)
HPMC	PVA, F_3O_4 water	Tumor therapy	Wang et al. (2017)
HPC	Water	Drug delivery	Baumgartner et al. (2002)
HEC	Water	Drug delivery	Baumgartner et al. (2002)
HEC	CS, glyccrophosphatc	Injectable delivery vehicle for tissue engineering application	Jihong et al. (2010)
EC	Water	Binder, dispersing agent, stabilizer and a slow releasing agent	Bruno et al. (2012)
CMC	Water	Hydrogel for skin burns, improve moisturizing effects, thickener, stabilizer, drug delivery	Gulrez et al. (2011), Calo and Khutoryanskiy (2015) and Benslimane et al. (2016)
CMC	Pectin, GP	Hydrogel for treatment wounds	Calo and Khutoryanskiy (2015)
CMC	Bentonite, water	In drilling fluids	Benslimane et al. (2016)
CMC	CS	Drug delivery carriers or microsensors and actuators	Kang et al. (2016)
CMC	PVA, water	Drug release carrier for oral administration	Congming and Yongkang (2007)
CMC	Pectin	Various processed foods	Bekkour et al. (2014)
Modified CMC	GP, water	Hydrogel for treatment wounds	Calo and Khutoryanskiy (2015)
CLMC	PVA, water	Carrier for drug controll release	Congming et al. (2012)

As for the pharmaceutical industry, the gels based on the cellulose derivatives are commonly applied as drug delivery systems (Kang et al. 2016). Their ability to cross-linking with increasing temperature has been used in tissue engineering. Cellulose derivative solutions are applied over the tissues during surgeries. After cross-linking due to an impact from increasing temperature they are characterized by mechanical properties similar to the properties of natural tissues (Dashtimoghadam et al. 2016; Ito et al. 2007). Thermal cross-linking has been taken advantage of during the treatment applied to eye diseases. A drop containing a cellulose derivative applied over the surface of an eyeball formulated gel (Winnicka 2008). Cellulose derivatives are also used to produce gel dressings (Sannino et al. 2009). In this case, the gels play the role of absorbing materials to remove water from the body during the treatment of edemas (Karpiński et al. 2015). Gel dressings containing glycerin and propylene glycol prevent the skin from drying out. A very interesting application of gels includes the treatment against obesity. Cellulose-based xerogel taken orally bulges in the stomach and causes the feeling of satiety (Sannino et al. 2009).

Cellulose-based gels may be applied in agriculture to optimise the usage of water resources. They are able to absorb big amounts of water, which is stored and released gradually to the soil during cultivation. Nutritious substances, used for soil fertilizing, may be introduced into the gels and later released gradually with water (Sannino et al. 2009).

5   Summary

The following paper presents the survey on cellulose derivatives which are capable of formulating the physical gels in aqueous solutions. The preparations based on the cellulose are biocompatible and biodegradable materials and hence they are widely used in chemical, food-processing, cosmetics and pharmaceutical industries. A noticeable part of cellulose derivatives are characterized by their capability to become gels when their solutions are heated. Solutions of individual cellulose derivatives differ in gelation temperatures. Its values may be modified by adding simple salts, surfactants, glycerin, etc. In ambient temperature, cellulose derivative-based physical gels may be obtained by changing pH, solvent composition or by using various additives. In the majority of cases, the formulated physical cross-linking results from intra- and interchain hydrogen bonds.

Acknowledgements This work was supported by PUT research grant No. 03/32/DSPB/0702.

References

Abdulmumin OA, Jianchun G, Shibin W et al (2016) Hydraulic fracture fluid for gas reservoirs in petroleum engineering applications using sodium carboxymethyl cellulose as gelling agent. J Nat Gas Sci Eng 32:491–500

Alamprese C, Mariotti M (2015) Modelling of methylcellulose thermogelation as a function of polymerconcentration and dissolution media properties. LWT-Food Sci Technol 60:811–816

Alvarez-Lorenzo C, Duro R, Gomez-Amoza JL et al (2001) Influence of polymer structure on the rheological behavior of hydroxypropylmethylcellulose–sodium carboxymethylcellulose dispersions. Colloid Polym Sci 279(11):1045–1057

Arvidson SA, Lott JR, McAllister JW et al (2013) Interplay of phase separation and thermoreversible gelation in aqueous methylcellulose solutions. Macromolecules 46:300–309

Bain MK, Bhowmick B, Maity D (2012) Effect of PVA on the gel temperature of MC and release kinetics of KT from MC based ophthalmic formulations. Int J Biol Macromol 50:565–572

Balaghi S, Edelby Y, Senge B (2014) Evaluation of thermal gelation behavior of different cellulose ether polymers by rheology. AIP Conf Proc 1593:755–761

Barros SC, Silva AA, Costa DB et al (2014) Thermo-sensitive chitosan–cellulose derivative hydrogels: swelling behaviour and morphologic studies. Cellulose 21:4531–4544

Barros SC, Silva AA, Costa DB et al (2015) Thermal–mechanical behaviour of chitosan–cellulose derivative thermoreversible hydrogel films. Cellulose 22(3):1911–1929

Baumgartner S, Kristl J, Peppas NA (2002) Network structure of cellulose ethers used in pharmaceutical applications during swelling and at equilibrium. Pharm Res-Dordr 19(8): 1084–1090

Bayer R, Knarr M (2012) Thermal precipitation or gelling behaviour of dissolved methylcellulose (MC) derivatives–behaviour in water and influence on the extrusion of ceramic pastes. Part 1: fundamentals of MC-derivatives. J Eur Ceram Soc 32(5):1007–1018

Bekkour K, Sun-Waterhouse D, Wadhwa SS (2014) Rheological properties and cloud point of aqueous carboxymethyl cellulose dispersions as modified by high or low methoxyl pectin. Food Res Int 66:247–256

Benslimane A, Bahlouli IM, Bekkour K (2016) Thermal gelation properties of carboxymethyl cellulose and bentonite carboxymethyl cellulose dispersions: Rheological considerations. Appl Clay Sci 132:702–710

Bodvik R, Dedinaite A, Karlson L et al (2010) Aggregation and network formation of aqueous methylcellulose and hydroxypropylmethylcellulose solutions. Colloid Surf A Physicochem Eng Asp 354:162–171

Bruno L, Kasapis S, Heng PWS (2012) Effect of polymer molecular weight on the structural properties of non-aqueous ethyl cellulose gels intended for topical drug delivery. Carbohyd Polym 88:382–388

Calo E, Khutoryanskiy VV (2015) Biomedical applications of hydrogels: a review of patents and commercial products. Eur Polym J 65:252–267

Chen CH, Tsai CC, Chen W et al (2006) novel living cell sheet harvest system composed of thermoreversible methylcellulose hydrogels. Biomacromol 7:736–743

Clark AH, Ross-Murphy SB (1987) Structural and mechanical properties of biopolymer gels. Biopolymers, vol 83. Advances in polymer science. Springer, Berlin, Heidelberg, pp 57–192

Congming X, Yongkang G (2007) Preparation and properties of physically crosslinked sodium carboxymethylcellulose/poly(vinyl alcohol) complex hydrogels. J Appl Polym Sci 107: 1568–1572

Congming X, Cunping X, Ma Y et al (2012) Preparation and characterization of dual sensitive carboxylatedmethyl cellulose/poly(vinyl alcohol) physical composite hydrogel. J Appl Polym Sci 127(6):4750–4755

Cuvelier G, Peigney-Nourry C, Launay B (1990) Viscoelastic properties of physical gels: critical behaviour at the gel point. In: Phillips GO, Wedlock DJ, Williams PA (eds) Gums and stabilisers for the food industry, vol 5. Oxford University Press, Oxford, New York, Tokyo, pp 549–552

Dashtimoghadam E, Bahlakeh G, Salimi-Kenari H (2016) Rheological study and molecular dynamics simulationof biopolymer blend thermogels of tunable strength. Biomacromol 17 (11):474–3484

Davidovich-Pinhas M, Barbut S, Marangoni AG (2015) The gelation of oil using ethyl cellulose. Carbohyd Polym 117:869–878

Desbrieres J, Hirrien M, Ross-Murphy SB (2000) Thermogelation of methylcellulose: rheological considerations. Polymer 41:2451–2461

Edelby Y, Balaghi S, Senge B (2014) Flow and sol-gel behavior of two types of methylcellulose at various concentrations. AIP Conf Proc 1593:750–754

Fahad M, Dickens P, Gilbert M (2013) Novel polymeric support materials for jetting based additive manufacturing processes. Rapid Prototyping J 19(4):230–239

Fahad M, Gilbert M, Dickens P (2017) Microscopy and FTIR investigations of the thermal gelation of methylcellulose in glycols. Polym Sci Ser A 59(1):88–97

Flory PJ (1953) Principles of polymer chemistry. Cornell University Press, Ithaca, New York

Ford JL (2014) Design and evaluation of hydroxypropyl methylcellulose matrix tablets for oral controlled release: a historical perspective. In: Timmins P, Pygall S, Melia CD (eds) Hydrophilic matrix tablets for oral controlled release, vol 16. Advances in pharmaceutical sciences. Springer, New York, pp 18–19

Funami T, Kataoka Y, Hiroe M et al (2007) Thermal aggregation of methylcellulose with different molecular weights. Food Hydrocolloid 21:46–58

Gosal WS, Clark AH, Ross-Murphy SB (2004) Fibrillar tl-lactoglobulin gels: part 2. Dynamic mechanical characterization of heat-set systems. Biomacromol 5:2420–2429

Gulrez SKH, Al-Assaf S, Phillips GO (2011) Hydrogels: methods of preparation, characterisation and applications. In: Capri A (ed) Progress in molecular and environmental bioengineering—from analysis and modeling to technology applications. InTech, Croatia, pp 117–150

Gyles DA, Castro LD, Silva JOC (2017) A review of the designs and prominent biomedical advances of natural and synthetic hydrogel formulations. Eur Polym J 88:373–392

Heng PWS, Chan LW, Chow KT (2005) Development of novel nonaqueous ethylcellulose gel matrices: rheological and mechanical characterization. Pharm Res-Dordr 22(4):676–684

Hoare T, Zurakowski D, Langer R et al (2010) Rheological blends for drug delivery. I. Characterization in vitro. J Biomed Mater Res A 92(2):575–585

Hoare T, Yeo Y, Bellas E (2014) Prevention of peritoneal adhesions using polymeric rheological blends. Acta Biomater 10:1187–1193

Hoffman AS (2012) Hydrogels for biomedical applications. Adv Drug Deliver Rev 64:18–23

Hussain S, Keary C, Craig DQM (2002) A thermorheological investigation into the gelation and phase separation of hydroxypropyl methylcellulose aqueous systems. Polymer 43(21):5623–5628

Ito T, Yeo Y, Highley CB (2007) The prevention of peritoneal adhesions by in situ cross-linking hydrogels of hyaluronic acid and cellulose derivatives. Biomaterials 28(6):975–983

Jaworska M, Vogt O (2013) Sorbitol and cellulose derivatives as gelling agents. Chemik 67 (3):242–249

Jihong Y, Liu Y, Guirong W et al (2010) Biocompatibility evaluationof chitosan-based injectable hydrogels for the culturingmice mesenchymal stemcells in vitro. J Biomater Appl 24(7):625–637

Joshi SC (2011) Sol-Gel Behavior of hydroxypropyl methylcellulose (HPMC) in ionic media including drug release. Materials 4:1861–1905

Joshi SC, Lam YC (2005) Modeling heat and degree of gelation for methyl cellulose hydrogels with nacl additives. J Appl Polym Sci 101:1620–1629

Joshi SC, Liang CM, Lam YC (2008) Effect of solvent state and isothermal conditions on gelation of methylcellulose hydrogels. J Biomater Sci Polymer Ed 19(12):1611–1623

Kang H, Liu R, Huang Y (2016) Cellulose-based gels. Macromol Chem Phys 217(12):1322–1334

Karlsson K, Schuster E, Stading M et al (2015) Foaming behavior of water-soluble cellulose derivatives: hydroxypropyl methylcellulose and ethyl hydroxyethyl cellulose. Cellulose 22 (4):2651–2664

Karpiński R, Górniak B, Maksymiuk J (2015) Biomedyczne zastosowania polimerów-materiały opatrunkowe. Nowoczesne Trendy w Medycynie, Lublin, pp 18–23

Kobayashi K, Huang C, Lodge TP (1991) Thermoreversible gelation of aqueous methylcellulose solutions. Macromolecules 32:7070–7077

Kuang Q, Cheng G, Zhao J et al (2006) Thermogelation hydrogels of methylcellulose and glycerol–methylcellulose systems. J Appl Polym Sci 100:4120–4126

Lam YC, Joshi SC, Tan BK (2007) Thermodynamic characteristics of gelation for methyl-cellulose hydrogels. J Therm Anal Calorim 87(2):475–482

Li L, Thangamathesvaran PM, Yue CY et al (2001) Gel network structure of methylcellulose in water. Langmuir 17(26):8062–8068

Li L, Liu E, Lim CH (2007) Micro-DSC and rheological studies of interactions between methylcellulose and surfactants. J Phys Chem B 111:6410–6416

Li H, Hao X, Xie Y et al (2012) Effect of β-cyclodextrin upon the sol–gel transition of methylcellulose solutions in the presence of sodium dodecyl sulfate. Chinese J Chem Phys 25 (2):242–248

Lopes da Silva JA, Rao MA (2007) Rheological behavior of food gels. In: Barbosa-Canovas GV (ed) Rheology of fluid and semisolid foods: principles and applications. Food engineering series, 2nd edn. Springer, Boston, pp 338–391

Lue A, Zhang L (2008) Investigation of the scaling law on cellulose solution prepared at low temperature. J Phys Chem 112:4488–4495

Luo K, Yang Y, Shao Z (2016) Physically crosslinked biocompatible silk-fibroin-based hydrogels with high mechanical performance. Adv Funct Mater 26:872–880

Mandala I, Savvas TP, Kostaropoulos AE (2004) Xanthan and locust bean gum influence on the rheology and structure of a white model-sauce. J Food Eng 64:335–342

Miura Y (2014) Solvent isotope effect on sol–gel transition of methylcellulose studied by DSC. Polym Bull 71:1441–1448

Murtaza G (2012) Ethylcellulose microparticles: a review. Acta Pol Pharm 69(1):11–22

Nasatto PL, Pignon F, Silveira JLM, Duarte M, Noseda M, Rinaudo M (2015) Methylcellulose, a cellulose derivative with original physical properties and extended applications. Polymers 7(5):777–803

Nijenhuis KT, Winter HH (1989) Mechanical properties at the gel point of a crystallizing poly (vinyl chloride) solution. Macromolecules 22(1):411–414

Nishinari N (2000) Rheology of physical gels and gelling processes. Rep Prog Polym Phys Jpn 43:163–192

Nishinari K (2009) Some thoughts on the definition of a gel. In: Tokita M, Nishinari K (eds) Gels: structures, properties, and functions. fundamentals and applications, vol 136. Progress in colloid and polymer science. Springer, Berlin, Heidelberg, pp 87–94

Osswald TA, Rudolph N (2014) Polymer rheology fundamentals and applications. Hanser, Cincinnati

Picout DR, Ross-Murphy SB (2003) Rheology of biopolymer solutions and gels. Sci World J 3:105–121

Pingping L, Ruigang L (2015) Cellulose gels and microgels: synthesis, service, and supramolecular interactions. In: Seiffert S (ed) Supramolecular polymer networks and gels, vol 268. Advances in polymer science. Springer, Cham, pp 209–238

Ramli S, Ja'afar SM, Sisak MAA et al (2015) Formulation and physical characterization of microemulsions based carboxymethyl cellulose as vitamin C carrier. Malaysian J Anal Sci 19(1):275–283

Razmkhah S, Razavi SMA, Mohammadifar MA (2017) Dilute solution, flow behavior, thixotropy and viscoelastic characterization of cress seed (*Lepidium sativum*) gum fractions. Food Hydrocolloid 63:404–413

Ren LZ, Ren PG, Zhang XL (2014) Preparation and mechanical properties of regenerated cellulose/poly(vinyl-alcohol) physical composite hydrogel. Compos Interface 21(9):853–867

Sannino A, Demitri C, Madaghiele M (2009) Biodegradable cellulose-based hydrogels: design and applications. Materials 2(2):353–373

Sarode A, Wang P, Cote P et al (2013) Low-viscosity hydroxypropylcellulose (HPC) grades SL and SSL: versatile pharmaceutical polymers for dissolution enhancement, controlled release, and pharmaceutical processing. AAPS Pharm Sci Tech 14(1):151–159

Scalan JC, Winter HH (1991) Composition dependence of the viscoelasticity of end-linked poly (dimethylsiloxane) at the gel point. Macromolecules 24:47–54

Schupper N, Rabin Y, Rosenbluh M (2008) Multiple stages in the aging of a physical polymer gel. Macromolecules 41:3983–3994

Schwittay C, Mours M, Winter HH (1995) Rheological expression of physical gelation in polymers. Faraday Discuss 101:93–104

Shifeng Y, Jingbo Y, Li T et al (2010) Novel physically crosslinked hydrogels of carboxymethyl chitosan and cellulose ethers: structure and controlled drug release behavior. J Appl Polym Sci 119(4):2350–2358

Siepmann J, Peppas NA (2012) Modeling of drug release from delivery systems based on hydroxypropyl methylcellulose (HPMC). Adv Drug Deliv Rev 64:163–174

Silva SMC, Pinto FV, Antunes FE et al (2008) Aggregation and gelation in hydroxypropylmethyl cellulose aqueous solutions. J Colloid Interf Sci 327(2):333–340

Silva SMC, Antunes FE, Sousa JJS et al (2011) New insights on the interaction between hydroxypropylmethyl cellulose and sodium dodecyl sulfate. Carbohyd Polym 86:35–44

Su JC, Liu SQ, Joshi SC et al (2008) Effect of sds on the gelation of hydroxypropylmethylcellulose hydrogels. J Therm Anal Calorim 93(2):495–501

Tang Y, Wang X, Li Y et al (2010) Production and characterisation of novel injectable chitosan/methylcellulose/salt blend hydrogels with potential application as tissue engineering scaffolds. Carbohyd Polym 82:833–841

Tung CYM, Dynes PJ (1982) Relationship between viscoelastic properties and gelation in thermosetting systems. J Appl Polym Sci 27(2):569–574

Veríssimo MIS, Pais AAC, Gomes MTS (2010) Following HPMC gelation with a piezoelectric quartz crystal. Carbohyd Polym 82:363–369

Wang Q, Li L, Liu E et al (2006) Effects of SDS on the sol–gel transition of methylcellulose in water. Polymer 47:1372–1378

Wang T, Chen L, Shen T et al (2016) Preparation and properties of a novel thermo-sensitive hydrogelbased on chitosan/hydroxypropyl methylcellulose/glicerol. Int J Biol Macromol 93:775–782

Wang F, Yang Y, Ling Y (2017) Injectable and thermally contractible hydroxypropyl methyl cellulose/Fe_3O_4 for magnetic hyperthermia ablation of tumors. Biomaterials 128:84–93

Winnicka K (2008) Leki okulistyczne w aptece—nowe technologie. Gaz Farm 7:30–32

Winter HH, Chambon F (1986) Analysis of linear viscoelasticity of a crosslinking polymer at the gel point. J Rheol 30(2):367–382

Wuestenberg T (2014) Cellulose and cellulose derivatives in the food industry: fundamentals and applications. Wiley-VCH Verlag GmbH & Co. KGaA, Germany, pp 345–376

Xia X, Yao Y, Gong M et al (2014) Rheological behaviors of cellulose/[BMIM]Cl solutions varied with the dissolving proces. J Polym Res 21:512–519

Zhang L, Wang Y, Yu L et al (2015) Rheological and gel properties of hydroxypropyl methylcellulose/hydroxypropyl starch blends. Colloid Polym Sci 293(1):229–237

Zhang L, Li J, Liu H (2016) Thermal inverse phase transition of azobenzene hydroxypropylcellulose in aqueous solutions. Cellulose 23:1177–1188

Zheng P, Li L, Hu X (2004) Sol–gel transition of methylcellulose in phosphate buffer saline solutions. J Polym Sci Pol Phys 42:1849–1860

The Characterization of the Residence Time Distribution in a Fluid Mixer by Means of the Information Entropy

Marian Kordas, Daniel Pluskota and Rafał Rakoczy

1 Residence Time Distribution

Tracer diagnostics of flow patterns is a traditional tool of chemical engineering. It is a highly developed system of tests which are widely applied to many mixing systems (Yablonsky et al. 2009). The main concept of this approach is based on the residence time distribution (RTD). The RTD method is one of the most informative characteristics to obtain hydrodynamic information and it is the main concept of the Danckwerts approach (Danckwerts 1953, 1958).

The typical experiment related to the stimulus response RTD approach is carried out with non-reactive tracers. The flow entering a mixer or reactor is marked by a substance that does not participate in the reaction and the mixer or reactor outlet stream is then monitored (Fig. 1).

The determination of mixing patterns for the tested apparatus may be developed by means of the RTD results (Levenspiel 1998). The shape of the normalized RTD function is an indication of the type of mixing or flow behavior (e.g. ideal mixing, plug flow, dead zones, stagnancy, channeling, etc.) occurring in mixers and reactors. The RTD technique may be also treated as the quantitative method to derive simple hydrodynamic models in the case of complex engineering systems. The diagnostics and the characterization of a mixing process may be directly gained form the interpretation of the RTD curves (Levenspiel 1998). Many experimental investigations and theoretical considerations are reported on the RTD of impinging chemical engineering devices (Melo et al. 2001; Harris et al. 2002; Zhang et al. 2005; García-Sera et al. 2007; Pröll et al. 2007; Christensen et al. 2008; Guo et al. 2008; Hornung and Mackley 2009; Madhurabthakam et al. 2009; Mizonov et al. 2009; Nikitine et al. 2009; Gao et al. 2011).

M. Kordas · D. Pluskota · R. Rakoczy (✉)
Faculty of Chemical Technology and Engineering, West Pomeranian
University of Technology, Szczecin, Poland
e-mail: rafal.rakoczy@zut.edu.pl

© Springer International Publishing AG, part of Springer Nature 2018
M. Ochowiak et al. (eds.), *Practical Aspects of Chemical Engineering*,
Lecture Notes on Multidisciplinary Industrial Engineering,
https://doi.org/10.1007/978-3-319-73978-6_14

Fig. 1 Scheme of a
pulse-response experiment

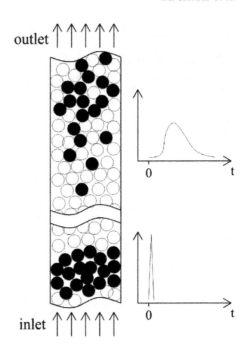

The RTD method allows to compare an obtained curve for a tested mixing device with ideal mixing models (plug flow reactor—PFR and a continuous stirred tank reactor CSTR) (Buso et al. 1997; Cocero and Garcia 2001; Znad et al. 2004; Yianatos et al. 2005). By analyzing the shape of a distribution curve, some non-ideal mixing properties inside the flow, such as dead zones, and bypassing paths, may be inferred (Martin 2000; Jafari and Soltan Mohammadzadeh 2005).

A mixing performance in continuous flow stirred tank reactors may be studied by using age distributions. The measure for the relative level of mixing of a molecular age in these systems was proposed by Danckwerts (1953, 1958). Zwietering further discussed the mixing degree based on the variance of the ages (Zwietering 1959). A method to compute the mixing degree proposed by Danckwerts and Zwietering as a measure of a mixing performance for a continuous flow reactor was presented by Liu (2012). The mean age theory is also developed to characterize a spatial non-uniformity in a CSTR. The effect of the device flow on the mixing process due to various designs of inlet and outlet locations is quantitatively measured (Liu 2012).

The RTD method is applied as an indicator of the type and extent of mixing. The RTD curves may be treated as the quantitative description of the mixing system. For the tracer input method (or pulse method) the RTD (or exit age-distribution function, $E(t)$, may be connected with the outlet concentration, $c(t)$, by the following relation

$$E(t) = \frac{c(t)}{\int_0^\infty c(t)\,dt} \qquad (1)$$

or

$$E(t) = \frac{c(t)}{\sum_{i=0}^{\infty} c(t_i)\,\Delta t_i} \qquad (2)$$

where: c—tracer concentration, e.g. kg tracer m^{-3} solvent; E—residence time distribution function, s^{-1}; t—time, s; t_i—time step for the measurements, s.

The fraction of whole material that resided for time t in mixing or reactor system in the range $(0, \infty)$ is equal to:

$$\int_0^\infty E(t)\,dt = 1 \qquad (3)$$

For any mixer or reactor system, $E(t)$ represents the volume fraction of fluid having a residence time distribution 0 and t. Based on this function, the cumulative and dimensionless F-curve may be obtained from:

$$F(t) = \int_0^t E(t)\,dt \qquad (4)$$

Both $E(t)$ and $F(t)$ are representation of RTD in the flow system. Function $E(t)$ can be obtained from $F(t)$ as follows:

$$E(t) = \frac{d\,F(t)}{dt} \qquad (5)$$

Taking into account the RTD function (Eq. 1), statistical parameter such as the mean residence time can be determined by using the following relationships (Rakoczy et al. 2013):

$$\tau = \frac{\int_0^\infty t\,E(t)\,dt}{\int_0^\infty E(t)\,dt} = \frac{\int_0^\infty t\,c(t)\,dt}{\int_0^\infty c(t)\,dt} \qquad (6)$$

or

$$\tau = \frac{\sum_{i=0}^{\infty} t_i\,c_i\Delta t_i}{\sum_{i=0}^{\infty} c_i\Delta t_i} \qquad (7)$$

where: τ—mean residence time distribution, s.

Moreover the variance can be calculated as follows (Jones et al. 2009):

$$\sigma^2 = \int_0^\infty (t - \tau)^2 E(t) \, dt \qquad (8)$$

or

$$\sigma^2 = \sum_{i=0}^\infty (t_i - \tau)^2 E(t) \, \Delta t_i \qquad (9)$$

where: σ^2—variance of time distribution, s^2.

Based on the definitions of the mean residence time and the variance, the coefficient of variation, CoV, can be defined:

$$CoV = \frac{\sqrt{\sigma^2}}{\tau} \qquad (10)$$

This coefficient may be applied for the assessment of the variability of the standard variation with the respect to the mean residence time distribution. In the context of the mixing process, the smaller the variance CoV, the closer the distribution to the mean residence time is and the better the mixing quality (Adeosun and Lawal 2010).

The RTD function may be also expressed in the dimensionless form:

$$E(\theta) = \tau \, E(t) \qquad (11)$$

where: θ—normalized time ($\theta = t/\tau$).

A normalized RTD function, $E(\theta)$, is used instead of $E(t)$ when the mixing performance of flow systems of different sizes or flow conditions is to be compared. In the case the standardized and restrictive conditions can be written as:

$$\int_0^\infty E(\theta) \, d\theta = 1 \qquad (12)$$

and

$$\int_0^\infty \theta \, E(\theta) \, d\theta = 1 \qquad (13)$$

The relationship between the RTD parameters (Eqs. 1–13) may be graphically presented in Fig. 2.

Fig. 2 The schematic
diagram of the certain
parameters of the RTD
measurements

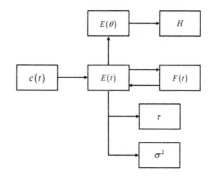

2 Information Mixing Capacity

One of the traditional approaches to the mathematical description of the chemical engineering processes is based on the statistical analysis. An interesting alternative to the classical methods is the mathematical modeling of these processes by means of the information theory and the Shannnon's entropy. According to Shannon, the information entropy is a function which measures the uncertainty of the even realization. If the event is certain, the value of this parameter is equal to zero. A higher uncertainty of event realization is connected with the higher value of entropy.

Generally, the informational entropy of a random variable is defined in terms of its probability distribution. A discrete random variable, X, taking values x_i ($i = 1,2, ..., n$) with probabilities $p_i = p(x_i)$ is given by:

$$\sum_{i=1}^{n} p_i = 1 \quad (p_i \geq 0, \ \forall i \in N)$$ (14)

The information entropy (Shannon's entropy) for discrete variables is defined as (Shannon 1948).

$$H = -\sum_{i=1}^{n} p(x_i) \ln p(x_i)$$ (15)

It should be noticed that the information associated with the receipt of signal x_i is equal to—$\ln[p(x_i)]$.

For a continuous random variable with the probability density $p(x)$ reads:

$$H = -\sum_{i=1}^{n} p(x_i) \ln p(x_i) = -\lim_{\Delta x \to 0} \sum_{i}^{\infty} p(x_i)\Delta x \ \ln\{p(x_i)\Delta x\}$$

$$= -\int_{0}^{\infty} p(x) \ \ln p(x) \, dx - \lim_{\Delta x \to 0} p(x)\Delta x \ \ln \Delta x$$ (16)

The information entropy can be applied for the description of the mixing efficiency in a reactor system. The informational theory can be also used to analyze the
RTD measurements. Taking into account the function $E(\theta)$ (Eq. 11) and the definition of the information entropy (Eq. 16) the uncertainty for the RTD measurements may be expressed as follows (Ogawa 2007):

$$H = - \int_0^\infty E(\theta) \ln E(\theta) \, d\theta \tag{17}$$

or

$$H = - \sum_{i=1}^n E(\theta_i) \Delta\theta \ln\{E(\theta_i)\Delta\theta\} \tag{18}$$

where: $\sum_{i=1}^n E(\theta_i) \Delta\theta$ is equal to 1 and n is the number of discrete times that is
sufficiently large to express the RTD.

From the practical point of view, an information mixing capacity can be defined
as:

$$M = \frac{H - H|_{min}}{H|_{max} - H|_{min}} \tag{19}$$

The proposed mixing capacity is using maximum, $H|_{max}$, and minimum, $H|_{min}$,
values of the information entropy. The condition under which $H|_{min}$ takes the
minimum value is realized when a plug flow is established in the equipment. It is
characterized by the fact the flow of fluid through a reactor is orderly with no
element of a fluid overtaking or mixing with any other element ahead or behind. In
this case the whole material processed through a reactor must have the same residence time and the material leaving such a reactor spends precisely the same
amount of time within a reactor. The condition of the minimum entropy corresponds to the ideal plug flow reactor (PFR). For the PFR it is shown that the
informational entropy is equal to 0 (Ogawa 2007). Also note that $H = 0$ if and only
if the probability of one outcome is certain and the probability of all outcomes is 0.

Taking into consideration the above definition the mixing capacity may be
defined as follows

$$M = \frac{H}{H|_{max}} \Rightarrow M = \frac{-\sum_{i=1}^n E(\theta_i)\Delta\theta \ln\{E(\theta_i)\Delta\theta\}}{H|_{max}} \tag{20}$$

Consequently, the informational entropy is a non-negative continuous function
which gradually increases from 0 to its maximum value at $H|_{max}$. The condition
under which H takes a maximum value is realized when the perfect mixing flow is
established in mixer or vessel content is completely homogeneous (no difference

exists between various portions of a vessel, and outlet stream properties are identical to vessel fluid properties). In an ideal, CSTR the concentration of any substance in the effluent stream is identical to a concentration throughout a mixer.

For a perfectly mixed CSTR the normalized RTD function, $E(\theta)$, is given as follows:

$$E(\theta) = e^{-\theta} \tag{21}$$

The maximum value of the informational entropy is defined by:

$$H|_{max(CSTR)} = -\int_0^\infty E(\theta) \ln E(\theta) d\theta \Rightarrow H|_{max(CSTR)}$$

$$= -\int_0^\infty e^{-\theta} \ln e^{-\theta} d\theta \Rightarrow H|_{max(CSTR)} = \ln(e) \int_0^\infty \theta e^{-\theta} d\theta \tag{22}$$

Therefore the informational entropy takes its maximum value as follows:

$$H|_{max} = \ln(e) \tag{23}$$

Taking into consideration the above assumptions the mixing capacity (Eq. 20) can be defined as the degree of approach from the plug flow to the perfect mixing flow:

$$M = \frac{H}{H|_{max}} \Rightarrow M = \frac{-\sum_{i=1}^{n} E(\theta_i)\Delta\theta \ln\{E(\theta_i)\Delta\theta\}}{\ln(e)} \tag{24}$$

The values of the mixing capacity, M, varies between 0 (for the plug flow) and 1 (for the perfect mixing flow). It should be noticed that the higher the value of the mixing capacity M, the closer the RTD measurement to the ideal mixing is and the better the mixing quality is.

3 Application of Information Mixing Capacity

3.1 Static Mixer

Mixing is one of the most popular operation used to reach desired physical conditions. Motionless mixers are able to provide reasonable mixing performance eliminating numerous issues of mechanical stirrers. Most of them are tubes equipped with internal elements modifying the flow-path of input streams in order to cause the dynamic effect applied to mixing operation. The process with static device is hustled by mechanical energy of fluid, which allows to design efficient equipment. Static mixers require less power and demand less maintenance care in comparison to actively stirred equipment involving energy loss on friction and service costs connected with moving elements. This is the main reason why interdisciplinary high-end technologies are focused on elimination of moving parts and replacing them with motionless units as far as it is possible. Providing high throughput, static mixers are very reasonable solution for continuous operations what is serious issue of actively stirred tanks. They require less space time to perform the process reaching approximate results in terms of homogenization and heat transfer. Active impellers in opposition to static units generate local zones of intensive action damaging fragile materials. Comparative research (Thakur et al. 2003) led to conclusions proposing the application of static mixers if their performance seem to meet process requirements.

Patents and commercial offers indicate a strong connection between the geometry and case specification such as flow character and physical properties of materials. Mixers designed for laminar flow contain highly developed interior elements filling bigger part of radial profile than the applied with the turbulent flow. It is an intuitively explainable design strategy caused by lower amount of mechanical energy of the flowing fluid and lack of self-inducted turbulence. Performance of laminar mixers is based on flow-division mixing mechanism. Turbulent devices take advantage of the flow character what require less intensive flow manipulation. Obviously, some motionless mixers can be applied with both turbulent and laminar flow, but specialized units remain the majority. For example, standard helical mixer developed by *Kenics* views good mixing performance with application of both flow characters (chemineer.com). Application of a wrong mixer at given process parameters may lead to poor performance or economical inefficiency caused by intensive friction.

In the case of this investigations the static mixer equipped with a new mixing insert, which was a modified version of the mixer described in the patent registration (Masiuk and Szymański 1997). The sketch of mixing device is presented in Fig. 3.

The single mixing element had two truncated cones connected with each other. The diameter of the bigger cone was equal to the inner diameter of the apparatus. On the side surface of the inlet cone the longitudinal slots were placed symmetrically along the circumference. The slots were equipped with steering blades placed on their longitudinal edges. The side surface of the inlet cone had an oval shape near its lower base. A stream of fluid flowing into the inlet cone was redistributed the slots on the surface of the cone were steering the flow outside, while the steering blades bestowed the arising streams with the rotational movement. A part of the stream leaving the inside of the insert through the outlet cone became turbulent due to the cone's oval shape. The inner and outer streams were mixed before entering the next mixing element where the streams were redistributed once again, rotated in the opposite direction, and reconnected. A novel mixing device was built up of five separate segments in the form of two top opened cones. The large cone had

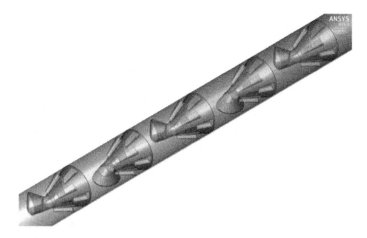

Fig. 3 View over the geometry of new construction for a static mixer

longitudinal holes on the conical face with motionless outside baffles. The experiments were carried out with the cascade including 5 mixing elements.

The mixing process in the tested static mixer was evaluated by the RTD measurements using the stimulus-response technique. This technique is the simplest and most direct way of finding the RTD curves. In the case of this experimental work, the RTD was obtained by injecting a tracer instantaneously (a pulse input) at the inlet of a flow system, and then measuring the tracer concentration, c(t), at the outlet as a function of time. The stimulus-response technique demands injecting the tracer within a very short time compared to a flow residence-time scale. This impulse can be described by the Dirac delta function (δ-Dirac function).

In this case the distribution function is a spike of infinite height and zero width, whose area is equal to 1. The Dirac delta function has the following properties:

$$\delta(x) = \begin{cases} 0 & for \quad x \neq 0 \\ \infty & for \quad x = 0 \end{cases} \tag{25}$$

and

$$\int_{-\infty}^{\infty} \delta(x)dx = 1 \tag{26}$$

The schematic diagram of experimental set-up is graphically presented in Fig. 4.

Experimental measurements were performed for tap water flow rates ranging between 0.028 and 0.25 kg s^{-1}. The RTD experiments were realized with a saturated solution of brine (25 wt% NaCl). The RTD measurements were determined by measuring the electrical conductivity of the liquid inside the static mixer. Samples

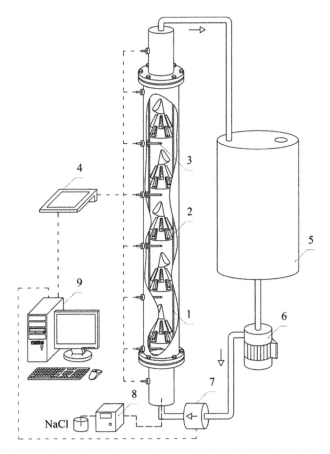

Fig. 4 The diagram of experimental set-up: 1 mixer; 2 mixing inert; 3 probe; 4 multifunction meter; 5 buffer; 6 circulation pump; 7 flow meter; 8 dosing pump; 9 computer

were then collected every 1 s until the tracer's disappearance in the tested magnetic mixer. Typical examples of conductivity measurements are presented in Fig. 5.

The experimental set-up was equipped with a measuring instrument which controlled the concertation of brine in the liquid and supervised the real-time acquisition of all the experimental data coming from the sensors. This device also measured the concentrations fluctuations inside the static mixer during the process. Electric signals were sampled by using the special sensors (CD-210, Elmetron, Poland) and were passed through the multifunction meter (CX-701, Elmetron, Poland) to a personal computer for further processing.

The conductivity was recalculated to concentration with the calibration curve (Rakoczy and Masiuk 2010).

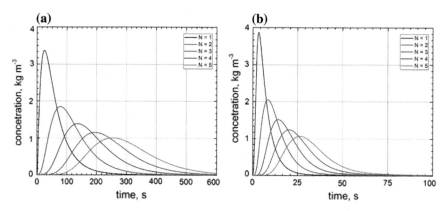

Fig. 5 Typical examples of RTD experimental results: **a** 0.028 kg s^{-1}; **b** 0.25 kg s^{-1} (where N is the number of mixing elements)

3.2 Information Mixing Capacity for Tested Static Mixer

In the following work, the electrical conductivity was measured at different longitudinal positions and was used to compute the RTD function (Eq. 11). These functions were calculated after 1, 2, 3, 4, and 5 incremental segments. The generalization of the results of the mixing process measurements may be done by using the proposed information mixing capacity (Eq. 20).

From the practical point of view the information mixing capacity, M, may be modeled by means of the following relation

$$M = f(\text{Re}) \Rightarrow \left(\frac{-\sum_{i=1}^{n} E(\theta_i)\Delta\theta \ \ln\{E(\theta_i)\Delta\theta\}}{\ln(e)} \right) = f\left(\frac{w_L D^2}{v_L}\right) \qquad (27)$$

where: w_L—disruptive velocity of liquid, m s^{-1}; D—diameter of static mixer, m; v_L—kinematic viscosity of liquid, m^2 s^{-1}.

In order to establish the effect of the Reynolds number on the information mixing capacity the calculated data obtained in this work are graphically illustrated in Fig. 6.

Results shown in Fig. 6 suggest that the mixing process in the tested static mixer may be analytically described by means of a unique monotonic function.

$$M = p_1 \text{Re}^{p_2} \qquad (28)$$

The constant, p_1, and exponent, p_2, were computed by employing the *Matlab* software and the principle of least square and the obtained values may be correlated as:

Fig. 6 The influence of
Reynolds number on the
information mixing capacity

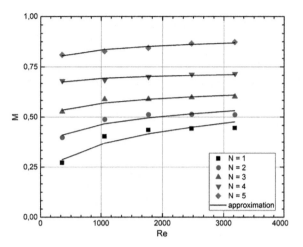

$$p_1 = 0.08\,N^{1.41} \tag{29}$$

and

$$p_2 = 0.25\,N^{-1.34} \tag{30}$$

Figure 6 presents results of the mixing capacity at different operating conditions. These results show that this parameter is strongly dependent on the flow rate of the continuous liquid phase and the number of mixing elements. The applied definition of the mixing capacity gives a useful compact way to describe the RTD measurements. In order to study the changes of the RTD curves with the increasing value of the Reynolds number, the mixing capacity basing on the informational entropy is determined for different operating conditions.

Figure 6 compares the mixing capacity obtained at the different number of mixing inserts for various flow rates. In the case of the experimental results, the calculated values of the mixing capacity show systematically increasing trends with the value of the dimensionless Reynolds number.

Based on the proposed relation (Eq. 28), the influence of the number of mixing inserts may be graphically presented. The influence of the number of mixing inserts on the information mixing capacity for the selected Reynolds numbers (Re = 357 and Re = 3185) is graphically presented in Fig. 7.

The analysis of the obtained results reveals that the increase in the number of the mixing inserts causes distinct modifications of the information mixing capacity. It can be seen that the values of the information mixing capacity for the higher values of the dimensionless Reynolds number are greater than the values of mixing capacity obtained in this work for lower values of this hydrodynamic parameter. It should be noticed that the enhancement of the information mixing capacity is connected with the flow regime. The observed discrepancies between the obtained

Fig. 7 Graphical presentation of the influence of the mixing inserts over the information mixing capacity

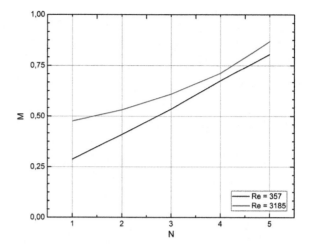

M-capacity decreases with the intensification of the hydrodynamic conditions (increase the number of mixing inserts).

Figure 8 shows cross-sectional vertical cut along the central axial axis of the modeled static mixer, for the different operating conditions and water as the working liquid. This figure was obtained by using the CFD technique. The numerical computations were carried out using the commercial CFD package *ANSYS Workbench 14.5* (*ANSYS, Inc.*, Canonsburg, USA). To calculate the liquid flow in the tested static mixer, geometry of static mixer's inner tube (of the same dimensions as the real mixer) with five inserts by *AutoCAD* software (*Autodesk, Inc.*, San Rafael, USA) was created. The created geometry was imported to Design Modeler and introduced to *ANSYS Meshing* software to generate the CFD mesh. Numerical mesh consists of over 1.5 million triangular elements of acceptable quality. The numerical computations of fluid flow process was performed by means of *ANSYS CFX* software. This solver uses a control-volume formulation for solving the conservation equations of mass, momentum and energy. Each simulation was assumed to converge when the residuals for equations of continuity and momentum were below 10^{-5}. The three-dimensional finite volume CFD code used here solves the steady, Reynolds-averaged Navier-Stokes (RANS) equations with the $k-\omega$ turbulence-closure model. This kind of numerical approach is allowed to compute complex laminar and turbulent flows in static mixers after applying appropriate corrections.

The basic, two equations $k-\omega$ model was proposed by Wilcox (Wilcox 1988). In the case of this paper the employed numerical model incorporates also modifications for low-Reynolds number effects, compressibility, and shear flow spreading. The applied $k-\omega$ model may be used to analyze the flow inside the static mixer with the complex construction (Jones et al. 2002). In this case the predicted flow may be extremely complicated and contains regions with low level of turbulence or regions with the local turbulence. Moreover, this model is employed with automatic

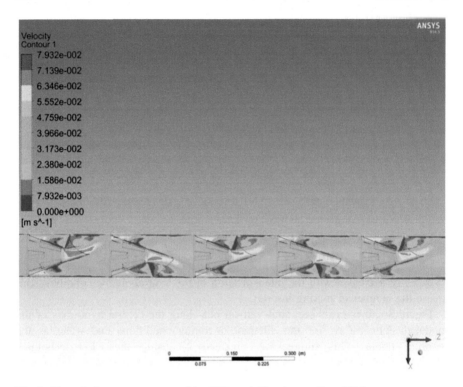

Fig. 8 The velocity contour generated by CFD modelling (water; $Re = 3185$)

near-wall treatment automatically switches from wall-functions to a low-Reynolds near wall formulation. In addition, the $k–\omega$ model was validated extensively in complex, three-dimensional shear flows (Lin and Sotiropoulos 1997). It should be emphasized that the $k–\omega$ model is expected to be able to reasonably capture the characteristics of the laminar flow because the ω-equation possesses a solution as the turbulent kinetic energy approaches zero (Peng et al. 1996). The application of the $k–\omega$ model for low-Reynolds regions is presented in the relevant literature (Togun et al. 2014; Stringer et al. 2014; Jones 2000).

The application of the motionless elements direct the flow from the center towards the wall. This flow improves the radial mixing eliminating velocity gradients and enhances the mixing process in the proposed construction of the static mixer. The tested mixing inserts give a modest improvement in the mixing process and it is achieved by lower values of the dimensionless Reynolds numbers. This leads to the conclusion that the mixing capacity is mainly dependent on the geometrical configuration of the motionless insert and the generated hydrodynamic conditions in the flowing fluid through the static mixer.

4 Conclusion

Using the RTD as a measure to indirectly evaluate a mixing behavior in a magnetic mixer, the calculations of the informational entropy were performed to obtain the mixing capacity. This study has emphasized the importance of the informational entropy as a critical complementary parameter for the experimental RTD studies in the evaluation of mixing in a flow system. It should be noticed that the RTD investigations were worked out for a static mixer. The information mixing capacity may be applied for a comparison between various mixing or reactor systems. This parameter reduces complex analysis of the RTD measurements to a single quantity measuring the progress of a mixing process. The proposed information mixing capacity may be applied instead of the approach applying well-knows theories (e.g. the coefficient CoV, which gives the ratio of the standard deviation to the mean and provides a measure of the relative variation of distribution).

References

Adeosun JT, Lawal A (2010) Residence-time distribution as a measure of mixing in T-junction and multilaminated/elongational flow micromixers. Chem Eng Sci 65:1865–1874

Buso A, Glomo M, Boaretto L (1997) New electrochemical reactor for wastewater treatment: mathematical model. Chem Eng Process 36:411–418

Christensen D, Nijenhuis J, van Ommen J et al (2008) Residence times in fluidized beds with secondary gas injection. Powder Technol 180:321–331

Cocero MJ, Garcia J (2001) Mathematical model of supercritical extraction applied to oil seed extraction by CO_2+ saturated alcohol—II. Shortcut methods. J Supercrit Fluid 20:245–255

Danckwerts PV (1953) Continuous flow systems. Chem Eng Sci 2:1–13

Danckwerts PV (1958) The effect of incomplete mixing on homogeneous reactions. Chem Eng Sci 8:93–99

Gao Y, Vanarase A, Muzzio F et al (2011) Characterizing continuous powder mixing using residence time distribution. Chem Eng Sci 66:417–425

García-Sera J, García-Verdugo E, Hyde JR et al (2007) Modelling residence time distribution in chemical reactors: a novel generalised n-laminar model. Application to supercritical CO_2 and subcritical water tubular reactors. J Supercrit Fluid 41:82–91

Guo Q, Liang Q, Ni J et al (2008) Markov chain model of residence time distribution in a new type entrained-flow gasifier. Chem Eng Process 47:2061–2065

Harris A, Thorpe R, Davidson J (2002) Stochastic modelling of the particle residence time distribution in circulating fluidised bed risers. Chem Eng Sci 57:4779–4796

Hornung C, Mackley M (2009) The measurements and characterisation of residence time distribution for laminar liquid flow in plastic microcapillary arrays. Chem Eng Sci 64:3889–3902

Jafari M, Soltan Mohammadzadeh JS (2005) Mixing time, homogenization energy and residence time distribution in a gas-induced contractor. Trans IChemE Part A 83:452–459

Jones SC (2000) Static mixers for water treatment: a computational fluid dynamics model. Ph.D. thesis. Georgia Institute of Technology, Source DAI-B 61/04, p 2132

Jones SC, Sotiropoulos F, Amirtharajah A (2002) Numerical modeling of helical static mixers for water treatment. J Environ Eng 128:431–440

Jones PN, Özcan-Taşkin NG, Yianneskis M (2009) The use of momentum ration to evaluate the performance of CSTRs. Chem Eng Res Des 87:485–491

Levenspiel O (1998) Chemical reaction engineering. Wiley, New York

Lin FB, Sotiropoulos F (1997) Strongly-coupled multigrid method for 3-D incompressible flows using near-wall turbulence closures. J Fluids Eng 119:314–324

Liu M (2012) Age distribution and the degree of mixing in continuous flow stirred tank reactors. Chem Eng Sci 69:382–393

Madhurabthakam C, Pan Q, Rempel G (2009) Residence time distribution and liquid holdup in kenics KMX static mixer with hydrogenated nitrile butadiene rubber solution and hydrogen gas system. Chem Eng Sci 64:3320–3328

Martin AD (2000) Interpretation of residence time distribution. Chem Eng Sci 55:5907–5917

Masiuk M, Szymański E (1997) Polish Patent No. 324,150. Polish Patent and Trademark Office, Warszawa

Melo PA, Carlos Pinto J, Jr Biscaia E (2001) Characterization of the residence time distribution in loop reactors. Chem Eng Sci 56:2703–2713

Mizonov V, Berthiaux H, Gatumel C et al (2009) Influence of crosswise non-homogenity of particulate flow on residence time distribution in a continuous mixer. Powder Technol 190:6–9

Nikitine C, Rodier E, Sauceau M et al (2009) Residence time distribution of a pharmaceutical grade polymer melt in a single screw extrusion process. Chem Eng Res Des 87:809–816

Ogawa K (2007) Chemical engineering. A new perspective. Elsevier

Peng SH, Davidson L, Holmberg S (1996) The two-equations turbulence k-omega model applied to recirculating ventilation flows. Rept. 96/13, Thermo and Fluid Dynamics, Chalmers University of Technology, Göteborg

Pröll T, Todinca T, Şuta M et al (2007) Acid gas absorption in trickle flow columns—modelling of the residence time distribution of a pilot plant. Chem Eng Process 46:262–270

Rakoczy R, Masiuk S (2010) Influence of transverse rotating magnetic field on enhancement of solid dissolution process. AIChE J 56:1416–1433

Rakoczy R, Kordas M, Grądzik P et al (2013) Experimental study and mathematical modeling of the residence time distribution in magnetic mixer. Polish J Chem Technol 15:53–61

Shannon CE (1948) A mathematical theory of communication. Bell Syst Techn J 27:379–423 and 623–656

Stringer RM, Zang J, Hillis AJ (2014) Unsteady RANS computations of flow around a circular cylinder for a wide range of Reynolds numbers. Ocean Eng 87:1–9

Thakur RK, Vial C, Nigam KDP et al (2003) Static mixers in the process industries—a review. Institution of Chemical Engineers

Togun H, Safaei MR, Sadri R et al (2014) Numerical simulation of laminar to turbulent nanofluid flow and heat transfer over a backward-facing step. Appl Math Comput 239:153–170

Wilcox DC (1988) Reassessment of the scale-determining equation for advanced turbulence models. AIAA J 26:1299–1310

Yablonsky GS, Constales D, Marin GB (2009) A new approach to diagnostics of ideal and non-ideal flow patterns: I. The concept of reactive-mixing index (REMI) analysis. Chem Eng Sci 64:4875–4883

Yianatos JB, Bergh LG, Díaz F et al (2005) Mixing characteristics of industrial flotation equipment. Chem Eng Sci 60:2273–2282

Zhang T, Wang T, Wang J (2005) Mathematical modelling of the residence time distribution in loop reactors. Chem Eng Process 44:1221–1227

Znad H, Báleš V, Kawase Y (2004) Modeling and scale up of airlift bioreactor. Comput Chem Eng 28:2765–2777

Zwietering TN (1959) The degree of mixing in continuous flow system. Chem Eng Sci 11:1–15

Selected Aspects of Dust Removal from Gas Stream for Chamber Separators

Andżelika Krupińska, Marek Ochowiak and Sylwia Włodarczak

1 Theoretical Fundamental of Dust Removal Process

The basic quantity required in the course of designing or selection of the relevant equipment for dust removal from the gas stream is the concentration of particles suspended in it. Other important parameter include the dimensions (their arrangement), shape, weight, density, adhesiveness as well as physical and chemical properties of such particles (Chen et al. 2017). Solid particles are usually characterized by irregular shapes. Most frequently we come across acicular, tabular, rod-like, flat or stellate particles that differ in sizes. The shape which the particles adopt has a big impact over the dust removal processes, as it determines the particle motions inside the gas stream. Irregular particles collide with one another more frequently than the particles having regular structure and they demonstrate a stronger tendency to join, i.e. agglomeration (Kabsch 1992). Table 1 lists the most important quantities used to characterize the irregular particles (Warych 1998; Janka 2014).

The size of a particle suspended in a gas stream remains a very important parameter as well. Depending on what sizes of dust particles we have to handle, the operational effectiveness of various mechanisms used for dust removal varies. In order to separate the particles with relatively big sizes, gravity settling chambers, inertial separators or cyclones are applied (Punmia and Ashok 1998).

While describing the particles suspended in the gas stream one must bear in mind that we deal with polydisperse systems in most cases. Each single particle may be assigned to a given set, also referred to as a class or a fraction. Characteristic quantities to describe a given set include: particle size diameter

A. Krupińska (✉) · M. Ochowiak · S. Włodarczak
Institute of Chemical Technology and Engineering, Poznan University
of Technology, Poznań, Poland
e-mail: andzelika.krupinska@doctorate.put.poznan.pl

© Springer International Publishing AG, part of Springer Nature 2018
M. Ochowiak et al. (eds.), *Practical Aspects of Chemical Engineering*,
Lecture Notes on Multidisciplinary Industrial Engineering,
https://doi.org/10.1007/978-3-319-73978-6_15

Table 1 The most important quantities describing the shape of an irregular particle (Warych 1998; Janka 2014)

Size	Equation	Description
Sphericity (shape coefficient)	$\psi = \frac{A_k}{A_p} \leq 1$	A ration of sphere surface with the same volume as the volume of a given particle to the particle surface; in case of a spherical particle it equals 1
Shape coefficient	$\varphi = \frac{1}{\psi} \geq 1$	The inverse of sphericity
Specific surface area	$a = \frac{A}{V}$	The surface attributed to a unit of volume, e.g. for a sphere: $a = \frac{6}{d_p}$
Mesh diameter	$d_p = L$	It refers to the size of a mesh through which a given particle still goes through

distribution (maximum, minimum and mean values), particle specific surface area, sphericity, porosity, cumulative/fractional fraction.

Cumulative fraction is described by the following dependency:

$$f_i = \frac{n_i}{\sum_{i=1}^{i=n} n_i} \tag{1}$$

where: n_i—the number of particles of i-fraction.

It means that a cumulative fraction is defined as a share of the particles of a fraction (i-fraction) in the entire set examined. From among numerous mean diameters, the most practical application is attributed to Sauter mean diameter. This quantity may be defined by the following equation:

$$d_{ps} = \frac{6 \cdot V_p}{A_p} = \frac{\sum_{i=1}^{i=n} n_i d_{pi}^3}{\sum_{i=1}^{i=n} n_i d_{pi}^2} \tag{2}$$

where: d_{ps}—Sauter mean diameter, V_p—particle volume, A_p—particle surface, d_{pi}—particle diameter.

Specific surface area also remains a very important factor, which is expressed as:

$$a = \frac{6}{d_{ps}} \tag{3}$$

During all of the aforementioned analyses it has been assumed that we deal with a set of particles with spherical shapes. During dust removal, the particles are exposed to the impact consisting of a set of forces. This means that in the course of the process particles may settle gravitationally, may be subject to the impact from inertial forces or even diffusion forces. Additionally, the suspended particles move with relation to the gas stream, which has a specified direction and a sense of the velocity vector. The difference between a particle velocity and the speed of surrounding gas stream brings about the formulation of a passive force of aerodynamic

Fig. 1 Movement of a single
particle in motionless gas

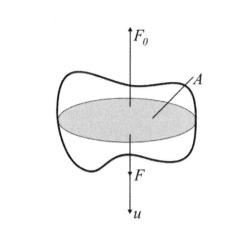

resistance (Ghadiri and Clift 2010). Figure 1 presents the movement of a single
aerosol particle in motionless gas (or in laminar gas streams).

Uniform motion of a single particle occurs when the force causing the motion
F will be balanced by the centre resistance force F_0. Force F_0 will have its sense
opposite to the velocity vector u. Force F_0 is described by the following
dependency:

$$F_0 = \zeta \cdot A \cdot \frac{\rho_g \cdot u^2}{2} \tag{4}$$

where: ζ—local resistance coefficient of the centre, A—particle surface perpendicular to the motion direction ρ_g—gas density.

In order to describe the particle motion with relation to the gas, the Reynolds
number is applied:

$$Re = \frac{u \cdot d_p \cdot \rho_g}{\mu_g} \tag{5}$$

where: μ_g—dynamic gas viscosity coefficient. It must be noted that $\zeta = f(Re, \psi)$.

Assuming that a particle analysed has a spherical shape (for individual established motion conditions), the values of centre resistance coefficient takes up the
form presented by Table 2 (Kabsch 1992).

Additionally in practice, in order to determine the centre resistance coefficient a
series of empirical formulas is used, e.g. Klaczka's formula, Schiller's formula and
Naumann's formula.

While examining the settling of a spherical particle with small dimensions,
moving in gas, within the laminar range, the resistance force amounts to:

Table 2 Values of the centre resistance coefficient (Kabsch 1992)

Motion range	Re	Centre resistance coefficient	Valid law
Laminar	$Re < 0.1$	$\zeta = \frac{24}{Re}$	Stokes law
Transient	$2 < Re < 500$	$\xi = \frac{18.5}{Re^{0.6}}$	Allen's law
Turbulent	$Re > 500$	$\zeta = 0.44$	Newton's law

$$F_0 = \frac{3 \cdot \pi \cdot d_p \cdot \mu_g \cdot u}{Cu} \qquad (6)$$

$$Cu = f(Kn) \qquad (7)$$

$$Kn = \frac{\lambda}{d_p} \qquad (8)$$

where λ is a mean length of gas particle free path, Cu remains the Cunningham's correction (slip ratio) and Kn remains Knudsen number.

In order for a particle to be separated, the following conditions must be fulfilled:

$$u_w \geq \frac{u_p \cdot D}{l} \qquad (9)$$

where: u_w—particle velocity along assumed straight trajectory, u_p—limit settling velocity, D—canal width (where a gas stream flows), l—length of a particle path along the flow direction (Janka 2014).

Three basic mechanisms for dust removal process can be distinguished: inertial, hitching and diffusion (Table 3) (Mueller 2007; Janka 2014).

In case of an inertial mechanism, exuding of solid particles occurs due to collision of a gas stream with suspended aerosol, with the collector's surface. A dependency is taken advantage of which results from increased inertia of dispersed phase particles than permanent phase ones. Such a collision brings about a rapid change of the flow direction.

For the particles characterized by big dimensions with regards to the size of the collector on which they are supposed to be placed, the interception mechanism is very important. It is very difficult to create such conditions to achieve the situation

Table 3 Mechanism of dust removal process

Dust removal mechanism	Description
Inertial	Inertial forces are used, intended especially for particles with diameters $d_p > 1$ μm
Interception	Crucial, for big particles with regards to the collector surface area
Diffusion	Occurs due to Brown's motions, thermo-phoresis, diffusion-phoresis, noticeable for particles $d_p < 1$ μm

in which the dust removal process shall run due to the said mechanism solely during the experiment (Janka 2014).

Emission of the particles with sub-micron dimensions (with the diameter below 1 μm) is connected predominantly with the diffusion mechanism. It occurs due to the Brown's motions, thermo-phoresis and diffusion-phoresis.

In practice, dust removing devices take advantage of the outcomes of several mechanism for particle separation from a gas stream at the same time. It has been demonstrated that efficiency of particle separation according to a given mechanism is proportional to the sizes of a particle and a collector as well as the particle velocity. Temperature, pressure, gas composition and properties also play some role and have an impact (Pell and Dunson 1997).

2 Parameters of the Dust Removing Process

Efficiency remains the most important parameter of the dust removing process. It may be defined as:

$$\eta = \frac{m_z}{m_0} \cdot 100\% \qquad (10)$$

where: m_z—weight of the particles stopped in the dust collector, m_0—weight of the introduced to the dust collector.

Determined efficiency values may vary for the same dust removing process. The efficiency is the function of the size and density of the particles in the dust settled. Depending on these values, the particles from the gas stream are settled in different manners. Hence, the general efficiency of dust removal remains the sum of efficiencies for individual particle sizes (Warych 1998).

In order to evaluate the quantities of dust leaving the settling chamber, the term penetration is used. Penetration may be described as the following dependence:

$$p = 1 - \eta = \frac{m_w}{m_0} \qquad (11)$$

where: m_w—weight of the particles leaving the settling chamber in purified gas stream. Another important parameter bearing importance during device exploitation and in the course of the designing is the gas pressure drop, and more precisely, the power which must be supplied in order to overcome the energy losses:

$$N = \dot{V} \cdot \Delta p \qquad (12)$$

where: N—power that must be supplied to the system in order to minimise the gas energy losses that are formulated when the gas flows through the settling chamber, $\dot{V}$—volumetric gas stream intensity, Δp—gas pressure drop during the dust removing process.

Other parameters for the dust removing process include: purifying device flow resistances, energy demand, costs and water consumption (refers to wet dust removal, i.e. scrubbers).

3   Dust Removing Devices

Figure 2 presents a basic division of dust removing devices (Błasiński and Młodziński 1976; Hao and Wang 2009). Table 4 presents gas purification manner (used mechanism) in individual types of apparatuses (dry dust collectors) (Płanowski et al. 1974; Mueller 2007; Mulyandasari 2011; Ghadiri and Clift 2010; Aranowski and Lewandowski 2016).

The simplest form of dust removal is to take advantage of the gravity (gravitational settling). In order to separate the solid particles from the gas stream several specified conditions must be met. First of all, the dimensions of the settling chamber, its length L and height H, must be selected in the way to allow the particles (dust) settle to the bottom, from where they will be irreversibly separated.

Gravitational settling chambers are divided into:

- chambers separators
- multiple tray settling chambers
- pipeline separators.

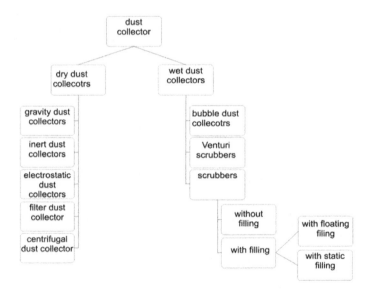

Fig. 2 Division of dust collectors (Błasiński and Młodziński 1976; Hao and Wang 2009)

Table 4 Ways to purify the gas stream in individual types of dust collectors

Apparatus	Gas purification manner
Dust chambers (settling chambers)	Dust particle settling by gravity—(using the difference in the specific gravity of the dust and the gas)
Cyclones, multi-cyclones	They use the centrifugal force (due to supplying of purified gas in a concentric manner, swirls are formed—centrifugal force; difference in specific gravity causes the situation in which forces acting upon the solid particles are noticeably higher than upon gas)
Inertial dust collectors	Dust removal by using inertia (dust particles move with high velocity and when they encounter an obstacle they collide with high inertia force losing their energy and speed in this way, later they settle by gravity)
Filters: bag, non-woven, grainy	Filtration (settling the solid particles on a partition—filter)
ESPs	Dust removal due to the impact of electrostatic forces (gas ionising by emission of electrons for the electrode, the charge transfers from ionised gas to the dust particles, attracting the dust by the settling electrode)

In case of centrifugal dust collectors, the phenomenon used for separation is the centrifugal force that acts in a different manner over the particles of gas and solid matter. Centrifugal force (F_r) is caused by the swirl motion of contaminated gas stream. Its strength is the function of the particle weight (m_p), particle location (distance) with regards to the motion axis (r) and its velocity (u):

$$F_r = \frac{m_p \cdot u^2}{r} \tag{13}$$

Two swirls are formed as a result of the action of this phenomenon: external one (where the dust concentration is very high) and internal one (with purified stream) (Solero and Coghe 2002).

Gravitational settling chambers are characterised by small flow resistances, simple design and, unfortunately, relatively low efficiency of particle separation. Usually, they are applied as preliminary settling chambers because they are not suitable for separating efficiently the particles with dimensions exceeding 40–50 μm. In case of centrifugal settling chambers, their assets include high efficiency in dust removal of particles with diameters 5–10 μm, low exploitation costs and small pressure losses. Their disadvantages include: exposure to excessive corrosion, low range efficiency in dust removal and the threat of explosion (Kabsch 1992; Elsayed and Lacor 2010).

Gravitational settling chambers and centrifugal dust collectors make use of basic physical phenomena in dust removal processes. In order to achieve such ones, easily accessible conditions must be met. Simple design and operating principle contribute to their popularity despite their numerous constraints (Wei et al. 2017).

There are attempts to modernize the apparatuses of this type, which has been presented in the papers published by Molerus and Glückler (1994), Elsayed and Lacor (2011), Ha et al. (2011) and Ochowiak et al. (2016).

Such undertakings aim to improve the operational parameters of chamber separators, to diminish the flow resistances and to increase the collection efficiency. Such modernisation intended to improving a classic chamber separator may be exemplified by an attempt to replace a cubical chamber with a cylindrical one. Inlet and outlet connectors had the same diameter and were placed along the axis of the apparatus. Figure 3 presents the discussed designs for the separators.

Owing to the modification applied, higher collection efficiency has been achieved (Ochowiak et al. 2016). The researching material included sands originating from the sandy sea beaches extending in Mielno, near Koszalin. Sea beach sands are well sorted and their grains are well dipped (have oval shapes). On average, they contain approximately 1–1.5% of gravimetric heavy minerals, i.e. such ones with the densities exceeding 2800 kg/m^3. Light fraction, constituting approximately 99% of the sand weight, consisted predominantly of the grains of quartz and feldspars. The first stage of the research included the analysis of sand according to the particle sizes made by mesh analysis. The research was performed by means of a set of sieves and a Retsch AS 200 sieve shaker. To measure the weight of solid particles stopped in the separator chamber, a precise scales Radwag PS 210/C/2 was used, as it measures the weight with the precision till ±0.001 g.

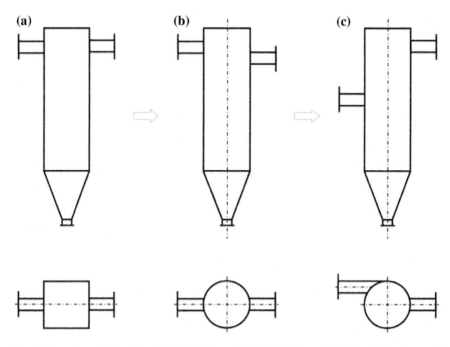

Fig. 3 Separators (Ochowiak et al. 2016; Ochowiak et al. 2017) **a** classic, **b** cylindrical, **c** swirl

Subsequently, the efficiency of the apparatuses for the purification of air stream from the grain with the diameters ranging from 100 to 600 μm was evaluated. Examinations were performed for various solid matter weights (2.5, 5, 10 g) as well as for various volumetric air flow intensities (from 1.5 to 3.15 m³/h). It has been acknowledged that the collection efficiency is the function of gas flow intensity, the sizes of particles subject to dust removal and dust concentration in the stream contaminated. The highest efficiency for dust removal has been achieved with the lowest values of gas flow intensity (Fig. 4). Collection efficiency decreases together with the increase in the flow intensity of contaminated air and along with the decreasing diameter (weight) of the dust particles. The weight of the dust introduced has a slight impact on the collection efficiency, whereby its increase contributes to the increased separator's efficiency. This impact is clearly visible with the small gas flow intensities (Ochowiak et al. 2016). Maximum efficiencies reached approximately 70%, while the minimum values 30%. The change in the chamber shape contributed to noticeable savings in materials and to the elimination of the so called "dead zones".

Figure 4 presents mean values for the collection efficiency with regards to a chamber separator which decouple the outcome from solid particle concentration in a gas stream. The obtained efficiency values are compliant to the data from the reference sources for this type of separators. The designed chamber separator with cylindrical shape contributes to the savings in the materials from which the device has been manufactured.

Fig. 4 Collection efficiency for a cylindrical chamber separator (Ochowiak et al. 2016)

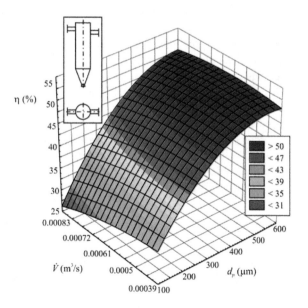

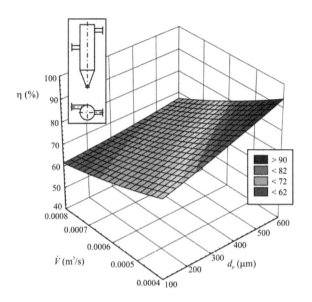

Fig. 5 Collection efficiency for a cylindrical chamber separator using the swirl motion phenomenon (Ochowiak et al. 2017)

Another attempt to improve the operation of a cylindrical chamber separator included the modification of its design. It encompassed the change of the inlet port location and taking advantage of the swirl motion phenomenon. For the flow intensities of contaminated gas below 4.2×10^{-4} m³/s, the separator acted as a classic gravitational separator (gravity force exceeds the value of the centrifugal force). When the outlet stream flow intensity exceeded 4.2×10^{-4} m³/s, the separator took advantage of the centrifugal force mainly. Based on the research conducted, it has been acknowledged that owing to the design modification for chamber separators, collection efficiency has been improved in comparison to the previously used solutions, i.e. a cylindrical chamber separator (Ochowiak et al. 2016). Figure 5 presents exemplary research outcomes (Ochowiak et al. 2017).

Figure 6 shows the values for collection efficiency obtained for the modified cylindrical chamber separators in case of the particles with diameters of 350 μm. Regardless of the inlet stream flow intensity, higher efficiency values were achieved for the filter using the swirl motion phenomenon. The lower the gas flow intensity, the higher efficiency values were recorded. This dependency is clearly visible in case of the second modification.

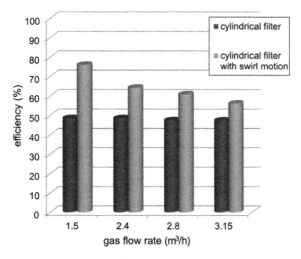

Fig. 6 Collection efficiency for solid particle fraction with diameter of 350 μm

Table 5 Applications for dust removing devices

Designing solution for a separator	Application
Settling chambers	Used mainly durian preliminary purification, on the boilers fired with coal dust or biomass
Inertial dust collectors	Collection of trick dry dusts, separation of waste dusts (ashes), compaction of dust being a product (powdered milk, pigments, medications)
Cyclones	Wood industry, plastics industry, paper manufacturing, power engineering—collection of boiler dusts
Filtration separators	Wood processing, metal processing, plastics processing, chemical and cement industry
ESPs	In power engineering—boilers fired with coal, coking, metallurgy (copper processing, material sintering), in the production of sulphuric acid

4 Applications for Dust Removal Devices

Dust removing devices are receiving more and more intensified interest nowadays, which is brought about by the enhanced awareness of the negative impact caused contaminated air and the increasing requirements with regards to the emission norms. The areas of application for dust removing devices include: environmental protection, production technologies, recycling of valuable substances emitted as pollutants, removal of harmful residual substances. Separators are applied in wood industry, furniture production, building material manufacturing, food processing industry, chemical industry, feed mix industry as well as plastics industry. Table 5 lists the exemplary applications for individual designing solutions (Schnelle and Brown 2001; Aranowski and Lewandowski 2016).

5 Summary

The aspect of health protection is tightly related to the advancements in technologies and enhanced awareness of mankind with regards to the air cleanness. Providing clean air is gaining importance and currently it remains one of the priority issues. This brings about an intensified interests in any solutions that may lead to effective improvement of air cleanness. Hence, the numerous research is being conducted intended to modernize the hitherto applied equipment, which will allow to obtain higher efficiency of dust removing processes. The following chapter presents two solutions which allow to achieve better outcomes for air stream purification by a chamber separator.

It has been acknowledged that by changing the design of the device one may not only increase its collection efficiency but also control its operational parameters. By replacing a cuboidal chamber with a cylindrical one, it was possible to get savings in the materials. The outcomes obtained may receive practical applications during the designing of chamber separators and chamber separators using the phenomenon of swirl motion.

Acknowledgements This work was supported by PUT research grant no. 03/32/DSPB/0702.

References

Aranowski R, Lewandowski WM (2016) Technologie ochrony środowiska w przemyśle i energetyce. PWN, Warszawa

Błasiński H, Młodziński B (1976) Aparatura przemysłu chemicznego. WNT, Warszawa

Chen L, Jin X, Yang L et al (2017) Particle transport characteristics in indoor environment with an air cleaner: the effect of nonuniform particle distributions. Build Simul 10:123–133. https://doi.org/10.1007/s12273-016-0310-7

Elsayed K, Lacor C (2010) Optimization of the cyclone separator geometry for minimum pressure drop using mathematical models and CFD simulations. Chem Eng Sci 65:6048–6058

Elsayed K, Lacor C (2011) The effect of cyclone inlet dimensions on the flow pattern and performance. Appl Math Model 35:1952–1968

Ghadiri M, Clift R (2010) Gas-solids separation, overview. http://www.thermopedia.com/content/48/. Accessed 2 June 2017

Ha G, Kim E, Kim Y et al (2011) A study on the optimal design of a cyclone system for vacuum cleaner with the consideration of house dust. J Mech Sci Technol 25:689–694

Hao J, Wang S (2009) Control techniques for particles. In: Yi Q (ed) Point sources of pollution: local effects and control. Eolss Publishers Company Limited, pp 194–222

Janka RM (2014) Zanieczyszczenia pyłowe i gazowe. Podstawy obliczania i sterowania poziomem emisji. PWN, Warszawa

Kabsch P (1992) Odpylanie i odpylacze. WNT, Warszawa

Molerus O, Glückler M (1994) Development of a cyclone separator with new design. Powder Technol 86:37–40

Mueller F (2007) Fundamentals of gas solids/liquids separation. Mueller Environmental Designs Inc. http://www.muellerenvironmental.com/res/uploads/media//200-059-GMRC-2004-Separation.pdf. Accessed 10 June 2017

Mulyandasari V (2011) Separator vessel selection and sizing (engineering design guideline). KLM Technology Group

Ochowiak M, Broniarz-Press L, Nastenko OV (2016) Oczyszczanie strumienia powietrza w cylindrycznym odpylaczu komorowym. Inż Ap Chem 2:66–67

Ochowiak M, Matuszak M, Włodarczak S et al (2017) Analiza wpływu wybranych parametrów na oczyszczanie powietrza w odpylaczu komorowym wykorzystującym zjawisko ruchu wirowego. Inż Ap Chem 3:96–97

Pell M, Dunson JB (1997) Gas-solid operations and equipment. In: Perry RH, Green DW (eds) Perry's chemical engineers handbook, 7th edn. McGraw Hill Company, New York

Płanowski AN, Ramm WM, Kagan SZ (1974) Procesy i aparaty w technologii chemicznej. WNT, Warszawa

Punmia BC, Ashok KJ (1998) Waste water engineering. Laxmi Publications (p) LTD, New Delhi

Schnelle KB, Brown CA (2001) Air pollution control technology handbook. CRC Press, Boca Raton

Solero G, Coghe A (2002) Experimental fluid dynamic characterization of a cyclone chamber. Exp Therm Fluid Sci 27:87–96

Warych J (1998) Oczyszczanie gazów. Procesy i aparatura. WNT, Warszawa

Wei J, Zhang H, Wang Y et al (2017) The gas-solid flow characteristics of cyclones. Powder Technol 308:178–192

Cleaning Porous Materials Using Supercritical Fluids

Jan Krzysztoforski and Marek Henczka

1 Introduction

Supercritical fluids (SCFs), especially supercritical carbon dioxide (scCO$_2$), have found numerous applications in processes involving porous materials. These processes include supercritical fluid extraction (SFE) of active substances from plant raw materials, wood impregnation, chemical modification of porous materials, sterilization of materials for biomedical applications, textile dyeing, polymer foaming and recovery of heavy metals from soils. One group of these processes is cleaning porous materials using supercritical fluids, during which the porous structure has to be cleaned by removal of a liquid or solid contaminant present inside the pores, or dissolved in the material of the porous structure. The replacement of liquid organic solvents used in standard methods for porous media cleaning with SCFs, especially scCO$_2$, may result in higher process safety and reduction of the amount of waste material produced during the cleaning processes. However, due to high investment costs of high pressure setups suitable for realization of processes involving SCFs, an extensive study of the novel process, including experimental investigation in lab scale, is usually necessary before the industrial implementation of the developed process. Such a study enables to analyze the course of the process, to determine the optimum process conditions, and to identify possible limitations and pitfalls.

The aim of this work is to present the potential of using supercritical fluids for cleaning porous materials. The work is divided into two parts. In the first part, a literature review of successful applications of supercritical fluids in cleaning porous materials is provided. The technical solutions utilizing SCFs are compared with standard techniques involving the use of liquid organic solvents. Moreover, the

J. Krzysztoforski · M. Henczka (✉)
Faculty of Chemical and Process Engineering,
Warsaw University of Technology, Warsaw, Poland
e-mail: marek.henczka@pw.edu.pl

© Springer International Publishing AG, part of Springer Nature 2018
M. Ochowiak et al. (eds.), *Practical Aspects of Chemical Engineering*,
Lecture Notes on Multidisciplinary Industrial Engineering,
https://doi.org/10.1007/978-3-319-73978-6_16

potential advantages and limitations of using SCFs are discussed. In the second part, guidelines for development of processes for cleaning porous materials using supercritical fluids are presented. Based on the guidelines, an exemplary process of porous media cleaning, namely microfiltration membrane cleaning using supercritical carbon dioxide, is analyzed and discussed.

2 Literature Review

A classic example of application of supercritical fluids for cleaning porous material is the coffee decaffeination process by supercritical fluid extraction, patented by Kurt Zosel from the Max Planck Institute (Zosel 1974). In this process, moist supercritical carbon dioxide was used for selective extraction of caffeine from water soaked coffee beans in a high pressure vessel. The $scCO_2$-caffeine mixture leaving the vessel was separated by extracting the caffeine with water. This method enabled to replace the organic solvents used in standard methods for coffee decaffeination. Although decaf coffee was originally regarded as the primary product of the decaffeination processes, pure caffeine being a "contaminant" in this process became a desirable product in the growing energy drink industry.

Supercritical carbon dioxide was also used for degreasing various metal parts, including geometries with holes of different shapes (Björklund et al. 1996). The oils used in the experiments, which were typical oils from the mechanical industry, were removed with a high efficiency, although some of the process runs were reported as slow. A strong influence of the hole geometry on the cleaning rate was observed. Moreover, a repeated pressure decrease during the cleaning process was found to be a viable method for increasing the process rate.

Another example are processes for cleaning and drying microelectronic structures, in which liquid or supercritical carbon dioxide is used (DeYoung et al. 2003). In these processes, carbon dioxide is used for removing water and solutes from the surface of a microelectronic substrate, which can have complex geometries including small cavities. The removal of the contaminants occurs through their immersion in the carbon dioxide drying composition. The process parameters have to be controlled so that the drying composition remains homogeneous during the cleaning process, which significantly reduces the amount of secondary deposits of the contaminants after completion of the process.

Supercritical carbon dioxide is also used in the process for drying silica alcogels in order to produce aerogels through removal of the contaminant—ethanol—present in the pores of the alcogel (Özbakır and Erkey 2015). The process was investigated both experimentally and numerically. The influence of various process parameters, including $scCO_2$ flow rate and alcogel thickness, on the course of the drying process was analyzed and the mass transfer mechanisms were identified. It was found that the mass transfer coefficients calculated based on experimental data were even higher than predicted by some correlations developed for supercritical fluid extraction.

Liquid/supercritical carbon dioxide is also an established medium in the dry cleaning process (Dewees et al. 1993). The presented dry cleaning equipment includes a high pressure vessel with an rotatable drum for the soiled fabrics, and with a system for recovery of energy associated with the CO_2 condensation and expansion. The system provides improved cleaning with reduced re-deposition of the contaminants of the soiled fabrics, as well as reduction of damage to polymer substrates.

There are many variations of this dry cleaning technique. For example, this approach was successfully applied for cleaning antique textiles, which require special care due to historic or artistic value (Sousa et al. 2007). In this work, supercritical and liquid CO_2 (pure or with isopropanol or isopropanol/water mixture as cosolvent) was used for cleaning old silk textiles. This cleaning method proved to be a safe method as the fibres and structures were not damaged and no loss material was observed.

Supercritical carbon dioxide can also be used for degreasing of sheepskins for the leather industry (Marsal et al. 2000). This technique enables to reduce emission of volatile organic compounds related to the conventional degreasing method utilizing organic solvents. A highly decreasing efficiency was observed and it was found that the degreasing efficiency decreased with increasing moisture content in the skin and increased with increasing CO_2 density, flow rate and cleaning time.

Another interesting example is the removal of 2,4,6-trichloroanisole (TCA) from cork stoppers using supercritical carbon dioxide (Taylor et al. 2000). TCA is the major compound associated with cork taint and using $scCO_2$ it can be almost completely removed from the cork stoppers in a fast and nearly solvent free process.

Supercritical fluids can also be applied as sterilizing agents (Kamihira et al. 1987). In this case, not physical removal, but biological inactivation of the contaminant is the main aim of the treatment. Moreover, for biomedical applications such as tissue scaffolds, residual amounts of potentially harmful sterilizing have to be avoided. Therefore, supercritical carbon dioxide is a promising medium for decontamination of biomaterials and tissue scaffolds (Tarafa et al. 2010). Supercritical carbon dioxide can be used for decontamination (which includes cleaning, disinfection and sterilization) of both metal and polymeric biomaterials.

Other applications of SCFs in cleaning porous materials include soil remediation (Cocero et al. 2000; Anitescu and Tavlarides 2006). Supercritical fluids (water or CO_2) are applied for the removal of various organic contaminants including hydrocarbons from a petrochemical plants, polycyclic aromatic hydrocarbons and pesticides. The extraction step can be coupled with adsorption on activated carbon.

Another group of applications of SCFs for cleaning porous materials is related to the production of microfiltration membranes using the Temperature Induced Phase Separation (TIPS) method. During this process, large amounts of organic solvents are usually used for removal of oil present in the membrane pores after the phase separation (Berghmans et al. 1996). The technical problems associated with these technical solutions include fire hazard, process costs related to high solvent usage and trace amounts of the solvent present in the porous structure after completion of the cleaning process. The alternative is to apply $scCO_2$ in the cleaning phase of the

Table 1 Advantages and disadvantages of using SCFs in porous media cleaning

Advantages	Disadvantages
No residual solvent inside the pores	High pressure equipment needed
Higher process safety (scCO$_2$)	Possible deterioration of materials
Easy separation of solvent-contaminant mixture through decompression	High energy costs in the solvent thermodynamic cycle
High penetration of porous material by solvent (removal of substances dissolved in the solid matrix of the porous structure)	Lower process rate than in standard techniques utilizing liquid organic solvent

TIPS method (Zhang et al. 2007). High process efficiency and a strong influence of numerous process parameters (time, pressure, temperature) on the course of the process was observed. It was found that the scCO$_2$-treated membrane is of superior quality as a lower tendency to shrinking and pore collapsing.

To sum up, the literature review shows, that the application of supercritical fluids (especially supercritical carbon dioxide) in many processes for cleaning porous materials has been proven to be a promising alternative to standard techniques utilizing organic solvents in liquid state. Various porous materials, which differ in size, shape, morphology of porous, chemical composition, type of contaminant, origin, and planned application field are successfully cleaned using supercritical carbon dioxide. Depending on the specific case, advantages in terms of product quality, process safety, and/or solvent usage can be achieved. However, the processes utilizing SCFs have their specific requirements, limitations and technical challenges. The main advantages and disadvantages of applying supercritical fluids for cleaning porous materials are summarized in Table 1.

3 Guidelines for the Design of Novel Cleaning Processes

As presented in the literature review, supercritical fluids show a great potential for application in processes for cleaning porous materials. However, the use of supercritical fluids involves some technical challenges and limitations, and usually requires higher investment costs due to the need of high pressure equipment.

Therefore, the developed process involving cleaning porous materials using supercritical fluid should be extensively studied before implementation in industry-scale plants. Moreover, various aspects of process development have to be covered by the study to ensure that the novel process is efficient, economically justifiable and safe both for the environment and the porous material itself. For this purpose, a set of practical guidelines for the design of novel cleaning processes involving porous materials and supercritical fluids has been formulated and is presented in this work.

The guidelines are grouped in five parts and they should be considered in the proposed order:

1. Identification of reference processes,
2. analysis of physical properties of the system,
3. assessment of possible destructive effects due to SCF treatment,
4. experimental investigation of the process, and
5. process optimization and scale-up.

The guidelines presented in this work are illustrated by an exemplary study of a novel process for cleaning porous media using supercritical carbon dioxide. As the exemplary process, the process of microfiltration membrane cleaning is considered:

Exemplary Process

Microfiltration membranes can be produced using the TIPS method (Berghmans et al. 1996). In this method, a liquid polymer/oil mixture is prepared by heating. Then, the mixture undergoes cooling under strictly defined conditions and temperature induced phase separation occurs. After this step, a porous polymer structure filled with oil is obtained. In order to complete the manufacturing procedure, the oil has to be removed from the membrane pores in a cleaning process. In standard realizations of the TIPS method, liquid organic solvents at elevated temperatures are used. In this work, the replacement of these organic solvents with supercritical carbon dioxide is considered as the exemplary cleaning process (Krzysztoforski 2016). The exemplary cleaning process is characterized below:

    Porous material: polypropylene capillary microfiltration membrane
    Contaminant: soybean oil
    Solvent: supercritical carbon dioxide
    Reference solvent: liquid isopropyl alcohol.

3.1 Identification of Reference Processes

The starting point of the development of a novel process for cleaning a porous material is a critical review of existing technical solutions which become the reference processes for the process to be developed. These solutions usually involve the use of organic solvents in the liquid state. This analysis of the state of the art enables to identify critical drawbacks related to standard solutions. Moreover, it helps to assess whether one should expect potential benefits from the replacement of organic solvents with supercritical fluids. Moreover, existing solutions often reveal technical limitations and minimum requirements related to the quality of the porous product, such as the purity, the mechanical strength, etc., which should be fulfilled by the novel process.

If no technical solutions for cleaning the said porous material using organic solvents are available in the literature, one should focus on literature data related to the processing of similar materials. The key questions to be answered are: What are the main causes that hinder the use of organic solvents for cleaning the porous material? Can these causes be eliminated by using supercritical fluids instead?

In the considered exemplary process, an established solution for cleaning the microfiltration membranes manufactured using the TIPS method is to use hot organic solvents (e.g. isopropyl alcohol). There are many drawbacks related to this approach. In this process, large amounts of isopropyl alcohol have to be used to obtain a high level of membrane purity. Moreover, the resultant oil/isopropyl alcohol mixture cannot be easily separated, which makes the reuse of the solvent impossible. Due to the usage of the organic solvent at elevated pressure, fire hazard exists, especially in the case of leakages. Another problem which arises in the process is the presence of residual solvent inside the porous material on completion of the cleaning process. Therefore, a secondary cleaning step must be conducted in order to remove the solvent from the membrane pores, e.g. by drying the porous material. Due to the presence of a air/solvent phase boundary, the evaporation of the organic solvent leads to collapsing of smaller pores inside the porous material, which is disadvantageous as it reduces the membrane permeability and it may result in minor amounts of organic solvent trapped inside the porous material. Regarding the requirements for the quality of the membrane, the pore size distribution should not be altered by the cleaning process and a high degree of purity of the membranes should be obtained.

The characteristics of the standard approach for cleaning microfiltration membranes reveal that there are many potential benefits in the replacement of isopropyl alcohol with supercritical carbon dioxide. First of all, the solvent usage can be reduced as the oil/solvent mixture can be separated downstream the high pressure extraction vessel by decompression, and the solvent can be recycled and reused in the process. Moreover, the process safety is considerably enhanced as the flammable organic solvent is replaced by inflammable $scCO_2$. Finally, as there is no CO_2/air phase boundary, the solvent can be removed from the membrane pores after the cleaning process by simple decompression of the system.

3.2   Analysis of Physical Properties of the System

After identification of reference processes, the physical properties of the multiphase and multi-component systems have to be studied both for the reference processes and for the novel process. This analysis enables to gather data on key properties affecting the course of the cleaning process, which are helpful for prediction of the expected cleaning rate and for preliminary comparison of the considered processes.

First of all, the type of the porous material has to be characterized, including its shape and size, porosity, pore size distribution, chemical composition, etc. The characteristics of the porous material determine the mechanisms of mass transfer in the cleaning process. Moreover, the properties of the contaminant have to be assessed, including its chemical composition (pure substance/mixture), thermodynamic state (liquid/solid), phase equilibria in the solvent-contaminant-porous material system, and transport properties. The location of the contaminant in the porous material (separate phase inside the pores, or molecules dissolved in the material of the porous matrix) is also a critical aspect. Finally, the physical properties of the solvent have to be determined.

In the exemplary process, the physical properties of the porous membrane, the soybean oil-isopropyl alcohol system and the soybean oil-scCO$_2$ system were studied based on literature data and product specification (Krzysztoforski 2016). Physical properties were analyzed for various process pressures (20–30 MPa) and process temperatures (313–343 K). Based on the product sheet of the manufacturer (Membrana GmbH), the geometry (outer diameter 2.8 mm, inner diameter 1.7 mm, porosity 0.7) and the morphology (porosity 0.7, mean pore size 0.45 µm) of the porous structure was determined. Soybean oil fills the membrane pores, but is not dissolved in the polypropylene itself. Based on the literature, the solubility of soybean oil in scCO$_2$ is 3–11 kg m^{-3}, while solubility of scCO$_2$ in soybean oil is 285–382 kg m^{-3}. In the reference process, soybean oil and isopropyl alcohol are fully miscible. Therefore, in the exemplary process, a smaller driving force for mass transfer can be expected than in the reference process. Moreover, the mechanisms of the processes differ significantly, as in the exemplary process a solvent-contaminant phase boundary arises due to the presence of the miscibility gap, while in the reference process, mass transfer takes place in a single fluid phase. Regarding the diffusivity values in the considered systems, diffusivity of soybean oil in scCO$_2$ is ca. one order of magnitude higher than in liquid isopropyl alcohol, which indicates lower mass transfer resistance when scCO$_2$ is used.

3.3 Assessment of Possible Destructive Effects Due to SCF Treatment

The primary aim of performing the cleaning process is to get rid of the contaminant present in the porous structure. Besides, no negative changes in the properties of the porous material should be induced by the SCF treatment. On the other hand, SCFs (and especially scCO$_2$) are known for their ability to penetrate into polymeric materials and to alter their structure at molecular scale. Other SCFs (such as supercritical water) may undergo chemical reactions with the porous structure. Processing of porous materials at elevated temperatures can also remain a cause for undesirable product alterations. Therefore, it has to be assessed whether a SCF can be used safely for cleaning a specific porous material without deteriorating its key properties, which in turn are related to the planned application of the porous material. This assessment can be performed theoretically (if literature data is available) or experimentally.

In the exemplary process, the $scCO_2$-polypropylene interaction has to be considered, as it is known that supercritical carbon dioxide interacts with numerous polymeric materials (Sawan et al. 1994). For microfiltration membranes, the two key properties which have to be preserved are the pore size distribution and the mechanical strength of the membranes, which should sustain a certain value of the transmembrane pressure. Therefore, the effect of $scCO_2$ treatment on physical properties of polypropylene microfiltration membranes was investigated (Tarabasz et al. 2016). The effect of treatment time, temperature, pressure and decompression rate was studied. Various analytical methods including scanning electron microscopy, tensile test, bubble point and wetting angle measurements were carried out. Untreated membranes were considered as reference samples. It was found that in the considered range of process parameters (process pressure 20–30 MPa and process temperature 313–343 K) no significant changes in the membrane properties were observed. Therefore, this range of process parameters can be regarded as safe in terms of possible deterioration of the porous material's key properties. However, for other values of process parameters and/or materials, additional tests would be necessary.

3.4 Experimental Investigation of the Process

The next stage is the lab-scale experimental investigation of the process for determination of the course of the process and the influence of various process parameters on the process rate. During this stage, cleaning tests utilizing the reference processes can be run as well for comparison.

The exemplary process was investigated experimentally in a lab-scale high pressure experimental system, in which a single membrane of 100 mm length, contaminated with pure soybean oil, was cleaned (Krzysztoforski 2016). The influence of process time, pressure, temperature, solvent mass flow rate, solvent type ($scCO_2$ or liquid isopropyl alcohol in the reference process) on the course of the process was investigated. The process efficiency was defined as relative reduction of oil mass present in the membrane pores. In Fig. 1, exemplary results of the experimental investigation of the exemplary process and reference process are shown.

Based on the experimental results, the influence of various process parameters on the course of the cleaning process was analyzed. Regarding the process time, it was found that the cleaning efficiency increases with increasing process time, while the process rate decreases, which suggests increasing mass transfer resistance during the process. In the exemplary process, a positive effect of increasing process pressure on the process rate was found. This is related to the known phenomenon of increased solvating power of supercritical fluids at higher process pressures. In the reference process, where liquid isopropyl alcohol is used, the pressure does not influence the course of the process. Hence, it can be carried out at ambient pressure. Regarding the process temperature, a negative effect of process temperature on the

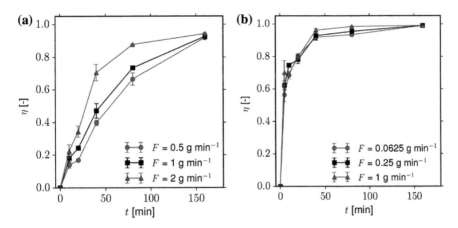

Fig. 1 Exemplary experimental results of membrane cleaning process. Cleaning efficiency η versus time t for different solvent mass flow rates F. Variants: **a** exemplary process with $scCO_2$ as solvent (30 MPa, 313 K), and **b** reference process with liquid isopropyl alcohol as solvent (10 MPa, 343 K)

process rate was observed in the exemplary process. However, the strength of this effect depends on process pressure and it becomes weaker and finally vanishes, when the process pressure is increased. In the reference process, a positive temperature effect was present. In none of the investigated processes any strong influence of solvent mass flow rate was observed. This may imply that the major mass transfer resistance is related to diffusive mass transfer inside the membrane pores and it is not affected by increasing the solvent flow rate. When the process rates of both processes are compared, the reference process turns out to be faster than the novel process utilizing $scCO_2$, which is also noticeable in Fig. 1. However, one has to bear in mind that replacement of the liquid organic solvent with $scCO_2$ contributes to lower solvent usage and higher process safety. Moreover, the process rate of the exemplary process might be enhanced by further increasing the process pressure (above 30 MPa) and/or by addition of minor amounts of organic cosolvents as solubility enhancers (Michałek et al. 2015).

3.5 Process Optimization and Scale-up

The final step of the process development is process optimization and scale up. Based on the experimental results, optimum process conditions can be determined. Additional experiments can be carried out, if necessary. Then, assumptions and guidelines for the process scale-up can be formulated and the final analysis regarding process economy can be performed. In this step of process development, a mathematical model of the process can be developed and validated using the experimental data. It can be used as a numerical tool for process investigation, optimization and scale-up.

Based on the experimental investigation of the exemplary process, the optimum process conditions were found (30 MPa and 313 K). Moreover, a mathematical model of the membrane cleaning process was developed, implemented into CFD software and validated using experimental data (Krzysztoforski et al. 2016; Krzysztoforski 2016). The mathematical model is based on the volume of fluid (VOF) multiphase model and includes mathematical representation of the porous zone, the transport properties of the fluid phases, as well as the phenomenon of mutual dissolution of soybean oil and $scCO_2$. The validation of the model confirmed a good agreement with experimental results. The model was used for further numerical investigation of the process, including simulations with process conditions which were not tested during the experimental investigation. From the simulation results it was confirmed that due to the gradually receding oil-solvent phase boundary, the diffusion path through the membrane becomes longer. This in turn results in increased mass transfer resistance and lower process rate, as the process advances. Due to high flow resistance in the porous material, the solvent flow around the porous membrane does not induce any convective flow inside the membrane pores. Therefore, diffusion is a predominant mass transfer mechanism inside the membrane pores. It was also found that due to the decreasing process rate, the soybean oil concentration at the extraction vessel outlet also decreases as the process advances. Therefore, for efficient solvent usage and reduction of energy consumption in the process, it might be advisable to run the cleaning process with a solvent mass flow rate decreasing in time. Moreover, the effect of oil phase swelling, due to the dissolution of CO_2 in the oil phase, on the process rate was investigated. It was found that it significantly contributes to a higher process rate in the first stage of the cleaning process. The validated mathematical model can be applied for process optimization and scale-up. Using CFD simulations, numerous variants of process realization can be tested without the need of performing costly experimental test in lab-scale or even larger scales.

4 Conclusions

Supercritical fluids, especially supercritical carbon dioxide, show a great potential in processes for cleaning porous media. In the literature, many successful applications of SCFs in such cleaning processes are known, such as removal of caffeine from coffee beans, production of aerogels, cleaning natural cork and microfiltration membranes. The advantages of these processes, when compared to standard cleaning methods utilizing liquid organic solvent, include reduction of solvent usage and waste material production, cleaner products and enhanced process safety. The major challenges are related to limited process rate and high investment costs due to the high pressure conditions which have to be provided. Practical guidelines for development a novel process for cleaning porous materials using supercritical fluids were formulated. The guidelines were demonstrated for an exemplary process, namely cleaning microfiltration membranes using supercritical carbon dioxide. It was shown that this process has the potential to replace the standard technique utilizing hot isopropyl alcohol in liquid state, which constitutes the reference process. Although the novel process is slower than the reference process, it is superior in terms of solvent usage, waste production and process safety. A mathematical model of the process developed can be applied in process investigation, optimization and scale-up.

Acknowledgements Research financed by the National Science Centre, Poland, Project No. 2014/15/N/ST8/01516.

References

Anitescu G, Tavlarides LL (2006) Supercritical extraction of contaminants from soils and sediments. J Supercrit Fluids 38(2):167–180

Berghmans S, Berghmans H, Meijer HEH (1996) Spinning of hollow porous fibres via the TIPS mechanism. J Membr Sci 116(2):171–189

Björklund E, Turner C, Karlsson L et al (1996) The influence of oil extractability and metal part geometry in degreasing processes using supercritical carbon dioxide. J Supercrit Fluid 9(1): 56–60

Cocero MJ, Alonso E, Lucas S (2000) Pilot plant for soil remediation with supercritical CO_2 under quasi-isobaric conditions. Ind Eng Chem Res 39(12):4597–4602

Dewees TG, Knafelc FM, Mitchell JD et al (1993) Liquid/supercritical carbon dioxide dry cleaning system. US Patent US 5267455 A

DeYoung JP, McClain JB, Gross SM (2003) Processes for cleaning and drying microelectronic structures using liquid or supercritical carbon dioxide. US Patent US 6562146 B1

Kamihira M, Taniguchi M, Kobayashi T (1987) Sterilization of microorganisms with supercritical carbon dioxide. Agric Biol Chem 51(2):407–412

Krzysztoforski J (2016) Transport phenomena in the process of porous membrane cleaning using supercritical fluids. PhD, Dissertation, Warsaw University of Technology, Faculty of Chemical and Process Engineering, Warsaw, Poland

Krzysztoforski J, Jenny P, Henczka M (2016) Mass transfer intensification in the process of membrane cleaning using supercritical fluids. Theor Found Chem Eng 50(6):907–913

Marsal A, Celma P, Cot J et al (2000) Supercritical CO_2 extraction as a clean degreasing process in the leather industry. J Supercrit Fluid 16(3):217–223

Michałek K, Krzysztoforski J, Henczka M et al (2015) Cleaning of microfiltration membranes from industrial contaminants using "greener" alternatives in a continuous mode. J Supercrit Fluid 102:115–122

Özbakır Y, Erkey C (2015) Experimental and theoretical investigation of supercritical drying of silica alcogels. J Supercrit Fluid 98:153–166

Sawan SP, Shieh Y-T, Su J-H (1994) Evaluation of the interactions between supercritical carbon dioxide and polymeric materials. Los Alamos National Laboratory

Sousa M, Melo MJ, Casimiro T et al (2007) The art of CO_2 for art conservation: a green approach to antique textile cleaning. Green Chem 9(9):943–947

Tarabasz K, Krzysztoforski J, Szwast M et al (2016) Investigation of the effect of treatment with supercritical carbon dioxide on structure and properties of polypropylene microfiltration membranes. Mater Lett 163:54–57

Tarafa PJ, Jiménez A, Zhang J et al (2010) Compressed carbon dioxide (CO_2) for decontamination of biomaterials and tissue scaffolds. J Supercrit Fluid 53(1):192–199

Taylor MK, Young T-M, Butzke CE et al (2000) Supercritical fluid extraction of 2, 4, 6-trichloroanisole from cork stoppers. J Agric Food Chem 48(6):2208–2211

Zhang CF, Zhu BK, Ji GL et al (2007) Supercritical carbon dioxide extraction in membrane formation by thermally induced phase separation. J Appl Polym Sci 103(3):1632–1639

Zosel K (1974) Process for recovering caffeine. US Patent US3806619 A

Large Eddy Simulations on Selected Problems in Chemical Engineering

Łukasz Makowski and Krzysztof Wojtas

1 Introduction

In recent years, with the development of computational capabilities, large eddy simulations (LES) have become increasingly common. LES remain an attractive alternative as they can simulate problems in a wide range of Reynolds and Schmidt numbers, which is often the case in engineering applications. Looking back on the development of CFD methods, the utility of LES models in solving engineering problems was quickly recognized.

The LES model is based on the observation that the kinetic energy of turbulence and flow anisotropy relates only to a large scale of motion depending on the environment in which they arise, and kinetic energy dissipation and isotropicity occur in the area of small scales. Large-scale eddies are directly simulated in LES, and small eddies are modeled using universal models, so-called subgrid scale (SGS) models. The accepted assumption makes it possible to use a coarser numerical grid than in direct calculations. However, the mutual influence of modeling on both scales should be emphasized. This is illustrated in Fig. 1.

In order to verify existing and new LES models, the jet reactors seem to be a proper choice. Impinging jet reactors are often used in practice because of the possibility of almost "instantly" mixing the contacted fluids. This is due to the formation of an area of high energy dissipation rate in the jet zone of the inlet streams. Each fluid element flowing to the system must pass through this area, without being able to bypass it. The area with high energy dissipation rates is due to the change of kinetic energy of the inlet streams in a highly turbulent flow, caused by collisions and a rapid change in the flow direction of the inlet streams in the a small space of the reactor. Very often in these systems co-exist the regions of

Ł. Makowski (✉) · K. Wojtas
Faculty of Chemical and Process Engineering,
Warsaw University of Technology, Warsaw, Poland
e-mail: lukasz.makowski.ichip@pw.edu.pl

© Springer International Publishing AG, part of Springer Nature 2018
M. Ochowiak et al. (eds.), *Practical Aspects of Chemical Engineering*,
Lecture Notes on Multidisciplinary Industrial Engineering,
https://doi.org/10.1007/978-3-319-73978-6_17

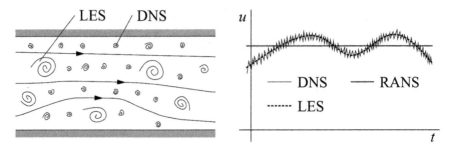

Fig. 1 Schematic representation of scales in turbulent flows and their relationship with modeling approaches (adapted from Ferziger and Perić 2002)

strictly laminar flow (for example inlet and outlet area) and the region where the highly turbulent flows occurs (impingement zone). Under such conditions RANS models very often lead to poor predictions. (Schwertfirm and Manhart 2010; Icardi et al. 2011; Makowski and Bałdyga 2011; Makowski et al. 2012; Wojtas et al. 2015b).

2 Mathematical Models

Transport equations in large eddy simulations are obtained using a filter procedure on the Navier-Stokes equations. This consists of separating eddies smaller than the width of the filter $(\overline{\Delta})$, which is the size of the numerical cells, from the directly solved flow field. Mathematically, this procedure, e.g., for velocity, takes the following form:

$$\overline{u_i}(\vec{x}, t) = \int G(\vec{x} - \vec{x}', t) u_i(\vec{x}', t) d^3\vec{x}' \tag{1}$$

where $G(\vec{x} - \vec{x}', t)$ is the filter function.

The momentum equation for incompressible flow with constant viscosity in the Cartesian coordinate system takes form:

$$\frac{\partial u_i}{\partial t} + \frac{\partial}{\partial x_j}(u_i u_j) = -\frac{1}{\rho}\frac{\partial p}{\partial x_i} + \frac{\partial}{\partial x_j}(2\nu S_{ij}) \tag{2}$$

where S_{ij} is the deformation tensor.

After applying the filtration procedure we receive:

$$\frac{\partial \overline{u_i}}{\partial t} + \frac{\partial}{\partial x_j}(\overline{u_i}\,\overline{u_j}) = -\frac{1}{\rho}\frac{\partial \overline{p}}{\partial x_i} + \frac{\partial}{\partial x_j}(2\nu\overline{S_{ij}} + T_{ij}) \tag{3}$$

where

$$T_{ij} = \overline{u_i}\,\overline{u_j} - \overline{u_i u_j} \tag{4}$$

is the subgrid tensor responsible for changing the momentum between the subgrid and the large scale zone.

Most subgrid models found in the literature, are models for subgrid stresses with a deformation tensor component for larger scales:

$$\tau_{ij} - \frac{\delta_{ij}}{3}\tau_{kk} = -2v_{sgs}\overline{S_{ij}} \tag{5}$$

where v_{sgs} is the subgrid viscosity.

In most cases, the subgrid viscosity is determined algebraically to limit additional equations that would increase the computation time. In addition, the correctness of the proposed solution can be justified by that small fluid motion scales are more homogeneous and isotropic than larger scales, so simple algebraic models should properly describe the motion in the subgrid scale. It is also known that subgrid stresses represent only a small fraction of total stresses, so any errors in the subgrid model should not significantly affect the accuracy of the obtained results (Piomelli 2000).

For modeling of scalar transported by a constant density fluid, one can write a transport equation:

$$\frac{\partial f}{\partial t} + \nabla \cdot (uf) = D_m \nabla^2 f \tag{6}$$

After the filtering procedure, the equation will take the following form:

$$\frac{\partial \overline{f}}{\partial t} + \nabla \cdot \left(\overline{uf}\right) = D_m \nabla^2 \overline{f} \tag{7}$$

Separating the non-linear term in subgrid and large scale regions one can obtain:

$$\frac{\partial \overline{f}}{\partial t} + \nabla \cdot (\overline{u}\overline{f}) = D_m \nabla^2 \overline{f} - \nabla \tau_f \tag{8}$$

where the subgrid scalar flux can be defined by equation:

$$\tau_f = \left(\overline{uf} - \overline{u}\overline{f}\right) \tag{9}$$

Most functional subgrid models are based on the subgrid diffusion paradigm, leading to the following general closure hypothesis:

$$\tau_f = D_{sgs}\nabla\bar{f} \tag{10}$$

The main mechanism of the process is considered to interpret the turbulent spectrum as a cascade of concentration observed for isotropic turbulence. This approach is analogous to the case of turbulent viscosity in the momentum balance.

In large eddy modeling of passive tracer mixing, a closure hypothesis based on the subgrid Schmidt number leading to:

$$D_{sgs} = \frac{\nu_{sgs}}{Sc_{sgs}} \tag{11}$$

The subgrid Schmidt number value in the literature ranges from 0.1 to 1.0, and the most often is equal to 0.4 (Fox 2003).

The use of the above-mentioned subgrid diffusion model in jet reactors has been verified in many of the authors' works (Makowski and Bałdyga 2011; Makowski et al. 2012; Wojtas et al. 2015b, 2017). The authors investigated the course of mixing process of fluids in two types of impinging jet reactors, which are shown in Fig. 2.

The symmetric T-mixer is the simplest geometry of this type of reactors. It can be characterized by symmetrically and coaxially arranged inlet pipes to the mixing chamber. The vortex mixer (V-mixer) has slightly modified geometry compared to the symmetric T-mixer, and has inlet pipes arranged tangentially to the outlet pipe. Inlet pipes diameter, d_{jet}, was equal to 7 mm, and outlet pipe diameter, d_{out}, was equal to 11 mm. The length of the inlet pipes was equal to 100 mm and as for the outlet pipes it was equal to 300 mm.

Particle image velocimetry (PIV) and planar laser induced fluorescence (PLIF) techniques were used to measure distributions of fluid velocity and tracer concentration respectively. A double-cavity Nd-YAG laser 532 nm with energy equal to 50 mJ per pulse was used. A laser beam was transformed to a collimated planar

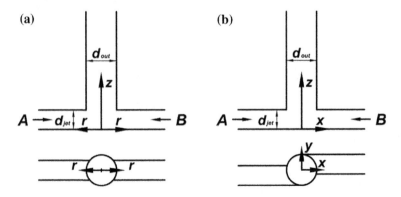

Fig. 2 Geometry of the mixers: **a** symmetric T-mixer (T-mixer); **b** vortex mixer (V-mixer)

laser sheet of thickness equal to 0.2 mm. The laser plane crossed the experimental system vertically through the axis of the outlet pipe.

Borosilicate glass particles of the average diameter equal to 10 μm were used as a seeding for PIV. Rhodamine B was used as a fluorescent tracer in PLIF, and its concentration in the inlet solution was equal to 0.2 g m^{-3}. Spatial resolution of the PLIF measurements results from a resolution of digital images and the thickness of the laser plane and in our case it was 9×9×200 μm. Spatial resolution of the PIV measurements depends also on size of a sampling window and in this work it was 144 × 144 × 200 μm. Simulations of hydrodynamics were carried out using ANSYS Fluent CFD software. The numerical meshes consisted of about 800,000 hexahedral cells for each mixer. More information about the measurement techniques and simulations one can find in the mentioned articles.

The experiments and calculations were performed for Re_{jet} equal to 1000, 2000 and 4000, where:

$$R_{jet} = \frac{u_{jet}\, d_{jet}\, \rho}{\mu} \tag{12}$$

and u_{jet} is the mean velocity at the inlet, ρ and μ are the density and dynamic viscosity of water respectively. Figures 3 and 4 show the measured and predicted contours of the dimensionless mean velocity magnitude and mean mixture fraction in the studied mixers, where U is a ratio of the local mean velocity to mean velocity in the inlet pipe, u_{jet}.

One can see that the agreement between the experimental and simulation results in the area of impingement, for all considered Re_{jet} numbers, is very good. Much better mixing is observed in the V-mixer, which can be explained by increased vorticity and turbulence in the impingement zone in this type of mixer.

3   SGS Variance Model

Knowledge of the variation of concentration in the subgrid scale plays a crucial role in the large eddy simulations of complex chemical processes. The course of chemical reactions in turbulent flow depends greatly on the local distribution of time-space scales occurring in the system. Transport equations, which describe the concentration of reagents after filtration, do not contain information on subgrid eddies. Therefore, an additional statistical equation describing the mixing of reagents in the subgrid scale is needed. Typically, a probability density function, determined by the filtered distribution of the passive tracer concentration and its subgrid variance, is used for this purpose.

The larger part of subgrid concentration variance value is found in the largest subgrid eddies. Therefore, in algebraic models the subgrid variance is estimated from the smallest scales, which are determined directly (filtered), using the test-filter

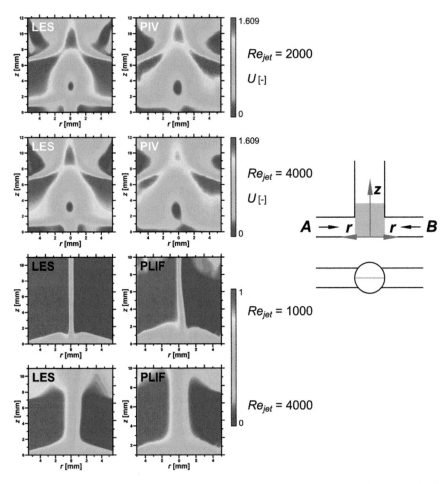

Fig. 3 Contours of the measured and predicted values of the dimensionless velocity magnitude and mean mixture fraction in the T-mixer in the injection zone

operation. An example of this model type is the scale-similarity model proposed by Cook and Riley (1994), in which the subgrid concentration variance is defined as:

$$\sigma_{sgs}^2(x,t) = \overline{(f')^2}(x,t) = \int \left[f(\vec{x}',t) - \bar{f}(\vec{x},t) \right]^2 G(\vec{x}' - \vec{x}) d^3\vec{x}'$$
$$= \overline{f^2}(x,t) - \bar{f}^2(x,t) \tag{13}$$

In the above equation, the term $\overline{f^2}(x,t)$ is unknown and requires modeling.

Several models for the SGS variance have been proposed in the literature (Cook and Riley 1994; Pierce and Moin 1998; Jiménez et al. 2001; Balarac et al. 2008; Ihme and Pitsch 2008). A comparison of most of these models and their application

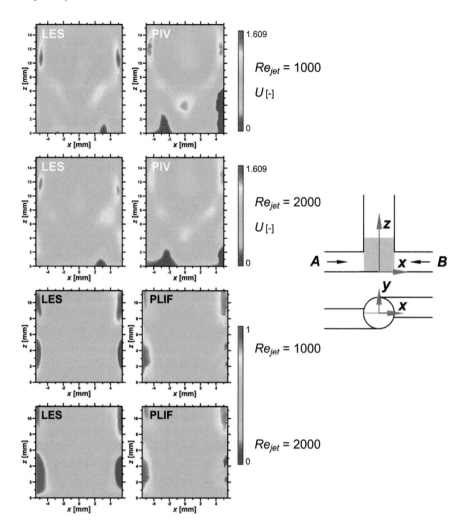

Fig. 4 Contours of the measured and predicted values of the dimensionless velocity magnitude and mean mixture fraction in the V-mixer in the injection zone

in predicting SGS concentration variance can be found in Wojtas et al. (2015a). The most commonly used model is the Cook and Riley model (1994), in which they proposed that the subgrid concentration variance can be predicted by assuming that the small scale statistics can be inferred from the large scale statistic:

$$\sigma_{sgs}^2 \approx c_f \widetilde{\sigma_{sgs}^2} = c_f \left(\widetilde{\bar{f}^2} - \tilde{\bar{f}}^2 \right) \tag{14}$$

where $\sim$ denotes test-filtered value, computed by applying the test filter (test-filter width $2\,\overline{\Delta}$ was used, where $\overline{\Delta}$ is the numerical grid size).

The above model contains the constant, which was determined by Cook (1997), for the mixing of gases. He obtained the values of the coefficient between 0.54 and 1.75. In later works, Michioka and Komori (2004), using direct numerical simulations, have identified that with the increase of the Schmidt number, the value of model constant increases reaching the value of 5. However, the constant factor value causes the model to ignore local changes of Reynolds and Schmidt numbers' values. The solution to this problem is the introduction of a new dynamic model coefficient to be used together with the scale similarity model (Wojtas et al. 2015a). In this approach, the concentration variance is calculated by the integration of the whole range of wave numbers of the energy spectrum of concentration fluctuations:

$$
\sigma_{sgs}^2 = \left\{ \left[1.70 \left(1 - \left(\frac{\lambda_K}{\overline{\Delta}} \right)^{2/3} \right) \right] + \frac{1.14}{Ba} \left[\frac{\varepsilon^{1/3}}{\overline{\Delta}^{2/3}} \left(\frac{\nu}{\varepsilon} \right)^{1/2} \ln Sc \right] \right\} \left(\widetilde{\overline{f^2}} - \widetilde{\overline{f}}^2 \right)
$$
$$
= c_{\widetilde{\Delta}} \left(\widetilde{\overline{f^2}} - \widetilde{\overline{f}}^2 \right)
\tag{15}
$$

where λ_K is the Kolmogorov scale, Ba is the Batchelor number, ε is the turbulence kinetic energy dissipation rate and $c_{\widetilde{\Delta}}$ is thus defined dynamic model coefficient for the scale similarity SGS concentration variance model.

An important feature of the model is the lack of the coefficient defined a priori or based on experimental data. What is more, the new model well predicts the value of the constant together with the change in the Schmidt number. For the Schmidt number equal to unity and for a high rate of energy dissipation, the constant $c_{\widetilde{\Delta}}$ takes values tending to 1.7, which is comparable to the result obtained by Cook (1997). For the Reynolds number tending to infinity, the viscosity term converges to zero, and for the same reasons, the ratio of the Kolmogorov microscale and the mesh grid size tends to zero as well. Along with the increase of the Schmidt number from 1 to infinity, the value of this constant increases and exceeds locally the constant value determined by Michioka and Komori (2004). Figure 5 shows instantaneous values of the new coefficient values in the T-mixer impingement zone. The obtained results change in space with the change of flow conditions and slightly differ from the Cook and Riley constant. Figure 6 shows contours of the mixture fraction variance calculated with and without an application of the new model coefficient. The obtained results are very similar and this is understandable because in this case, the model constant in the Cook and Riley model was estimated for water. More significant differences between models should be observed with increasing Sc values.

In order to investigate the effect of viscosity and diffusivity on the concentration variance distribution in jet reactors, the solutions of diethanolamine were used. Viscosity varied in the range form 0.89×10^{-6} to 2.70×10^{-6} m^2 s^{-1}, while diffusivity from 1.0×10^{-9} to 0.45×10^{-9} m^2 s^{-1}. This made possible to achieve a change in the Sc value from 890 to 6000.

Figure 7 shows the measured and predicted contours of the mean mixture fraction variance along the outlet pipes in smaller T-mixer.

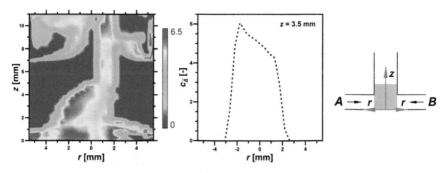

Fig. 5 Instantaneous distribution of the dynamic model coefficient (Wojtas et al. 2015a) in the T-mixer, $Re_{jet} = 4000$

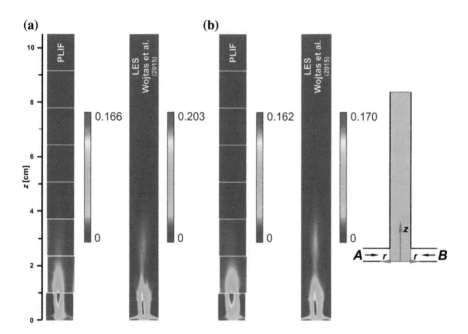

Fig. 6 Contours of the mean mixture fraction variance in the T-mixer: **a** $Re_{jet} = 2000$; **b** $Re_{jet} = 4000$

In this case, the inlet pipe diameter was equal to $d_{jet} = 4.6$ mm. The outlet diameter was the same as in the larger T-mixer.

The concentration variance distribution is almost identical in the injection zone, while increased variance values can be observed at higher parts of the reactor. The results of the Cook model are insensitive to the increase of Sc, whereas the model with the dynamic constant shows tendencies observed in the experiments, that is a small increase of the variance value above the injection zone. Detailed analysis of

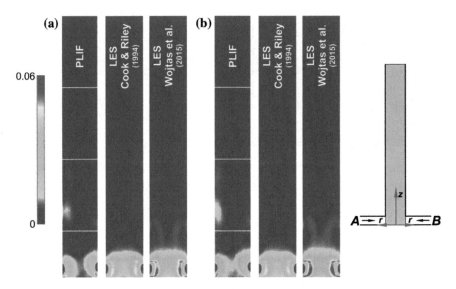

Fig. 7 Contours of the mixture fraction variance in the smaller T-mixer, $d_{jet} = 4.6$ mm, $Re_{jet} = 2000$: **a** $Sc = 2500$; **b** $Sc = 6000$

the results shows that there is an increase of the concentration variance with an increase of the Sc values, what the author's model predicted. This should have an influence on the course of complex chemical processes. More information on application of the model with the dynamic coefficient for fluids with a variable Schmidt number, one can find in the work of Wojtas et al. (2017).

4 Large Eddy Simulations of Parallel Chemical Reactions

Multiple chemical reactions are influenced by kinetics of a chemical reaction, as well as by a type of contact between reagents, mixing intensity, operation mode of the process. This allows for achieving a product of desired type and quality, not only with chemical methods (concentration, temperature, pH, etc.), but also with physical conditions (feeding point, method of dosing, stirring speed, type of reactor). Interpretation and understanding of the correlation between mixing and chemical reaction allow for the formulation of an appropriate theoretical model, which, when verified by laboratory results, can be successfully used in industrial applications for both assessing the influence of process parameters on product properties, as well as for scaling-up of chemical reactors. Such course of action allows to achieve better productivity and quality of obtained products as well as to limit quantity of side products (including products harmful to the environment). Additionally, unnecessary stages in production process can be eliminated.

To evaluate predictions of models describing reactive mixing, one can employ specially designed mixing sensitive test reactions (Bałdyga and Bourne 1999). In this work the parallel reaction system was used which includes the competitive neutralization of hydrochloric acid and the alkaline hydrolysis of ethyl chloroacetate (Bałdyga and Bourne 1990; Bourne and Yu 1994; Bałdyga et al. 1998; Bałdyga and Makowski 2004):

$$\underset{A}{\underline{NaOH}} + \underset{B}{\underline{HCl}} \xrightarrow{k_1 \to \infty} \underset{P}{\underline{NaCl}} + H_2O \tag{16}$$

$$\underset{A}{\underline{NaOH}} + \underset{C}{\underline{CH_2ClCOOC_2H_5}} \xrightarrow{k_2} \underset{S}{\underline{CH_2ClCOONa}} + C_2H_5OH \tag{17}$$

An application of Large Eddy Simulation CFD method to model mixing effects on the course of parallel chemical reaction requires modeling of the subgrid phenomena for the flow, mass transfer and chemical reactions (Makowski and Bałdyga 2011).

The subgrid model for parallel reactions was based on the application of the probability density function (PDF) of a chemically passive scalar tracer (Bałdyga and Bourne 1999) and can be expressed using the Beta probability distribution of the mixture fraction, f:

$$\Phi(f) = \frac{f^{v-1}(1-f)^{w-1}}{\int_0^1 y^{v-1}(1-y)^{w-1} dy} \tag{18}$$

where

$$v = \bar{f}\left[\frac{\overline{\bar{f}(1-\bar{f})}}{\sigma^2} - 1\right] \tag{19}$$

$$w = (1-\bar{f})\left[\frac{\overline{\bar{f}(1-\bar{f})}}{\sigma^2} - 1\right] \tag{20}$$

The concentrations of A and C reactants can be defined using a method of interpolation of local instantaneous reactant concentrations values, $c_i(f)$, between the concentrations for the instantaneous, $c_i^\infty(f)$, and infinitely slow reaction, $c_i^0(f)$:

$$c_i(f) = c_i^\infty(f) + \frac{\overline{c_i} - \overline{c_i^\infty}}{\overline{c_i} - \overline{c_i^\infty}}\left[c_i^0(f) - c_i^\infty(f)\right] \tag{21}$$

Obtained information about the local average subgrid values of the reaction kinetics terms, allows to solve a set of filtered differential balance equations:

$$\frac{\partial \overline{c_i}}{\partial t} + \overline{u_j} \frac{\partial \overline{c_i}}{\partial x_j} = \frac{\partial}{\partial x_j} \left[(D_m + D_{sgs}) \frac{\partial \overline{c_i}}{\partial x_j} \right] + \overline{r_i} \tag{22}$$

where the rate of the second reaction reads:

$$\overline{r_2} = k_2 \overline{c_A} \, \overline{c_C} = k_2 \int_0^1 c_A(f) c_C(f) \Phi(f) df \tag{23}$$

Figure 8 shows the comparison of the model predictions of jet Reynolds number influence on the selectivity, X_S, for inlet concentrations $c_{A0} = c_{B0} = c_{C0} = 100 \text{ mol m}^{-3}$. The final selectivity was expressed by:

$$X_S = \frac{c_{C,inlet} - c_{C,outlet}}{c_{A,inlet}} \tag{24}$$

The LES results are in good agreement with the experimental data. One can see that in each system the increase in the Reynolds number results in a decrease in final selectivity, which corresponds to a smaller amount of by-product. In the V-mixer predicted selectivity values were lower and the reason for this behavior lies in better mixing conditions inside the impingement zone, that is increased vorticity and turbulence intensity.

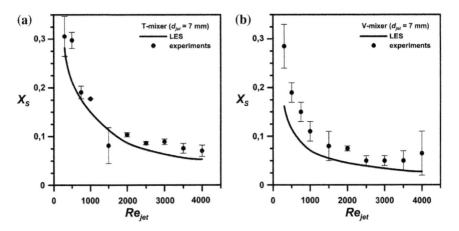

Fig. 8 Effects of Re_{jet} on the final selectivity of chemical reaction: **a** T-mixer; **b** V-mixer

5   Large Eddy Simulations of Precipitation Process

The precipitation process, understood as a rapid and irreversible crystallization characterized by high saturation, runs almost always in turbulent flows. Crystallization occurs when the reagents, mixed on the molecular scale, are rapidly reacting, which generates high values of the local supersaturation. In case of the rapid precipitation (as in the case of chemical reactions), the formation and growth of crystals begin when the solutions are not perfectly mixed. For this reason, the size and shape of the precipitated particles can be affected by changing the mixing method. An application of proper numerical simulation methods can significantly simplify and reduce costs in a design process of chemical reactors, in which chemical processes like precipitation occur.

The well-known and described process was chosen, precipitation of a sparingly soluble test material (barium sulfate), obtained from two aqueous ionic solutions, barium chloride and sodium sulfate:

$$\underbrace{BaSO_4}_{D} + \underbrace{Na_2SO_4}_{E} \xrightarrow{k_3} \underbrace{BaSO_4}_{F} + 2NaCl \tag{25}$$

In the case of barium sulfate precipitation simulations, local instantaneous rates of nucleation and crystal growth depended on the values of the activity coefficient, which was calculated according to Bromley method (Bromley 1973). Nucleation kinetics for $BaSO_4$ was described by the following empirical equations for nucleation rate R_N (Bałdyga and Orciuch 2001):

$$R_N = 1.06 \times 10^{12} \exp\left(\frac{-44.6}{\ln^2 S_a}\right) + 1.5 \times 10^{45} \exp\left(\frac{-3020}{\ln^2 S_a}\right) \tag{26}$$

$$S_a = \left(\frac{c_D c_E}{K_{Sa}}\right) \gamma_{DE} \tag{27}$$

where K_{Sa} is the thermodynamic solubility product of $BaSO_4$, γ_{DE} is the activity coefficient.

The rate of crystal growth, G depended on the ions concentrations in the solution and on the ions concentrations at the crystal surface, which results in the following expression for G:

$$G = k_g \left[\left(\frac{\left(c_D - \frac{G}{k_D}\right)\left(c_E - \frac{G}{k_D}\right)}{K_{Sa}}\right)^{0.5} \gamma_{DE} - 1\right]^2 \tag{28}$$

where $k_g = 4.0 \times 10^{-11}$ m s^{-1} (Wei and Garside 1997) and $k_D = 10^{-4}$ (m s^{-1})(m^3 kmol^{-1}) (Bałdyga and Orciuch 1997).

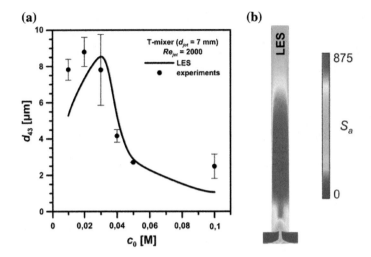

Fig. 9 **a** the effects of inlet concentration on mean particle diameter, d_{43}, in the T-mixer, $Re_{jet} = 2000$; **b** contours of saturation ratio in the T-mixer, $c_{D0} = c_{E0} = 0.1\,\mathrm{M}$, $Re_{jet} = 2000$

Microphotographs of the product showed no aggregates in the samples. For this reason in the population balance of dispersed phase, only the nucleation and crystal growth were considered, while aggregation and breakage were neglected. The balance was solved using the standard method of moments (Hulburt and Katz 1964) and the closure procedure proposed for RANS simulations by Bałdyga and Orciuch (1997). In the closure procedure the concept of the mixture fraction has been used.

In the case of large eddy simulation of precipitation process, the subgrid fluctuations were neglected. Based on the time constant analysis, Makowski et al. (2012), have shown that if the time constant for inertial-convective process and characteristic time constant for nucleation process are comparable and the subgrid mixing time is smaller than the nucleation time, one is just on the limit when subgrid closure will be necessary (Fig. 9).

One can see that the agreement between the numerical results and experimental data is good. The model correctly predicts the increase of the particles diameter, d_{43}, with the decrease of the inlet concentrations.

6   New Model for SGS Diffusivity

As was mentioned in Sect. 1, most functional subgrid models are based on the subgrid diffusion paradigm described by Eq. 11. As a consequence, the main mechanism of the process is considered to interpret the turbulent spectrum as a concentration cascade observed for the isotropic turbulence. So far, the simple model based on the subgrid Schmidt number was applied. A significant disadvantage of the model is that for fully resolved velocity field, the value of subgrid diffusivity is zeroed, while some subgrid scalar fluctuations could exist. A typical case of this will be the placement of the cut-off scale within the inertial-convective range.

In general, the equation describing the subgrid diffusivity should be a function of the fundamental quantities (Sagaut 2006):

$$D_{sgs} = D_{sgs}\left(Sc, \overline{\Delta}, |\overline{S}|, k_{sgs}, \varepsilon, |\nabla \overline{f}|, \sigma^2_{sgs}, \varepsilon_C\right) \tag{29}$$

where $|\overline{S}|$ is the strain rate tensor, k_{sgs} is the subgrid kinetic energy of turbulence, ε_C is the concentration variance dissipation rate.

An example of a model that is independent of the subgrid Schmidt number is Schmidt and Schumann (1989) model. It is a second order model, which assumes the concept of gradient transport:

$$D_{sgs} = C_D \overline{\Delta} \sqrt{k_{sgs}} \tag{30}$$

where C_D is the model coefficient and its value was determined by the authors of subgrid models. The value is in range from 0.161 to 0.204 (Deardorff 1974; Sommeria 1976; Schemm and Lipps 1976; Schmidt and Schumann 1989).

In the analysis of Eq. 30, the subgrid energy of kinetic turbulence becomes important. Using the Yoshizawa model (1988), the kinetic energy can be expressed by the following equation:

$$k_{sgs} = 2C_I \overline{\Delta}^2 |\overline{S}|^2 \tag{31}$$

The value of constant C_I ranges from 0.01 to $1/\pi^2$ (Moin et al. 1991; Fureby et al. 1997). Substituting Eq. 31 into Eq. 30, one can obtain:

$$D_{sgs} = C_{sgs} \overline{\Delta}^2 |\overline{S}| \tag{32}$$

where the constant C_{sgs} is derived from the C_D and C_I coefficients and in our calculations it was equal to 0.092.

Another interesting but more complex model was proposed by Yoshizawa (1988):

$$D_{sgs} = C \frac{\left(\sigma^2_{sgs}\right)^2 \varepsilon}{\varepsilon_C^2} \tag{33}$$

where $C = 0.446$.

The concentration variance σ_{sgs}^2 can be calculated by integrating over whole range of wave numbers of the concentration spectrum, consisting of the inertial-convective and viscous-convective subranges. The SGS variance can be expressed either by using the scale-similarity model (Eq. 14) or as a difference between $\overline{\sigma_{\approx}^2}_{k_\Delta}$ and $\overline{\sigma_{k_\Delta}^2}$, where $k_\Delta = 1/\overline{\Delta}$ is the wave number corresponding to the reciprocal of the size of the numerical grid and $\tilde{k}_\Delta = 1/\tilde{\overline{\Delta}} = 1/2\overline{\Delta}$ is expressed by size that is two times larger than the size of a numerical cell (Wojtas et al. 2015a):

$$\overline{\sigma_{\approx}^2}_\Delta = c_{\approx}\left(\overline{\tilde{\bar{f}}^2} - \overline{\tilde{f}^2}\right) = c_{\approx}\left(\overline{\sigma_{\approx}^2}_{k_\Delta} - \overline{\sigma_{k_\Delta}^2}\right) \tag{34}$$

$$\overline{\sigma_{k_\Delta}^2} = \int_{k_\Delta}^{k_B} E_c(k) = \frac{3}{2} Ba\varepsilon_c\varepsilon^{-1/3}\left[\frac{1}{k_\Delta^{2/3}} - \frac{1}{k_K^{2/3}}\right] + \varepsilon_c\left(\frac{\nu}{\varepsilon}\right)^{1/2}\ln Sc \tag{35}$$

$$\overline{\sigma_{\approx}^2}_{k_\Delta} = \int_{\tilde{k}_\Delta}^{k_B} E_c(k) = \frac{3}{2} Ba\varepsilon_c\varepsilon^{-1/3}\left[\frac{1}{\tilde{k}_\Delta^{2/3}} - \frac{1}{k_K^{2/3}}\right] + \varepsilon_c\left(\frac{\nu}{\varepsilon}\right)^{1/2}\ln Sc \tag{36}$$

where k_K is the Kolmogorov wave number.

This way one can determine the relation describing the concentration variance dissipation rate (Wojtas et al. 2015a):

$$\varepsilon_c = \left(\overline{\tilde{\bar{f}}^2} - \overline{\tilde{f}^2}\right)\frac{2}{3}\frac{\varepsilon^{1/3}k_\Delta^{2/3}}{Ba \cdot 0.587} \tag{37}$$

Now, one can insert Eqs. 15 and 37 into Eq. 33 and get expression for new subgrid diffusivity coefficient, which takes into account the local variation of the turbulence Reynolds number and depends on both Schmidt number and numerical cell size.

Figure 10 shows the contours of measured and predicted values of the mean mixture fraction in the T-mixer in the injection zone. The experimental results were obtained using PLIF method. Model I in the figure means application of turbulent Schmidt number equal to 0.4 (Eq. 11), model II - Schmidt and Schumann model (Eq. 32), while model III modified Yoshizawa model (Eq. 33). All models well predicted the mixing process of non-reactive tracer in the impingement zone for higher tested value of Re_{jet}, however, the best results overall were obtained for the model III, which correctly predicts the range of the non-mixing zone for both of presented jet Reynolds number. This may be particularly important in predictions of processes carried out in reactors with weaker mixing conditions.

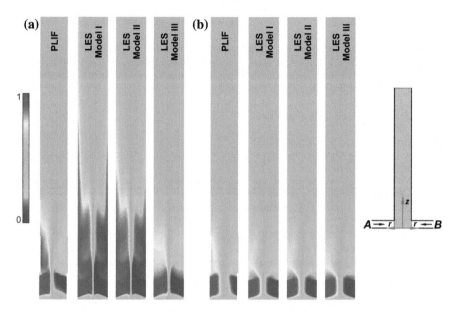

Fig. 10 Contours of the mean mixture fraction in the T-mixer: **a** $Re_{jet} = 1000$; **b** $Re_{jet} = 4000$

7 Conclusions

The results of large eddy simulation predictions of complex chemical processes were presented. LES modeling seems to be a very promising method for obtaining very accurate results, in particular to describe processes, in which the character of the flow changes from laminar to turbulent. For this reason, mathematical models describing the course of processes, especially in the subgrid scale, should be constantly developed. The work presented the new models for subgrid scale concentration variance and diffusivity coefficient, as well as the closure methods for parallel chemical reactions and precipitation. An important advantage of the presented models is a possibility of their wide application, lack (or small number) of model constants and the fact that they take into account a local and time variation of the turbulence Reynolds number, the Schmidt number and the numerical cell size. In the considered test processes carried out in jet reactors, a very good agreement between the experimental and numerical results was obtained.

References

Balarac G, Pitsch H, Raman V (2008) Development of a dynamic model for the subfilter scalar variance using the concept of optimal estimators. Phys Fluids 20:035114

Bałdyga J, Bourne JR (1990) The effect of micromixing on parallel reactions. Chem Eng Sci 45:907–916

Bałdyga J, Bourne JR (1999) Turbulent mixing and chemical reactions. Wiley, New York

Bałdyga J, Makowski Ł (2004) CFD modelling of mixing effects on the course of parallel chemical reactions carried out in a stirred tank. Chem Eng Technol 27:225–231

Bałdyga J, Orciuch W (1997) Closure problem for precipitation. Chem Eng Res Des 75:160–170

Bałdyga J, Orciuch W (2001) Barium sulphate precipitation in a pipe - an experimental study and CFD modelling. Chem Eng Sci 56:2435–2444

Bałdyga J, Rożeń A, Mostert F (1998) A model of laminar micromixing with application to parallel chemical reactions. Chem Eng J 69:7–20

Bourne JR, Yu S (1994) Investigation of micromixing in stirred tank reactors using parallel reactions. Ind Eng Chem Res 33:41–55

Bromley LA (1973) Thermodynamic properties of strong electrolytes in aqueous solutions. AIChE J 19:313–320

Cook AW (1997) Determination of the constant coefficient in scale similarity models of turbulence. Phys Fluids 9:1485–1487

Cook AW, Riley JJ (1994) A subgrid model for equilibrium chemistry in turbulent flows. Phys Fluids 6:2868–2870

Deardorff JW (1974) Three-dimensional numerical study of turbulence in an entraining mixed layer. Bound-Layer Meteorol

Ferziger JH, Perić M (2002) Computational methods for fluid dynamics. Springer, Berlin

Fox RO (2003) Computational models for turbulent reacting flows. Cambridge University Press, Cambridge

Fureby C, Tabor G, Weller HG, Gosman AD (1997) A comparative study of subgrid scale models in homogeneous isotropic turbulence. Phys Fluids 9:1416–1429

Hulburt HM, Katz S (1964) Some problems in particle technology. Chem Eng Sci 19:555–574

Icardi M, Gavi E, Marchisio DL et al (2011) Validation of LES predictions for turbulent flow in a Confined Impinging Jets Reactor. Appl Math Model 35:1591–1602

Ihme M, Pitsch H (2008) Prediction of extinction and reignition in nonpremixed turbulent flames using a flamelet/progress variable model. Combust Flame 155:90–107

Jiménez C, Ducros F, Cuenot B, Bédat B (2001) Subgrid scale variance and dissipation of a scalar field in large eddy simulations. Phys Fluids 13:1748–1754

Makowski Ł, Bałdyga J (2011) Large eddy simulation of mixing effects on the course of parallel chemical reactions and comparison with k–ε modeling. Chem Eng Process Process Intensif 50:1035–1040

Makowski Ł, Orciuch W, Bałdyga J (2012) Large eddy simulations of mixing effects on the course of precipitation process. Chem Eng Sci 77:85–94

Michioka T, Komori S (2004) Large-eddy simulation of a turbulent reacting liquid flow. AIChE J 50:2705–2720

Moin P, Squires K, Cabot W, Lee S (1991) A dynamic subgrid-scale model for compressible turbulence and scalar transport. Phys Fluids Fluid Dyn 3:2746–2757

Pierce CD, Moin P (1998) A dynamic model for subgrid-scale variance and dissipation rate of a conserved scalar. Phys Fluids 10:3041–3044

Piomelli U (2000) Large-eddy and direct simulation of turbulent flows, pp G1–G70

Sagaut P (2006) Large eddy simulation for incompressible flows: an introduction. Springer, New York

Schemm CE, Lipps FB (1976) Some results from a simplified three-dimensional numerical model of atmospheric turbulence. J Atmospheric Sci 33:1021–1041

Schmidt H, Schumann U (1989) Coherent structure of the convective boundary layer derived from large-eddy simulations. J Fluid Mech 200:511–562

Schwertfirm F, Manhart M (2010) A numerical approach for simulation of turbulent mixing and chemical reaction at high Schmidt numbers. In: Bockhorn H, Mewes D, Peukert W, Warnecke H-J (eds) Micro and macro mixing. Springer, Berlin, pp 305–324

Sommeria G (1976) Three-dimensional simulation of turbulent processes in an undisturbed trade wind boundary layer. J Atmospheric Sci 33:216–241

Wei H, Garside J (1997) Application of CFD modelling to precipitation systems. Chem Eng Res Des 75:219–227

Wojtas K, Makowski Ł, Orciuch W, Bałdyga J (2015a) Comparison of subgrid closure methods for passive scalar variance at high Schmidt number. Chem Eng Technol 38:2087–2095

Wojtas K, Orciuch W, Makowski Ł (2015b) Comparison of large eddy simulations and k-ε modelling of fluid velocity and tracer concentration in impinging jet mixers. Chem Process Eng 36:251–262

Wojtas K, Orciuch W, Wysocki Ł, Makowski Ł (2017) Modeling and experimental validation of subgrid scale scalar variance at high Schmidt numbers. Chem Eng Res Des 123:141–151

Yoshizawa A (1988) Statistical modelling of passive-scalar diffusion in turbulent shear flows. J Fluid Mech 195:541–555

Practical Aspects of Settling Tanks Design

Małgorzata Markowska, Szymon Woziwodzki, Magdalena Matuszak
and Marek Ochowiak

1 Introduction

Purification of surface water remains an extremely important question as it allows to provide proper living conditions for human beings and adequate conditions for industrial purposes. New technologies frequently contribute to the formation of excessive quantities of impurities which mix with water and may lead to the disturbance of water management and of the balance in the natural environment. For this reason, the modernisation works are being carried out in wastewater purification plants and water treatment systems.

The apparatuses for water purification with simple design are being developed as well. It allows to separate impurities from of solid particles of light and heavy fractions the fluids and at relatively low costs. Solutions of this type are under more demand as they do not require for complicated constructions to be applied or rare or hard-to-reach materials to be used.

2 Fundamentals of the Sedimentation Process

Slight suspensions occurring in precipitation wastewaters are separated in settling tanks through sedimentation, which is based on the settling of solid particles in a liquid. During the said process, a movement of each particle influences the adjacent particles (Darby et al. 1940; Bandrowski et al. 2001). Particle settling occurs most frequently through the use of gravity. However, under some circumstances and in order to intensify the settling process centrifugal force is taken advantage of, which

M. Markowska (✉) · S. Woziwodzki · M. Matuszak · M. Ochowiak
Institute of Chemical Technology and Engineering,
Poznan University of Technology, Poznań, Poland
e-mail: malgorzata.markowska@doctorate.put.poznan.pl

© Springer International Publishing AG, part of Springer Nature 2018
M. Ochowiak et al. (eds.), *Practical Aspects of Chemical Engineering*,
Lecture Notes on Multidisciplinary Industrial Engineering,
https://doi.org/10.1007/978-3-319-73978-6_18

is the case in settling centrifuges (Orzechowski et al. 1997). Sedimentation may be divided into periodical and permanent ones (Dorr et al. 1941). Periodical sedimentation refers to a precisely determined quantity of suspension, where an inflow of fresh mixture and sediment discharge do not occur. Permanent sedimentation is characterized by the lack of constraints with regards to the suspension inflowing, discharging of purified liquid and sediment removal, which is used in water and waste water purification systems (Camp 1945; Garcia 2008).

2.1 Characteristics of Suspensions

A suspension is a two-phase system consisting of solid particles (dispersed phase) suspended in a liquid, i.e. in dispersing phase (Bieszk 2016). A suspension may be differently referred to as a volatile dispersive system, which makes it different from colloids, which remain durable dispersive systems. When solid particles are dispersed evenly within the entire volume, we come up with a homogenous suspension. When it is not the case, we come up with an inhomogeneous suspension. What is more, solid particles may be characterised with different shapes, such as a ball, a cuboid, a cylinder or other solid figures. If the shape of particles in a suspension is mutually commensurate along three mutually perpendicular directions as far as the size is concerned, then such particles are described as isotropic ones. However, in case of an opposite situation they are anisotropic (Chen et al. 1993; Bandrowski et al. 2001; Garcia 2008).

In order to describe a suspension in a proper manner, one must take into consideration such parameters as: solid particle equivalent diameter, suspension viscosity and density as well as the relevant quantitative ratios between a fluid, a solid and a suspension (Dietrich 1982; Bandrowski et al. 2001).

In case of water system management, sediment and suspension constitute the major threats which disturb the functioning of precipitation water sewer system. Impurities which get into the network bring about silting and, subsequently, improper functioning of the systems (Wavin 2015).

2.2 Settling Velocity for Solid Particles

The process of free settling remains the simplest case of solid particle settling in fluids under the influence of the gravity field. The process runs uniformly inside a motionless centre, where a slight particle share occurs. This, in turn, brings about the lack of mutual interactions of the particles. Free settling velocity depends on the size, shape and properties of a particle and the properties of the centre itself (Serwiński 1971; Khan 1987). Physical properties of a fluid and a particle which are taken into consideration include particle density ρ_R and fluid density ρ_c as well as its viscosity η_c (Concha et al. 1979; Orzechowski et al. 1997). During free settling two forces act upon a particle: gravity G and buoyancy W. The difference between these forces causes the resistance of the centre, which, depending on the centre type, may be referred to as hydraulic or aerodynamic resistance. By taking advantage of the force balance, the general equation for free settling velocity is obtained for a single spherical particle u_s, depending on the particle diameter d, acceleration of gravity g and the density module Γ_ρ, having the form of:

$$u_s = \sqrt{\frac{4 \cdot g \cdot d}{3 \cdot \xi} \cdot \frac{\rho_R - \rho_C}{\rho_C}} = \sqrt{\frac{4 \cdot g \cdot d}{3 \cdot \xi} \cdot \Gamma_\rho} \tag{1}$$

The value of resistance coefficient ξ, on which the free settling velocity depends, is determined by the value of Reynolds number (Karamanev 1992).

Depending on the amount of the aforementioned criterion number, which determines the ratio between inertial force and viscosity force, the character of the movement may be established for a freely settling particle. We can distinguish between stratified movement, transition movement and turbulent movement. By determination of the function of the resistance coefficient for single settling particle from the Reynolds number and sphericity shape factor, one may distinguish between three subsequent flow rages: Stokes', Allen's and Newton's ones. Within the laminar range, i.e. within the Stokes' range, the diagram demonstrates a straight line that meets the condition of a linear dependency of the resistance coefficient on the Reynolds number (Orzechowski et al. 1997; Broniarz-Press et al. 2008). For the laminar range ($Re < 2$), the equation determined by Stokes was the equation to describe the theoretical settling velocity for a single particle u_0, depending on the fluid dynamic viscosity factor η_c, the density of the fluid dispersed phase ρ_R and fluid density ρ_c, particle diameter d and constant gravity acceleration g (Bandrowski et al. 2001; Broniarz-Press et al. 2013), also referred to as the Stokes' equation:

$$u_0 = \frac{d^2 \cdot (\rho_R - \rho_C) \cdot g}{18 \cdot \eta_C} \tag{2}$$

There also exists the settling occurring within the centrifugal forces field, however then separation of solid particles from a liquid makes sense if centrifugal acceleration is comparable to the gravity acceleration. Then, the equation analogical to the one for the settling within the laminar range (Eq. 2) takes advantage of centrifugal acceleration, not the gravity acceleration.

Sedimentation velocity is determined by a similar dependency, with relation to a theoretical settling velocity of a single particle within the laminar range. The difference lies in the fact that parameters of a suspension, not pure water, are taken into consideration, i.e. suspension density ρ_m and its viscosity η_m, which is dependent on the suspension porosity (Stafford 1994). However, all the aforementioned considerations refer to the particles having the same shapes. It is known that real suspensions contain the particles of various shapes and sizes, which hampers the application of generalized dependencies for all particles of a solid fraction (Orzechowski et al. 1997).

3 Settling Tanks

Separation of impurities from waste water and removal of concentrated suspensions occurs in the devices referred to as settling tanks (Dallas 1955). They stop and separate easily settling suspensions, natural suspension or the ones formulated in the process of coagulation and chemical precipitation of sparingly water soluble compounds into light and heavy fractions and purified liquid (Dorr et al. 1941; Bieszk 2016). They are characterized by a wide spectrum of applications, e.g. in surface and underground water purification systems. What is more, their operating mode may be periodical, semi-continuous and permanent (Darby et al. 1940; Kowal et al. 2009).

The system of settling tanks in water purification systems may vary and is dependent on the actual demand (Darby et al. 1940). It is influenced by the types and the sequence of unit processes used in order to purify a stream of a liquid (Forster 2003). Owing to their simple design and non-complicated operating principle, it is possible to locate settling tanks freely within the composition of any wastewater purification system (Dallas 1955).

Several types of settling tanks can be distinguished due to the direction of a liquid flow. Rectangular are those in which single-direction horizontal flow occurs.

In vertical ones, the direction remains vertical, from the feeding place at the bottom towards the settling tank periphery. Round settling tank are characterized by radial flow from the centre towards the device periphery (Dallas 1955; Ambler 1959; Filipowicz 1980). Apart from such ones, we can distinguish the settling tanks with suspended sediment and multi-stream settling tanks intended for mechanical purification of precipitation wastewaters without hydraulic by-passing (Orzechowski et al. 1997; Separator Service 2017). Swirl settling tanks constitute the youngest group of devices. They are intended to pick up solid impurities, sediments and suspensions from rain and precipitation wastewaters as well as from technological wastewaters flowing by gravity in the sanitary system (Wavin 2015; Eco-Tech 2017). They remain a proper solution for urban areas due to their high purification efficiency and small sizes (Wavin 2015). Their unique design allows to achieve high efficiency for suspension separation at high hydraulic loads (Ecol-Unicon 2017; Pur Aqua 2017).

Operational efficiency of settling tanks is measured by the contents of suspensions in water after sedimentation purification process. Process efficiency depends on types of sedimentation particles and the process duration and is expressed in percentages. This efficiency is the ration between the weight of the suspensions picked up in the settling tank and the weight of suspensions in water prior to sedimentation process (Kowal et al. 2009).

3.1   Operational Parameters for Settling Tanks

There exist the relevant values which describe operations of the settling tanks. The first parameter is the volume of the sediment stored. It is determined by the volume of the sedimentation part of the device. This value is very important due to a possible life time and to determine the multiplicity of sediment removal. The quantity of the sediment picked up must be taken into consideration as well, i.e. the dependence of suspension concentration at the inlet and outlet, the surface of reduced catchment area and the annual amount of precipitations (Darby et al. 1940; Svarovsky 2001). Catchment area is the surface from which precipitations are received and it may include the tarmac, a pavement or a park. However, the surface of reduced catchment area is the total surface of the catchment area changed by the permeability factor (Ecol-Unicon 2015). Another parameters to describe the functioning of the settling tank are its efficiency and flow resistances occurring during the fluid motions at the inlet to the device, within its volume and at the outlet. It is related to the diameters of the pipes supplying and discharging fluids or wastewaters. A crucial issue is the ratio of sizes, i.e. are they the same, or do they have a totally different size (Mcgivern et al. 1967). On top of that, flow resistances inside the device may be determined by means of the relevant analysis of fluid damming for individual zones of the device (Dallas 1955). Values of such damming depend on the volumetric flow intensity, i.e. hydraulic loads.

3.2   Selection of Settling Tanks

The first criterion to select a settling tank is a correct evaluation of the place where the apparatus is supposed to be used. Landform, available space for land development and the values of hydraulic intensity remain some of the most important questions that determine the selection of the type and size of the apparatus needed (Dallas 1955). Additionally, physical and chemical parameters of the suspension which will be flowing into the tank must be taken into consideration They include viscosity and density of dispersed and dispersing phase, their volumes, sediment hydration, porosity and the value of settling velocity in the aspect of sediment storage as well as the multiplicity of its removal from the device (Mcgivern et al. 1967). Determination of the relevant average parameters for the sizes of solid particle picked up at the bottom of the settling tank also influences its proper selection in line with the current needs. In order to ameliorate the choice, it is recommended to use the catalogues of the settling tank designing companies existing on the market as they offer a wide range of device types (Ecol-Unicon 2016).

3.3 Settling Tanks Design

3.3.1 Settling Tanks with Horizontal Flow

Settling tanks with horizontal flow demonstrate the floating of solid particles depending on the water flow velocity in the tank and the settling velocity (Darby et al. 1940). The value that describes the horizontal settling tank is its active length, which is dependent on the surface of the device section, its width and the number of devices, as they may operate in serial arrangement. The total length of horizontal settling tank remains the sum of its active length and the length taken by the overflows. A very important question is to maintain the proportions between the settling tank length and its width, as it is important to achieve operational stability and small disturbance zone (Dallas 1955; Hirsch 1967; Kowal et al. 2009).

Horizontal settling tanks are characterized by higher stability and smaller disturbance area in comparison to other types of such devices. In Poland, covered settling tanks are generally constructed due to climatic conditions. Otherwise, the stream might get frozen which could result in diminishing of its intensity or the total lack of flow (Kowal et al. 2009).

3.3.2 Vertical Flow Settling Tanks

Settling tanks with vertical flow are used to purification of small quantities of liquids. Their design is predominantly characterized by circular cross-section, less frequently by square cross-section (Bieszk 2016). The peculiarity of their design boils down to the fact that fluid is supplied to a central pipe or a reaction chamber and subsequently to a clarification chamber or the suspended sediment layer (Brix et al. 2005).

The fundamental parameter to describe this type of settling tank remains the velocity of vertical liquid (suspension) flow. We distinguish many designing solutions for settling tanks with vertical flow. The simplest one is a cylindrical apparatus with conical bottom. Along its vertical axis there is a central pipe ending with a supporting construction, and an overflowing trough discharging the purified liquid is located on top (Kowal et al. 2009).

3.3.3 Dorr Sedimentation Tank

Sedimentation tanks is a variation of vertical flow settling tank characterised by gravity action to separate the suspensions and the presence of a scraper along the device axis (Dorr et al. 1941; Bieszk 2016). In case of the sedimentation tanks operating in permanent mode, the inflow of suspension, discharge of clarified fluid and discharge of silt (solid particle sediment) occurs in permanent mode and they are equipped with a cylindrical, conical bottom with a large angle of flare (Dorr et al. 1941). There is a rectangular trough in the highest zone of the cylindrical part. Its role is to discharge the clarified fluid from the apparatus by overflowing. Whereas, suspension is supplied through a pipe located coaxially with the scraper shaft and it flows into the settling tank volume below the upper liquid level, which allows for it dispersion within the entire surface of the apparatus cross-section (Bieszk 2016).

3.3.4 Swirl Settling Tank

Swirl settling tank is another group of cylindrical devices taking advantage of centrifugal forces apart from the force of gravity. However, in this case vortex motion is observed very clearly (Ambler 1959; Veerapen 2005). Consequently to that, high separation efficiency is achieved even for high hydraulic loads (Sullivan 1978). Vortex motion may be caused by means of a directional deflector, located near the wastewaters inlet into the device chamber. The outlet is located in the central section of the tank and may be protected with a siphon or a grid made of stainless steel (Vokes 1943; Ecol-Unicon 2017; Eco-Plast 2017). Supplying the settling tanks should be carried out through designed inlets only, wastewaters discharging may not occur directly from the surface (Eco-Plast 2014). Swirl settling tanks provide effective pre-treatment of wastewaters from general suspensions to a standardized level, protect against excessive quantities of suspensions flowing into the devices in the systems with the separators or in wastewater treatment plants and provide effective separations of petroleum substances within the system with an integrated lamella section (Nixor 2012; Ecol-Unicon 2017; Eco-Tech 2017). For swirl settling tanks the characteristic parameters for the apparatuses must be determined in order to select proper series of swirl settling tank. Hydraulic values include effective separation of light and solid fractions as well as solid particle size distribution. Additionally, physical and chemical parameters are taken into account, like: viscosity and density of dispersed and dispersions phases, their volumes, sediment hydration, influence from porosity and settling velocity (Veerapen 2005; Pur Aqua 2017).

The main advantages of swirl settling tanks include high efficiency of purification, smaller gross covered surface compared to traditional settling tanks, a possibility to locate the wastewater inlet to the tank at any angle, which ameliorates connection to the sanitary system, easy exploitation, a possibility for installation along the road lanes and regulation of placing depth (Ecol-Unicon 2017; Eco-Plast 2017).

Application range for swirl settling tanks is very wide, however it depends on the ratio between the settling tank volume and anticipated quantity of sludge (Sullivan 1978). For small quantities of sludge, the settling tanks are used within the areas collecting precipitation waters with sediment originating from traffic or collecting pools within the areas of petrol tanks and roofed filling stations (Nixor 2012). On the other hand, for the medium quantities of sludge, we apply the said settling tanks in filling stations, manual car washes, washing areas for buses, for the sewers from garages and parking spaces as well as in power stations and mechanical plants (Wavin 2015). With regards to high quantities of sludge, there is a need to use a swirl settling tank with proper dimensions near the washing devices for all-terrain vehicles, construction equipment, agricultural machines, near the washing areas for trucks (Eco-Plast 2014).

There are two major designing solutions for swirl settling tanks: a single-chamber and a two-chamber one. In the two-chamber swirl settling tank, the first chamber serves to separate the heavy fraction, while the second one is usually divided into two sections, where the first one constitutes a trap for light fraction particles with the density below the density of a liquid (also petroleum substances), while the second one functions as a discharge chamber (Ecol-Unicon 2017; Pur Aqua 2017).

The research over the single-chamber swirl settling tanks is intended to determine the designing solution with the highest efficiency of purification of liquid from light and heavy fraction particles (Ochowiak et al. 2017b). Sea sand remains a researching material used in the systems of settling tanks used for purification of liquids from heavy fraction. It was divided into fractions due to the diameter by using screening analysis. The given fractions are 100–150, 150–200, 200–300 and >300 μm (Ochowiak et al. 2016). In case of the system to purify from light fraction, the researching materials is mineral oil 20–70 supplied by the Institute for Kerosene an Gas based in Cracow. The viscosity was 0.060 Pa s, and the density was 865 kg/m^3 (Ochowiak et al. 2017a). The device parameters that were taken into consideration were location and the shape of inlet connector and the arrangement of the outlet connector as well as the location of internal cylinder along the apparatus axis. Constructions under examination included a standard swirl settling tank (OW), a settling tank with a directional U-bend (OWK), a settling tank with a profiled pipe (OWR), a settling tank with a reversed outlet connector (OWKR) and a settling tank with internal cylinder (OWC) all of which are presented by Fig. 1 (Ochowiak et al. 2016).

Examinations demonstrated that the settling tank with a profiled pipe (OWR) is the most effective. Inlet connector has a shape of a profiled pipe directed vertically downwards and finished with a U-bend with 45 degree of inclination relative to the horizontal and tilted from the vertical by 30 degrees. The outlet pipe, located 0.1 m lower than the orifice for inlet connector, consists of a ninety degree U-bend directed upwards and a horizontal pipe. This design is characterized by the highest efficiency of purifying a liquid stream from heavy fraction and lowest values for hydraulic damming to occur inside the chamber (Ochowiak et al. 2017b). In order to separate a fluid and light fraction (oil fraction), a settling tank that is characterized by a smaller cylinder located inside the external cylinder has been chosen. Pre-treatment of a fluid stream from solid fraction particles takes place within the entire volume of the apparatus, while petroleum substances are stopped in a ring

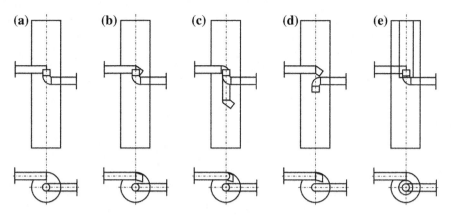

Fig. 1 Designs of examined settling tanks (Ochowiak et al. 2016, 2017a): **a** OWKR, **b** OWK, **c** OWR, **d** OW, **e** OWC

zone, i.e. the space between the cylinders. It has been detected that the internal cylinder functions as a deflector at the outlet of the settling tank, stops the floating substances and pacifies a fluid stream (Ochowiak et al. 2017a).

The examinations also demonstrated that the fluid damming increases along with the increase in fluid flow intensity. The values for fluid damming height depending on the fluid stream intensity have been presented in Fig. 2. The local resistance during fluid flow though the apparatus has also been taken into consideration. For a given flow intensity and the linear speed of a permanent phase w and the fluid damming height value ΔH, a local resistance coefficient ξ may be calculated for all types of obstacles:

$$\Delta H = \xi \frac{w^2}{2g} \tag{3}$$

For the examined OWC system, the local resistance coefficient is constant and amounts to 3.06 on average.

Subsequently, the efficiency in purification of fluid stream from heavy and light fraction particles was determined. In both cases, the value is determined as a degree to which the solid particles or oil are stopped inside the apparatus and is expressed as a percentage. Figures 3 and 4 show the efficiencies of the systems examined depending on the fluid flow intensity at the inlet to the apparatus. The aforementioned modified designing solutions serve to adjust the settling tank design in order to achieve the best possible purification of a liquid stream from heavy and light fractions. This leads to the amelioration of purification efficiency for the already existing solutions, which is extremely important for environmental issues.

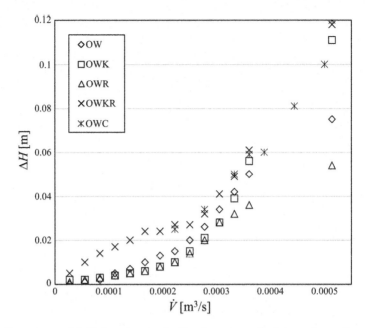

Fig. 2 Dependency of fluid damming on the flow intensity at the inlet to the settling tank for various designing solutions of a swirl settling tank

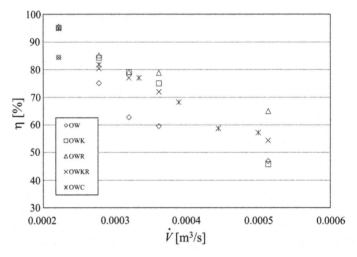

Fig. 3 Degree of stopping sand grains with the diameter of 100–150 μm in swirl settling tanks: OW, OWK, OWR, OWKR, OWC

Fig. 4 Dependence of purification for fluid contaminated with light fraction on the volumetric fluid stream for a OWC swirl settling tank

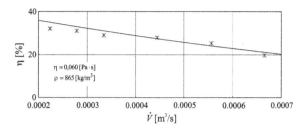

4 Summary

Information in the reference sources with regards to sedimentation and fluid stream purification from solid particles shows that settling phenomenon remains an effective method for fluid purification and suspension compaction. This method is characterized by simplicity and a wide spectrum of possible applications in various branches of economy. The devices applied, including the settling tanks and sedimentation tanks, may be modified. Designing and construction of the new solutions will contribute noticeably to improved efficiency in removing the impurities from liquids and to reducing the sizes of the devices themselves. Their simple design and non-complicated purification process as well as easy exploitation make the settling tank an excellent solution to the problem of purifying precipitation waters from usable surfaces and wastewaters from various branches of the industry. This will provide us with a possibility to take care of our natural environment.

Acknowledgements This work was supported by PUT research grant no. 03/32/DSPB/0702.

References

Ambler CM (1959) The theory of scaling up laboratory data for the sedimentation type centrifuge. Biotechnol Bioeng 1(2):185–205

Bandrowski J, Merta H, Zioło J (2001) Sedymentacja zawiesin. Zasady i projektowanie. Wydawnictwo Politechniki Śląskiej, Gliwice

Bieszk H (2016) Wybrane aparaty do rozdzielania zawiesin. Odstojniki. Katedra Aparatury i Maszynoznawstwa Chemicznego. Politechnika Gdańska, Gdańsk

Brix H, Arias CA (2005) The use of vertical flow constructed wetlands for on-site treatment of domestic wastewater. Ecol Eng 25(5):491–500

Broniarz-Press L, Dobrychłop J (2013) Opór opadania pojedynczej cząstki w wodnych roztworach soli sodowej karboksymetylocelulozy. Inż Ap Chem 3:155–156

Broniarz-Press L, Różański J, Różańska S et al (2008) Opadanie swobodne cząstek kulistych w płynach rozrzedzanych ścinaniem. Inż Ap Chem 6:13–14

Camp Thomas R (1945) Sedimentation and the design of settling tanks. Proc ASCE 71(4):445–486

Chen S, Timmons MB, Aneshansley DJ et al (1993) Suspended solids characteristics from recirculating aquacultural systems and design implications. Aquaculture 112(2–3):143–155

Concha F, Almendra ER (1979) Settling velocities of particulate systems. Int J Miner Process 5 (4):349–367

Dallas JL, U.S. Patent 2,708,520, 17 May 1955

Darby GM, Dorr JVN, Roberts EJ et al (1940) U.S. Patent 2,185,785

Dietrich WE (1982) Settling velocity of natural particles. Water Resour Res 18(6):1615–1626

Dorr JVN, Weber WC (1941) U.S. Patent 2,263,168

Eco-Plast (2017) Osadniki http://eco-plast.pl/oferta/osadniki. Accessed 29 April 2017

Eco-Plast (2014) Katalog osadniki piasku. Pleszew

Eco-Tech (2017) Settling tanks, http://www.eco-tech.pl/index1.php?lang=pl&grupa=1. Accessed 28 April 2017

Ecol-Unicon Sp. z o.o. (2015) Katalog projektanta. Gdańsk

Ecol-Unicon Sp. z o.o. (2016) Product catalogue. Gdańsk

Ecol-Unicon Sp. z o.o. (2017) Vortex settling tanks EOW, http://www.ecol-unicon.com/urzadzenia/osadniki/osadniki-wirowe. Accessed 28 April 2017

Filipowicz M, Filipowicz P (1980) U.S. Patent 4,199,459

Forster C (2003) Wastewater treatment and technology. Thomas Telford, London

Garcia MH (2008) Sedimentation engineering: processes, measurements, modeling and practice. ASCE 978–0–7844–0814–8:1132

Hirsch AA (1967) U.S. Patent 3,353,676

Karamanev DG, Nikolov LN (1992) Free rising spheres do not obey newton's law for free settling. AIChE 38(11):1843–1846

Khan AR, Richardson JF (1987) Fluid-particle interactions andflow characteristics of fluidized beds and settling suspensions of spherical particles. Chem Eng Commun 78(1):111–130

Kowal AL, Świderska-Bróż M (2009) Oczyszczanie wody. Podstawy teoretyczne i technologiczne, procesy i urządzenia. Wydawnictwo naukowe PWN, Warszawa

Mcgivern RF, Ucker LJ (1967) U.S Patent 3,333,704

Nixor Polska Sp. z o.o. (2012) Katalog separatory, osadniki, filtry antyodorowe. Kowale

Ochowiak M, Matuszak M, Włodarczak S et al (2016) Badania sprawności oczyszczania strumienia wód opadowych w osadnikach-piaskownikach wirowych. Inż Ap Chem 5:199–200

Ochowiak M, Matuszak M, Włodarczak S et al (2017a) Ocena pracy zmodyfikowanego osadnika wirowego do oczyszczania strumienia wody zanieczyszczonej frakcją lekką. Inż Ap Chem 4:132–133

Ochowiak M, Matuszak M, Włodarczak S et al (2017b) The modified swirl sedimentation tanks for water purification. J Environ Manage 189:22–28

Orzechowski Z, Prywer J, Zarzycki R (1997) Mechanika płynów w inżynierii środowiska. Wydawnictwo Naukowo-Techniczne, Warszawa

Pur Aqua Sp. z o.o. (2017) Osadniki wirowe, http://www.puraqua.pl/osadniki-wirowe.html. Accessed 29 April 2017

Separator Service (2017) Osadnik wielostrumieniowy STOPPOL, http://separator.pl/stoppol-htm. Accessed 19 June 2017

Serwiński M (1971) Zasady inżynierii chemicznej. Operacje jednostkowe. Wydawnictwo Naukowo-Techniczne, Warszawa

Stafford WF III (1994) Boundary analysis in sedimentation velocity experiments. Method Enzymol 240:478–501

Sullivan RH, Cohn MM, Ure JE et al (1978) The swirl primary separator: development and pilot demonstration. APWA 600/2–78–122

Svarovsky L (2001) Hydrocyclones. Elsevier, Butterworth-Heinemann

Veerapen JP, Lowry BJ, Couturier MF (2005) Design methodology for the swirl separator. Aquacult Eng 33:21–45

Vokes FC, Jenkins SH (1943) Experiments with a circular sedimentation tank. ICE 19(3):193

Wavin Polska SA (2015) Systemy do zagospodarowania wód deszczowych. Katalog produktów, Buk

Aerosol Therapy Development and Methods of Increasing Nebulization Effectiveness

Magdalena Matuszak, Marek Ochowiak and Michał Doligalski

1 Introduction

Aerosol therapy is a modern method of instrumental therapy and represents a very fast developing area of knowledge, which is predominantly the effect of significant technological progress (Sosnowski 2012). Drugs atomization may aim at delivering drugs of local effect (Bisgaard 1998; Khilnani and Banga 2008; Ari et al. 2009; Elliott and Dunne 2011) or drugs of system effect (Khilnani and Banga 2008; Sosnowski 2012) to a patient's respiratory tract.

Drug atomization during aerosol therapy is conducted in devices referred to as inhalers or nebulizers. Such devices belong to the oldest and the most frequently used in medicine (Petersen 2004; Pilcer and Amighi 2010; Dolovich and Dhand 2011). Considering mechanisms used to produce aerosol, medical inhalers can be divided into: classical and new generation devices. Classical devices include pressurized metered dose inhalers (pMDI) and dry powder inhalers (DPI) as well as pneumatic and ultrasonic nebulizers (Sosnowski 2012; Chan et al. 2014).

In effect of atomization an aerosol is produced, which consists a disperse system of solid object small particles suspended in a gas (continuous phase, most often it is air or oxygen) or of a liquid (dispersed phase). Produced medical aerosols are classified considering size and type of the particles suspended in gas, and finally we can distinguish: monodisperse aerosols, where the particles suspended in gas are of the same size, and polydisperse aerosols, where the particles are of a different size (Ochowiak et al. 2016).

M. Matuszak (✉) · M. Ochowiak
Institute of Chemical Technology and Engineering,
Poznan University of Technology, Poznań, Poland
e-mail: magdalena.matuszak@put.poznan.pl

M. Doligalski
Institute of Metrology, Electronics and Computer Science,
University of Zielona Góra, Zielona Góra, Poland

© Springer International Publishing AG, part of Springer Nature 2018
M. Ochowiak et al. (eds.), *Practical Aspects of Chemical Engineering*,
Lecture Notes on Multidisciplinary Industrial Engineering,
https://doi.org/10.1007/978-3-319-73978-6_19

Drug delivery via the respiratory tract has a lot of advantages and is of increasing interest. Methods of inhaled drug delivery show significant advantage over other methods, such as: vaccine or oral (Cipolla and Gonda 2011). The main advantages of the inhaled therapy include: relatively easy and comfortable construction of inhaling devices, low invasiveness, a possibility of selective, painless and quick delivery of medicine to required areas of the respiratory tract, a possibility of individual drug selection and its dose, a possibility of conducting both oxygen- and pharmaco-therapy (Ari et al. 2009; Elliott and Dunne 2011; Sosnowski 2012). It must be underlined that the possibility of direct drug delivery to chosen patient's respiratory tract areas allows to decrease a drug dose in order to achieve the same therapy effect as with other methods (Ari et al. 2009; Elliott and Dunne 2011; Sheth et al. 2015).

It should be underlined that aerosol therapy effectiveness depends on the knowledge and keeping the necessary rules of using inhaling methods under various conditions. Moreover, lots of factors such as: physicochemical properties of the atomized solution, device construction and its proper use and also individual anatomy of a patient's respiratory tract as well as inhaling dynamics (breathing frequency and volume of inhaled air) influence the aerosol therapy effectiveness (McCallion et al. 1995, 1996).

It is important to mention that contemporary respiratory tract diseases belong to one of the most frequent and the fastest developing diseases in the world which people of wide age range fight with (Amirav 2004; Marianecci et al. 2011; Pahwa et al. 2012). Consequently, a visible progress and increase in the number of new or improved pharmaceuticals and devices for inhalations occur. Liquid atomization is a complex process and its effectiveness is described by micro- and macro-parameters involving, among others, mean drop size and drop size histogram. In the literature you can find the most often used mean diameters as the mean volume-surface drop diameter D_{32} (*SMD, Sauter Mean Diameter*) and diameter $D_{0.5}$ (*MMD, Mass Median Diameter* and *CMD, Count Median Diameter*) (McCallion et al. 1995; Sosnowski 2012; Broniarz-Press et al. 2014). The diameter $D_{0.5}$ states exactly 50% of drop size distribution (mass-, volume- and number-based) (Lefebvre 1989). Diameters $D_{0.1}$ and $D_{0.9}$ are worth mentioning–they state respectively that 10% or 90% of liquid volume consists of drops with lower diameters than $D_{0.1}$ and $D_{0.9}$. The difference between the aforementioned diameters determines the homogeneity level of size distribution (Lefebvre 1989). It is worth underlining that the analysis of aerosol drop sizes plays a significant role in defining the effectiveness of size distribution and drug delivery via inhalation.

2 Dynamic Development of Aerosol Therapy

Consequently to dynamic technological progress, a visible development of diagnostic methods and medical therapies (Sosnowski 2012) accompanied by a growing interest in drug delivery via inhaled methods (Khilnani and Banga 2008; Ochowiak et al. 2016) have appeared. Continuous progress and a big innovation index with regards to inhaled devices and medicines enable to extend the area of possible applications for aerosol therapy. Because of that, you can see new pharmaceuticals on the market to be delivered via inhaling, such as:

- vaccines (Licalsi et al. 1999; Dilraj et al. 2000; Dolovich and Dhand 2011; Chan et al. 2014; Zhou et al. 2014; Berkenfeld et al. 2015; Wanning et al. 2015)
- nebulized antibiotics (Bisgaard 1998; Vaughan 2004; Vecellio et al. 2011; Olveira et al. 2014; Zhou et al. 2014)
- inhaled steroids (Bisgaard 1998; Terzano and Allegra 2002; Cipolla and Gonda 2011; Olveira et al. 2014)
- inhaled biopharmaceuticals (Shoyele and Cawthorne 2006)
- anticancer drugs (Brown et al. 2001; Willis et al. 2012; Chan et al. 2014)
- anesthetics (Isaev and Morozov 2005; Cipolla and Gonda 2011; Berkenfeld et al. 2015)
- antifungals (Willis et al. 2012; Olveira et al. 2014)
- anti-infective drugs (Chan et al. 2014; Zhou et al. 2014)
- anti-tuberculosis drugs (Chow et al. 2007; Willis et al. 2012; Andrade et al. 2013; Park et al. 2013)
- anticoagulants (Olveira et al. 2014)
- nebulized mucolytics (Odziomek et al. 2012; Olveira et al. 2014)
- anticholinergics (Cipolla and Gonda 2011)
- immunosuppressive drugs (Wang et al. 2014).

Nebulized drugs are also used in sinuses cure in patients with chronic sinuses inflammation who suffer acute, recurrent infections, which constitutes an important medical problem (Kamijyo et al. 2001; Vaughan 2004).

Shah (2011) in turn, describes in his paper the possibility of inhaled dosing of exogenous surfactants as the alternative to their delivery during intubation. This method enables to decrease side effects resulting from mechanical ventilation and leads to improved treatment effectiveness. Soll et al. (1994), proved that using the substitute therapy with surfactants decreases sick rate and death number among the newborns with respiratory distress syndrome. In the paper by Dijk et al. (1997), it was shown that inhaled surfactants delivery to lungs (especially in newborns) can be a good alternative to less effective methods. This method can contribute to improved uniform distribution and also minimize and exclude negative hemodynamics effects (Lewis et al. 1991, 1993). It is also important that conducted research did not demonstrate any changes in surfactants structure in distribution process. However quite a big volume loss of surfactant aerosol was observed, so an attempt to increase the nebulization effectiveness should be undertaken in the future (Dijk et al. 1997).

Moreover, Argent et al. (2008), conducted initial nebulization research of adrenaline to treat airway obstruction in patients with acute severe croup.

An innovative approach to the treatment was also described by Salama et al. (2014). The article refers to concurrent combining of the two drug delivery ways to a body, i.e. oral and inhalation. The authors proved that the suggested method is very effective and efficient in treatment (Salama et al. 2014).

The next innovative implementation of aerosol therapy is to deliver peptides and therapeutic proteins to the body. The process of distribution includes, among others, insulin, calcitonins, growth hormone and parathormone (parathyroid hormone–substance produced and secreted by the parathyroid glands) (Bosquillon et al. 2004; Shoyele and Cawthorne 2006; Shoyele and Slowey 2006; Depreter and Amighi 2010; Cassidy et al. 2011; Wanning et al. 2015). The experimental data shown in Cefalu's paper (2007) acknowledge the treatment effectiveness via inhalation of diseases such as diabetes. It is worth stressing that implementation of the inhaled insulin in aerosol therapy is considered by the regulations office in the United States and Europe (Shoyele and Slowey 2006).

Aerosol gene therapy and new generation pharmaceuticals (called recombinant medicines) represent quite an important issue (Kleemann et al. 2004; Lentz et al. 2005, 2006; Johnson et al. 2008; Mohajel et al. 2012). These medicines represent endogenous copies of biologically active proteins which contain hormones, vaccines, antibodies and diagnostic means. They are created in the effect of several-stages biotechnological processes conducted and genetic engineering techniques (Shoyele and Cawthorne 2006). They are applied in treatment of respiratory tract, various infections, genetic, auto-immune, neurological and blood diseases and also cancer (Robbins et al. 2003; Gill et al. 2004; Bivas-Benita et al. 2005; Lentz et al. 2006; Zou et al. 2007; Luisettia et al. 2011; Marianecci et al. 2011).

Dynamics of technological progress regarding aerosol therapy influenced not only modifications and improvements of new pharmaceuticals but also numerous trials were undertaken to improve devices used for drug atomization. These devices were significantly modified with regards to operation and construction. They are called new generation inhalers and show fewer faults than conventional devices. Figure 1 presents the division of unconventional inhalers (Sosnowski 2012). Moreover, they are characterized by higher effectiveness, precision and recurrence of a dose and enable quicker pulmonary drug delivery. New generation inhalers enable inhalation of drugs containing sensitive agents and such ones susceptible to degradation (Sosnowski 2012). The main obstacle to use them commonly by patients is their high cost (Kesser and Geller 2009).

Mesh nebulizers belong to a group which is the most often described and used devices of the new generation. These devices are equipped with vibrating mesh or "micropump" plate in the form of metal perforated membrane (Dhand 2002; Lass et al. 2006; Sosnowski 2012). Liquid atomization occurs in effect to work of vibrating piezoelectric crystal, which, in turn, makes the perforated membrane vibrating. The produced aerosol is of monodisperse character and the size of drops is in submicrones and can be regulated by the selection of "micropump" size (Ashgriz 2011; Longest et al. 2012; Sosnowski 2012). These nebulizers can be divided into active and passive ones (Dhand 2002; Ashgriz 2011; Najlah et al. 2013, 2014; Olveira et al. 2014). Nebulizers of this type have lots of advantages in comparison to conventional ultrasonic nebulizers, which include: no heating of

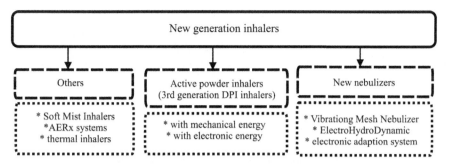

Fig. 1 The general classification of new generation inhalers used in aerosol therapy (Sosnowski 2012)

atomized liquid or heating is slight, low energy-absorptiveness, quiet work, low residual volume, uniform construction and shorter inhalation time (Dhand 2002; Lass et al. 2006; Ashgriz 2011; Beck-Broichsitter et al. 2012; Sosnowski 2012; Kwok and Chan 2014; Olveira et al. 2014). The main advantage of vibrating mesh nebulizers is that they are suitable for the distribution of drugs susceptible to degradation (Shoyele and Slowey 2006; Elhissi et al. 2007; Johnson et al. 2008; Dolovich and Dhand 2011; Elhissi et al. 2011; Beck-Broichsitter et al. 2012; Willis et al. 2012).

The next devices of the new generation are electrohydrodynamic atomizers (EHDA), which use electrohydrodynamic atomization to produce aerosol (electro-spray). The liquid disintegration into drops happens as a result of hydro-dynamic instabilities in a liquid stream during the outflow from the capillary (Gomez 2002; Jaworek 2007; Sosnowski 2012). The advantages of these devices are: possibility to distribute suspensions and drugs susceptible to degradation (e.g. proteins, liposomes), monodisperse character of aerosol (effective targeted therapy), minimum risk of nozzle clogging, particle size can range from nanometers to millimeters (Gomez 2002; Sosnowski 2012). The disadvantage is the possibility to distribute drugs only in a form of liquid and relatively low capacity, which makes this method suitable mainly to distribute new generation drugs of highly active character (Sosnowski 2012).

In the group of unconventional aerosol devices there is also the adaptative aerosol delivery (AAD). These are innovative nebulizers of advanced and modern technology enabling to register a patient's lungs work (air flow rate during inhale and exhale), which ensures adjusting the right dose to individual breathing way to achieve as high treatment effectiveness as possible (Denyer and Dyche 2010; Yeo et al. 2010; Dolovich and Dhand 2011; Sosnowski 2012; Zhou et al. 2014). The additional advantage of these devices is a minimum drug quantity getting to sur-roundings and a minimum risk of mistakenly conducted inhalation and lack of patient's age limits.

Soft mist inhaler (SMI) also belongs to the group of modern inhalers. The inhaler Respimat SMI by Boehringer Ingelheim is the example of such a device (Iacono et al. 2000; Dalby et al. 2004; Sosnowski 2012). Aerolization of a drug, being in a liquid form, takes place by pressure produced by using the mechanical energy to force drug solution through the inhaler nozzle (Dalby et al. 2004; Sosnowski 2012; Kwok and Chan 2014). The most important advantage of this inhaler is producing aerosol containing very high *Fine Particle Fraction* (*FPF*)–about 65% and *MMD* of 2 μm (for liquid solution) (Sosnowski 2012). Moreover, soft mist inhaler shows a very big effectiveness of drug deposition in lungs and it is simpler for a patient to use it properly (Dalby et al. 2004; Hess 2008; Hodder and Price 2009; Hodder et al. 2009; Sosnowski 2012).

Inhaler AERx of Aradigm also belongs to the new generation inhalers and enables to achieve a better capacity of inhalation than conventional devices. This inhaler (like soft mist inhaler) also generates aerosol in a form of mist. It means liquid flowing through a perforated plate by pressure (Sosnowski 2012; Chan et al. 2014). The advantages of this device are: monodisperse aerosol character, possibility to adjust aerosol drops size through size of hole diameters in the plate and also high efficiency of distribution of anesthetics and medicines in a form of proteins and colloids (Sosnowski 2012). It must be underlined that accuracy in holes making exerts influence on the character and monodisperse level of the generated aerosol. AREx inhalers are divided into electronic and mechanical ones.

The last inhaler belonging to unconventional devices is thermal inhaler which operation is based on innovative technology connected with condensation systems. The aerosol generating results from steam condensation in ambient air. The important feature of them is that they are good for atomization of drug both in liquid form (CAG, Capillary Aerosol Generator), and in solid (Staccato inhaler of Alexca Pharmaceuticals using sublimation/resublimation phenomena) (Sosnowski 2012). Inhalers of this type enable to produce monodisperse aerosol containing drops of very small sizes–1 μm. Device operational efficiency depends on several factors, heating power, flow rate of drug and capillary diameter, which gives the possibility of adjusting drop sizes, being among of them (Sosnowski 2012).

3   Methods of Increased Atomization Process Effectiveness

As mentioned before, for the last few decades a great growth in the number of new drugs and medical inhalers has taken place. Numerous research published referred to the improvement of atomization efficiency and increased treatment effectiveness in aerosol therapy. Some works were devoted to modifications and searched for new spray devices while the other referred to design engineering and testing of new drugs or research aiming at better understanding of complicated spray process.

One of the most important factors influencing effectiveness of aerosol therapy and the size of drops and drop size histogram is the temperature of the produced aerosol. The majority of available pneumatic nebulizers generate aerosol which temperature at the nozzle outlet is decreased to the temperature ranging between 10 and 15 °C. Much lower aerosol temperature in relation to human body temperature can arise side effects during inhalation having the form of the disturbance of air-passages mucus membrane homeostasis or bronchospasm. It is especially dangerous mainly for the newborns, babies and patients with hypersensitiveness, allergies and bronchial tree overreaction (Ferron and Roth 1998; Steckel and Eskandar 2003; Ochowiak et al. 2016). Steckel and Eskandar (2003) prove that, both, in the case of pneumatic nebulizer (PariBoy) and the ultrasonic one (Multisonic), you can see a visible temperature change of the sprayed liquid (lyophylisate resolved in water solution of 0.9% sodium chloride), which is presented in Fig. 2. During nebulization in Multisonic nebulizer the temperature increase of sprayed liquid is noticed from 24.53 to 45.9 °C, but in the case of PariBoy nebulizer the temperature decrease of liquid atomized takes place by about 8 °C.

The temperature decrease of the sprayed liquid in effect of pneumatic nebulizer was also registered by Ferron et al. (1976), Clay et al. (1983), Taylor et al. (1992), McCallion et al. (1996) and Beck-Broichsitter et al. (2012). Moreover, McCallion et al. (1996) described temperature influence on the size of drops and total volume of produced aerosol. Consequently to the aerosol temperature decrease, solubility of some drugs decreases (e.g. nebulized mucolytics or antimicrobials) with the growth of liquid viscosity and liquid surface tension present (Taylor et al. 1992).

In connection with that, aerosol with the temperature similar to human body temperature should be produced. According to Lewis (1983), the risk of bronchoconstriction disappears when aerosol temperature is 37 °C. Thermostating technique can be a solution to this problem. It allows to achieve aerosol of optimal temperature and, what is more, it contributes to decrease size of the created aerosol particles. It is worth stressing that smaller particles of thermal aerosol facilitate drug

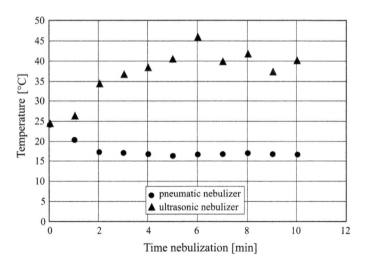

Fig. 2 The change in temperature of liquid during atomization by medical nebulizers (PariBoy and Multisonic) (Steckel and Eskandar 2003)

delivery to lower parts of respiratory tract (Lyutov 2006). Because of that, Ochowiak et al. (2016) designed, produced and tested a pneumatic nebulizer Microlife NEB 100 equipped with Philips Jet Pro nebulizer cup of a modified construction. The nebulizer cup was equipped with a heating element (resistance wire roll) and a temperature regulator enabling steering of the produced aerosol temperature. Following the research it was proven that growth of the sprayed liquid in the range from 20 to 70 °C contributes to a noticeable *SMD* decrease (from 16.3 to 2 μm) and to the growth of drops number with the diameter ≤ 5 μm from 34.8 to 100% (Table 1). Following the temperature increase, the drop size distribution has narrowed and moved towards lower values of drop diameters (Ochowiak et al. 2016).

Lots of research was conducted to estimate the impact of different factors on spray effectiveness in pneumatic nebulizers which proved, among others, that gas flow rate shows a noticeable impact on the size of formed drops and drop diameter distribution (Clay et al. 1983; Newman et al. 1986; Niven and Brian 1994; MacCallion et al. 1996; Flament et al. 1997; O'Callaghan and Barry 1997; Matuszak et al. 2017; Ochowiak and Matuszak 2017). Newman et al. (1986) conducted an analysis of atomization process of hypertonic saline solution (7%) for eleven pneumatic nebulizers of different constructions. In the paper, it was shown that gas flow rate increase (in range 6–12 l/min) contributes to *MMD* value decrease and an increase in the percentage of small particles with diameter not bigger than 5 μm. The dependence between aerosol drop size and volumetric gas flow rate for five different nebulizers is shown in Fig. 3. Moreover, a big differentiation is observed in case of achieved *MMD* values for particular nebulizer types, which probably results from different diameters of the air inlet of these devices. Then this fact results in different values of pressure decreases for the given gas flow rate.

Flament et al. (1997) proved that during solution atomization of α_1 protease inhibitor using the pneumatic nebulizer the increase of inhaled particle fraction volume is observed, and also the shortening of nebulizing time as a result of gas flow rate increase from 6 to 16 l/min and gas pressure increase from 0.5 to 2.5 bar. Clay's researches (1983) also show that gas flow rate increase (from 4 to 8 l/min) used in liquid atomization conducted in four pneumatic nebulizers caused decrease of the *Mass Median Aerodynamic Diameter* value, which corresponds to *Mass Median Diameter* by 50%.

Table 1 The values of *SMD* and $D_{\leq 5m}$ (Ochowiak et al. 2016)

Temperature, °C	*SMD*, m	$D_{\leq 5m}$, %
20	16.3	34.8
30	13.1	43.9
40	9.5	73.5
50	4.6	95.7
60	3.0	99.7
70	2.0	100

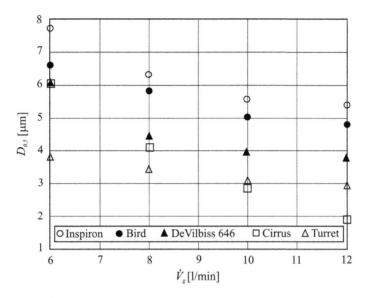

Fig. 3 Drop size dependence on gas flow rate (Newman et al. 1986)

Petersen (2004) and Markuszewska (2011) conducted researches to state impact air injection construction on atomization process effectiveness. To do that, the atomizer was equipped with an aerator (of different construction and holes number), which let deliver to the system a jet of gas aerating the liquid. The experimental data showed that additional liquid aeration causes loss of relatively big drops (diameter > 12 μm), and drop size distribution becomes more homogeneous.

In connection with that the authors Matuszak et al. (2017) and Ochowiak and Matuszak (2017) made a trial to increase effectiveness of pneumatic nebulization process (MedelJet Family nebulizer) by additional liquid aeration and regulating gas flow rate aerating the liquid. To implement it, four holes of the same diameters equal to 0.0014 m were made in the nebulizer cup, and four hoses were put into the holes to deliver to the system the additional gas source. The construction of the experimental place was described in Ochowiak and Matuszak's work (2017). The analysis of the obtained experimental results showed that the growth of volumetric gas flow rate in range of 0 to 8.33×10^{-5} m^3/s contributed to reduction of mean drop size values (D_{32} and $D_{0.5}$) (Fig. 4), and drop size histogram is narrowed and moved to lower values.

What is more, in the paper by Matuszak et al. (2017) it was proven that simultaneous implementation of additional liquid aeration and liquid heating (in temperature ranging from 20 to 60 °C) results in a decrease in the diameters of forming drops, the aerosol produced has monodisperse character and a drop size histogram becomes more and more homogeneous (Fig. 5). The distribution was made in MedelJet Family pneumatic nebulizer together with MedelJet Basic nebulizer cup and water or water solutions of 0.9% NaCl were used for atomization.

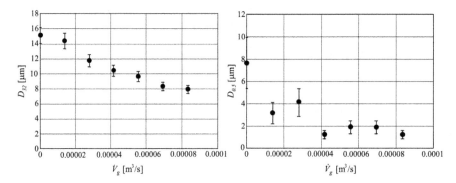

Fig. 4 Relationship between mean drop size values and volumetric gas flow rate (Ochowiak and Matuszak 2017)

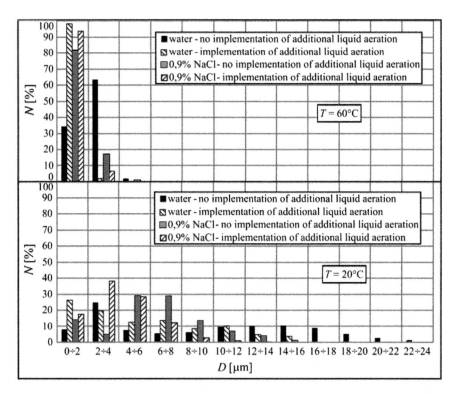

Fig. 5 The distributions of liquid at simultaneous implementation of additional liquid aeration and liquid heating (Matuszak et al. 2017)

4 Summary

Taking into consideration literature reports regarding aerosol therapy, you can come to the conclusion that nebulization constitutes an effective method in drug delivery to the respiratory system and can be widely implemented in medicine. This method has a lot of advantages and thanks to that it becomes more and more frequent often alternative to the treatment of diseases others than respiratory system diseases. It is worth stressing that aerosol therapy belongs to constantly developing areas, which results in the advent of new and more effective pharmaceuticals as well as modern devices used in drug atomization. It has been proven that factors having a considerable impact on the atomization process include: physicochemical properties of the atomized liquid, gas and liquid flow rate, liquid aeration and temperature. The attempts presented in the article and regarding nebulization effectiveness increase showed a visible decrease of drop size of the generated aerosol while drop size histograms became more homogenous. You must remember that atomization process is complex in character and the effectiveness of inhale therapy depends on many factors. Therefore further studies and improvement of atomization methods are necessary.

Acknowledgements This work was supported by PUT research grant no. 03/32/DSPB/0702.

References

Amirav I (2004) Aerosol therapy (Aerosol terapia). Ital J Pediatr 30:147–156

Andrade F, Rafael D, Videira M (2013) Nanotechnology and pulmonary delivery to overcome resistance in infectious diseases. Adv Drug Deliv Rev 65:1816–1827

Argent AC, Hatherill M, Newth CJL (2008) The effect of epinephrine by nebulization on measures of airway obstruction in patients with acute severe croup. Intensive Care Med 34:138–147

Ari A, Hess D, Myers TR (2009) A guide to aerosol delivery devices for respiratory therapists, 2nd edn. Am Assoc Resp Care, Texas

Ashgriz N (2011) Handbook of atomization and sprays. Theory and applications. Springer Science Business Media, LLC, New York

Beck-Broichsitter M, Kleimann P, Schmehl T (2012) Impact of lyoprotectants for the stabilization of biodegradable nanoparticles on the performance of air-jet, ultrasonic, and vibrating-mesh nebulizers. Eur J Pharm Biopharm 82:272–280

Berkenfeld K, Lamprecht A, McConville JT (2015) Devices for dry powder drug delivery to the lung. AAPS Pharm Sci Tech 16(3):479–490

Bisgaard H (1998) Targeting drugs to the respiratory tract. Res Immunol 149:229–331

Bivas-Benita M, Ottenhoff TH, Junginger HE et al (2005) Pulmonary DNA vaccination: concepts, possibilities and perspectives. J Control Release 107:1–29

Bosquillon C, Preat V, Vanbever R (2004) Pulmonary delivery of growth hormone using dry powders and visualization of its local fate in rats. J Control Release 96:233–244

Broniarz-Press L, Ochowiak M, Matuszak M et al (2014) The effect of shear and extensional viscosity on atomization in medical inhaler. Int J Pharm 468:199–206

Brown JS, Zeman KL, Bennett WD (2001) Regional deposition of coarse particles and ventilation distribution in healthy subjects and patients with cystic fibrosis. J Aerosol Med 14:443–454

Cassidy JP, Amin N, Marino M et al (2011) Insulin lung deposition and clearance following Technosphere®Insulin inhalation powder administration. Pharm Res 28:2157–2164

Cefalu WT (2007) The new diabetes inhalers: new tools for the clinician. Curr Diabetes Rep 7:165–167

Chan JGY, Wong J, Zhou QT et al (2014) Advances in device and formulation technologies for pulmonary drug delivery. AAPS Pharm Sci Tech 15(4):882–897

Chow AHL, Tong HHY, Chattopadhyay P et al (2007) Particle engineering for pulmonary drug delivery. Pharm Res 24(3):411–437

Cipolla DC, Gonda I (2011) Formulation technology to repurpose drugs for inhalation delivery. Drug Discov Today Ther Strateg 8(3–4):123–130

Clay MM, Pavia D, Newman SP (1983) Assessment of jet nebulisers for lung aerosol therapy. Lancet 322(8350):592–594

Dalby R, Spallek M, Voshaar T (2004) A review of the development of Respimat® Soft MistTM Inhaler. Int J Pharm 283:1–9

Denyer J, Dyche T (2010) The adaptive aerosol delivery (AAD) technology: past, present, and future. J Aerosol Med Pulm Drug Deliv 23(1):1–10

Depreter F, Amighi K (2010) Formulation and in vitro evaluation of highly dispersive insulin dry powder formulations for lung administration. Eur J Pharm Biopharm 76:454–463

Dhand R (2002) Nebulizers that use a vibrating mesh or plate with multiple apertures to generate aerosol. Resp Care 47:1406–1416

Dijk PH, Heikamp A, Piers DA et al (1997) Surfactant nebulisation: safety, efficiency and influence on surface lowering properties and biochemical composition. Intensive Care Med 23:456–462

Dilraj A, Cutts FT, de Castro JF et al (2000) Response to different measles vaccine strains given by aerosol and subcutaneous routes to schoolchildren: a randomised trial. Lancet 355:798–803

Dolovich MB, Dhand R (2011) Aerosol drug delivery: developments in device design and clinical use. Lancet 377:1032–1045

Elhissi AMA, Faizi M, Naji WF et al (2007) Physical stability and aerosol properties of liposomes delivered using an air-jet nebulizer and a novel micropump device with large mesh apertures. Int J Pharm 334:62–70

Elhissi A, Gill H, Ahmed W et al (2011) Vibrating-mesh nebulization of liposomes generated using an ethanol-based proliposome technology. J Liposome Res 21:173–180

Elliott D, Dunne P (2011) A guide to aerosol delivery devices for physicians, nurses, pharmacists, and other health care professionals. Am Assoc Resp Care, Texas

Ferron GA, Roth C (1998) A heat conservation model to estimate the temperature of aerosols from a jet nebulizer. J Aerosol Sci 29(1):763–764

Ferron GA, Kerrebijn KF, Weber J (1976) Properties of aerosols produced with three nebulizers. Am Rev Respir Dis 114:899–908

Flament MP, Leterme P, Burnouf T et al (1997) Jet nebulisation: influence of dynamic conditions and nebuliser on nebulisation quality. Application to the $\alpha 1$ protease inhibitor. Int J Pharm 148:93–101

Gill DR, Davies LA, Pringle IA et al (2004) The development of gene therapy for diseases of the lung. Cell Mol Life Sci 61:355–368

Gomez A (2002) The electrospray and its application to targeted drug inhalation. Respir Care 47:1419–1433

Hess DR (2008) Aerosol delivery devices in the treatment of asthma. Respir Care 53(6):699–725

Hodder R, Price D (2009) Patient preferences for inhaler devices in chronic obstructive pulmonary disease: experience with Respimat Soft Mist Inhaler. Int J Chron Obstructive Pulmon Dis 4:381–390

Hodder R, Reese PR, Slaton T (2009) Asthma patients prefer Respimat Soft Mist Inhaler to Turbuhaler. Int J Chron Obstructive Pulmon Dis 4:225–232

Iacono P, Velicitat P, Guemas E et al (2000) Improved delivery of ipratropium bromide using Respimat® (a new soft mist inhaler) compared with a conventional metered dose inhaler: cumulative dose response study in patients with COPD. Respir Med 94:490–495

Isaev IV, Morozov VY (2005) Dosing of low concentration anesthetics with evaporators used in inhalation anesthesia apparatuses. Biomed Eng 39(6):299–300

Jaworek A (2007) Micro- and nanoparticle production by electrospraying. Powder Technol 176:18–35

Johnson JC, Waldrep JC, Guo J et al (2008) Aerosol delivery of recombinant human DNase I: in vitro comparison of a vibrating-mesh nebulizer with a jet nebulizer. Respir Care 53 (12):1703–1708

Kamijyo A, Matsuzaki Z, Kikushima K et al (2001) Fosfomycin nebulizer therapy to chronic sinusitis. Auris Nas L 28:227–232

Kesser KC, Geller DE (2009) New aerosol delivery devices for cystic fibrosis. Respir Care 54 (6):754–767

Khilnani GC, Banga A (2008) Aerosol therapy. Indian J Chest Dis Allied Sci 50(2):209–220

Kleemann E, Dailey LA, Abdelhady HG et al (2004) Modified polyethylenimines as non-viral gene delivery systems for aerosol gene therapy: investigations of the complex structure and stability during air-jet and ultrasonic nebulization. J Control Release 100:437–450

Kwok PCL, Chan HK (2014) Delivery of inhalation drugs to children for asthma and other respiratory diseases. Adv Drug Deliv Rev 73:83–88

Lass JS, Sant A, Knoch M (2006) New advances in aerosolised drug delivery: vibrating membrane nebuliser technology. Expert Opin Drug Deliv 3:693–702

Lefebvre AH (1989) Atomization and sprays. Hemisphere Publishing Corporation, New York

Lentz YK, Worden LR, Anchordoquy TJ (2005) Effect of jet nebulization on DNA: identifying the dominant degradation mechanism and mitigation methods. Aerosol Sci 36:973–990

Lentz YK, Anchordoquy TJ, Lengsfeld CS (2006) Rationale for the selection of an aerosol delivery system for gene delivery. J Aerosol Med 19:372–384

Lewis RA (1983) Nebulisers for lung aerosol therapy. Lancet 322(8354):849

Lewis JF, Ikegami M, Jobe AH (1991) Aerosolized surfactant treatment of preterm lambs. J Appl Physiol 70(2):869–876

Lewis JF, Tabor B, Ikegami M et al (1993) Lung function and surfactant distribution in saline lavaged sheep given instilled vas nebulized surfactant. J Appl Physiol 74(3):1256–1264

Licalsi C, Christensen C, Bennett T et al (1999) Dry powder in-halation as a potential delivery method for vaccines. Vaccine 17:1796–1803

Longest PW, Spence BM, Holbrook LT et al (2012) Production of inhalable submicrometer aerosols from conventional mesh nebulizers for improved respiratory drug delivery. J Aerosol Sci 51:66–80

Luisettia M, Kroneberg P, Suzuki T et al (2011) Physical properties, lung deposition modeling, and bioactivity of recombinant GM-CSF aerosolised with a highly efficient nebulizer. Pulm Pharmacol Ther 24:123–127

Lyutov GP (2006) Methods for increasing the efficiency of aerosol inhalers. Biom Eng 40(1):1–3

Marianecci C, Di Marzio L, Rinaldi F et al (2011) Pulmonary delivery: innovative approaches and perspectives. J Biomater Nanobiotechnol 2:567–575

Markuszewska M (2011) Projekt dyszy pęcherzykowej. Poznan University of Technology, Poznan

Matuszak M, Ochowiak M, Włodarczak S (2017) Metody poprawy efektywności procesu rozpylania w nebulizatorach pneumatycznych. Inż Ap Chem 56(3):90–91

McCallion ONM, Taylor KMG, Thomas M et al (1995) Nebulization of fluids of different physicochemical properties with air-jet and ultrasonic nebulizers. Pharm Res 12(11):1682–1688

McCallion ONM, Taylor KMG, Bridges PA et al (1996) Jet nebulisers for pulmonary drug delivery. Int J Pharm 130:1–11

Mohajel N, Najafabadi AR, Azadmanesh K et al (2012) Optimization of a spray drying process to prepare dry powder microparticles containing plasmid nanocomplex. Int J Pharm 423:577–585

Najlah M, Vali A, Taylor M et al (2013) A study of the effects of sodium halides on the performance of air-jet and vibrating-mesh nebulizers. Int J Pharm 456:520–527

Najlah M, Parveen I, Alhnan MA et al (2014) The effects of suspension particle size on the performance of air-jet, ultrasonic and vibrating-mesh nebulisers. Int J Pharm 461:234–241

Newman SP, Pellow PGD, Clarke SW (1986) Droplet size distributions of nebulised aerosols for inhalation therapy. Clin Phys Physiol Meas 7(2):139–146

Niven RW, Brain JD (1994) Some functional aspects of air-jet nebulizers. Int J Pharm 104:73–85

O'Callaghan C, Barry PW (1997) The science of nebulised drug delivery. Thorax 52(2):31–44

Ochowiak M, Matuszak M (2017) The effect of additional aeration of liquid on the atomization process for a pneumatic nebulizer. Eur J Pharm Sci 97:99–105

Ochowiak M, Doligalski M, Broniarz-Press L et al (2016) Characterization of sprays for thermo-stabilized pneumatic nebulizer. Eur J Pharm Sci 85:53–58

Odziomek M, Sosnowski TR, Gradoń L (2012) Conception, preparation and properties of functional carrier particles for pulmonary drug delivery. Int J Pharm 433:51–59

Olveira C, Muñoz A, Domenech A (2014) Nebulized therapy. SEPAR Year. Arch Bronconeumol 50(12):535–545

Pahwa R, Pankaj G, Singh M et al (2012) Nebulizer therapy: a platform for pulmonary drug delivery. Der Pharmacia Sinica 3(6):630–636

Park JH, Jin HE, Kim DD et al (2013) Chitosan microspheres as an alveolar macrophage delivery system of ofloxacin via pulmonary inhalation. Int J Pharm 441:562–569

Petersen FJ (2004) A new approach for pharmaceutical sprays. Effervescent atomization. atomizer design and spray characterization. Ph.D. thesis, Department of Pharmaceutics, The Danish University of Pharmaceutical Sciences, Copenhagen

Pilcer G, Amighi K (2010) Formulation strategy and use of excipients in pulmonary drug delivery. Int J Pharm 392:1–19

Robbins PD, Evans CH, Chernajovsky Y (2003) Gene therapy for arthritis. Gene Ther 10:902–911

Salama RO, Young PM, Traini D (2014) Concurrent oral and inhalation drug de-livery using a dual formulation system: the use of oral theophylline carrier with combined inhalable budesonide and terbutaline. Drug Deliv and Transl Res 4:256–267

Shah S (2011) Exogenous surfactant: intubated present, nebulized future? World J Pediatr 7(1): 11–15

Sheth P, Stein SW, Myrdal PB (2015) Factors influencing aerodynamic particle size distribution of suspension pressurized metered dose inhalers. AAPS Pharm Sci Tech 16(1):192–201

Shoyele SA, Cawthorne S (2006) Particle engineering techniques for inhaled bio-pharmaceuticals. Adv Drug Deliv Rev 58:1009–1029

Shoyele SA, Slowey A (2006) Prospects of formulating proteins/peptides as aerosols for pulmonary drug delivery. Int J Pharm 314:1–8

Soll RF, Merritt TA, Hallman M (1994) Surfactant in the prevention and treatment of respiratory distress syndrome. In: Boynton BR (ed) New therapies for neonatal respiratory failure. Cambridge University Press, Cambridge, pp 49–80

Sosnowski TR (2012) Aerozole wziewne i inhalatory. Warsaw University of Technology, Warsaw

Steckel H, Eskandar F (2003) Factors affecting aerosol performance during nebulization with jet and ultrasonic nebulizers. Eur J Pharm Sci 19:443–455

Taylor KMG, Venthoye G, Chawla A (1992) Pentamidine isethionate delivery from jet nebulisers. Int J Pharm 85:203–208

Terzano C, Allegra L (2002) Importance of drug delivery system in steroid aerosol therapy via nebulizer. Pulm Pharmacol Ther 15:449–454

Vaughan WC (2004) Nebulization of antibiotics in management of sinusitis. Curr Infect Dis Rep 6:187–190

Vecellio L, Abdelrahim ME, Montharu J et al (2011) Disposable versus reusable jet nebulizers for cystic fibrosis treatment with tobramycin. J Cyst Fibros 10:86–92

Wang YB, Watts AB, Peters JI et al (2014) In vitro and in vivo performance of dry powder inhalation formulations: comparison of particles prepared by thin film freezing and micronization. AAPS Pharm Sci Tech 15(4):281–993

Wanning S, Süverkrüp R, Lamprecht A (2015) Pharmaceutical spray freeze drying. Int J Pharm 488:136–153

Willis L, Hayes DJr, Mansour HM (2012) Therapeutic liposomal dry powder inhalation aerosols for targeted lung delivery. Lung 190:251–262

Yeo LY, Friend JR, McIntosh MP et al (2010) Ultrasonic nebulization platforms for pulmonary drug delivery. Exp Opin Drug Deliv 7(6):663–679

Zhou QT, Tang P, Leung SSY et al (2014) Emerging inhala-tion aerosol devices and strategies: where are we headed? Adv Drug Deliv Rev 75:3–17

Zou Y, Tornos C, Qiu X et al (2007) p53 aerosol formulation with low toxicity and high efficiency for early lung cancer treatment. Clin Cancer Res 13:4900–4908

Hydraulic Mixing

Piotr Tomasz Mitkowski, Waldemar Szaferski and Mariusz Adamski

1 Introduction

Although mixing is one of the basic unit operations in process industries there is still a need for more economical and safer alternatives tailored to the specific requirements. Mixing is usually performed with the use of various agitators in stirred tanks, pipelines, packed beds, fluidized beds, circulation loops, columns with various internals, etc. (Smith 2005). Apart from mechanical mixing, by using different agitators, more and more varied kinds of static mixers or other kinds of internals are used in the industries (Thakur et al. 2003; Ghanem et al. 2014). The static mixers constitute a part of the broader group of motionless mixers, which can be defined as unit operation devices which are able to mix fluids without parts moving within the mixed medium (Mitkowski et al. 2016). On other hand, there is a concept of passive micromixers, which are defined as micromixers utilizing no energy input other than the pressure head used to drive the fluid at a constant rate (Lee et al. 2016).

In general, the motionless mixing technologies are desired when mild sheering forces (e.g. biologically active or fragile ingredients), restriction to generation of static electricity, exhaust gases and/or aerosol occur. In many cases the motionless

P. T. Mitkowski (✉) · W. Szaferski
Institute of Chemical Technology and Engineering, Faculty of Chemical Technology,
Poznan University of Technology, Poznań, Poland
e-mail: piotr.mitkowski@put.poznan.pl

M. Adamski
Institute of Biosystems Engineering, Faculty of Agriculture and Bioengineering,
Poznan University of Life Sciences, Poznań, Poland

© Springer International Publishing AG, part of Springer Nature 2018
M. Ochowiak et al. (eds.), *Practical Aspects of Chemical Engineering*,
Lecture Notes on Multidisciplinary Industrial Engineering,
https://doi.org/10.1007/978-3-319-73978-6_20

or passive mixing technologies can be treated as inherently safer processes. Within that group of mixing technologies three main groups can be classified: static mixers, pneumatic mixers and jet mixers which were widely discussed by various researchers (Bałdyga et al. 1995; Forney and Nafia 1998; Patwardhan and Gaikwad 2003; Espinosa-Solares et al. 2008; Katoh and Yoshida 2010; Xu and Huang 2012; Karcz 2013; Woziwodzki 2013).

This work presents a state-of-the-art of group of mixing devices named hydraulic mixers (HM) which belong to motionless mixing technologies. Therefore, the basis of hydraulic mixing process and assessment of its performance are presented.

2 Materials and Methods

Visualization technique based on the dilution of dye is used in assessment of mixing performance. The use of various dyes for visualization of mixing has been widely accepted as the technique for flow analysis (Hardt and Schönfeld 2003; Schönfeld et al. 2004; Ghanem et al. 2014). In the cases discussed in this chapter, the 3 ml of aqueous solution of malachite green dye containing 20 w% of malachite green in distilled water was injected with the use of a surgical needle. The dye solution was delivered at either one of the two places in the bottom of HM, i.e. in the centre of the HM bottom(abbreviated as CoM in Fig. 1) and in the half-distance between the outer surface of inner cylinder and the inner surface of outer cylinder (abbreviated as HoM, Fig. 1). The mixing of dye solution in HM has been recorded with the use of digital camera (Canon EOS 60D). The movies had a rate of 50 fps and the resolution of 1280×720 pixels and were saved in *.mov files. The analysis of the movies recorded was performed with the assistance of the following software: FFmpeg (Bellard 2015) and ImageJ 1.48 (National Institutes of Health 2015) with Bio-Formats 5.1.0 (Open Microscopy Environment 2015) plug-in. More details about experimental procedure are discussed in (Mitkowski et al. 2016).

The visual assessment of mixing performance remains troublesome because of various effects, which are discussed in the following sections. Two characteristic measures have been introduced: process time and height of dye. The process time is the time at which dye reaches the gas-liquid interface and the liquid has a uniform colour throughout the liquid zones. The height of dye represents the height of uniform colour zone counted from the bottom of hydraulic mixer. The height of dye reported here, if not stated otherwise, represents the arithmetical average value of the measured heights of dye in six places: next to the outer walls of outer compartment, in the middle between walls of outer compartment and next to the inner walls of outer compartment.

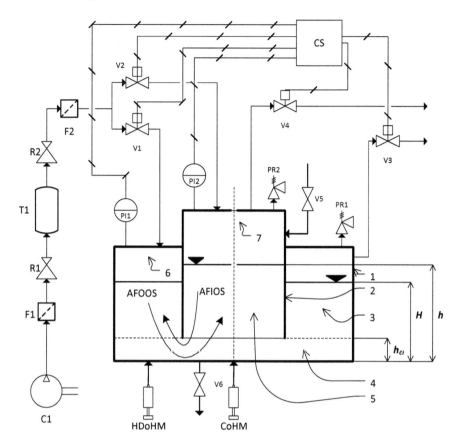

Fig. 1 Experimental set-up of hydraulic mixer set-up (based on Mitkowski et al. 2016). *C1* compressor; *F1*, *F2* air filter; *T1* gas reservoir (27.2 dm^3); *R1*, *R2* pressure reducing valves; *V1* air inlet valve to outer compartment; *V2* air inlet to inner compartment; *V3* air outlet valve from outer compartment; *V4* air outlet valve from inner compartment; *V5*, *V6* communicated valves; *PR1* pressure relieve valve in outer compartment; *PR2* pressure relieve valve in inner compartment; *PI1* pressure meter in outer compartment; *PI2* pressure meter in inner compartment; *CS* control system connected with PC; *1* hydraulic mixer casing; *2* casing of inner compartment; *3* liquid volume in outer compartment with liquid height *H*; *4* common volume with height h_{cl}; *5* liquid volume in inner compartment with liquid height *h*; *6* gas volume in outer comportment; *7* gas volume in inner compartment; *AFOOS* annular flow out of outer space; *AFIOS* annular flow into outer space, *CoHM* centre place of dye injection, *HDoHM* half-distance of dye injection

3 Experimental Set-up and Configuration

The hydraulic mixer (HM) is a tank-based equipment in which three characteristic volumes can be distinguished, namely outer and inner compartments and common volume as it is schematically presented in Fig. 1. The distinction of two compartments is made because of the concentrically located inner tube in outer tube.

Due to that, the HM has three defined cross-sections: (1) the annular cross-section in outer compartment, (2) the circular cross-section in inner compartment, and (3) the tube-like cross-section of common circular space. The common circular space has 194 mm in diameter and default height of 20 mm. That height is called the clearance height (h_c). The clearance can be increased stepwise before performing experiments by 20 mm from 20 mm up to 60 mm. The set of feeding and releasing valves (V1, V2 and V3, V4, respectively) is controlled by computer-aided system (CS) which utilize pressure measurements from pressure indicators (PI1 and PI2) to open and close valves in order to obtain:

- annular flow into outer space (AFIOS), and
- annular flow out of outer space (AFOOS).

The AFIOS flow is obtained when the high pressure is applied to inner compartment (7) by delivering gas through valve V2. That increase causes the liquid flow from inner compartment (5) to outer compartment (3). It is also manifested by a decrease of liquid height h in inner compartment (5) and increase of liquid height H in outer compartment (3). The AFOOS flow is obtained when higher pressure is applied in outer space (6), causing the decrease of liquid height H in outer compartment (3) and increase of h in inner compartment (5). It has to be kept in mind that superficial flows in inner and outer compartments are perpendicular to the gas-liquid interface whereas in common volume (4) is perpendicular to the clearance.

4 Parameters Affecting Hydraulic Mixing

Parameters influencing the efficiency of mixing in HM can be divided into three groups: (1) geometric invariants, (2) fluid properties and (3) process parameters. The first group consists of internal tank diameter (D_{in}), outer diameter of inner compartment (d_{out}), height of clearance (h) and available height of compartments for liquid (h_{av}). In general, the second group consists of set of physical properties such as density and viscosity. The parameters gathered in the third group are related to the inducted pressure changes such as inlet pressure, relaxation pressure, release pressure, relaxation time, liquid height, pressure in inner and outer compartments and temperatures of liquid and inlet gas. In general, the hydraulic mixing is a result of periodic pressure changes in outer and inner compartments, caused by inlet and outlet flows of gas, causing changes of liquid height in inner (h) and outer compartment (H). The periodic changes of liquid levels in conjunction with pressure changes are schematically presented in Fig. 2.

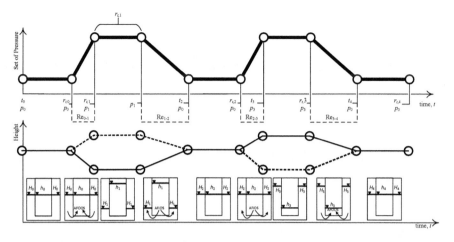

Fig. 2 Schematic representation of pressure set changes and accompanied variation of liquid level

5 Local, Oscillatory and Overall Reynolds Numbers Versus Experimental Results

The classical definition of Reynolds number was coined in 1908 by Arnold Sommerfeld (Rott 1990) who probably based on the work published by Reynolds (1883) which was, on the other hand, inspired by (Stokes 1851) and is represented by Eq. (1). In that equation w is linear superficial net flow velocity, d_e is characteristic dimension for the flow, ρ is density and η is dynamic viscosity. In general, it can be stated that the Reynolds number provides the dimensionless characterization of the flow in a continuous process.

$$Re = \frac{w d_e \rho}{\eta} \tag{1}$$

Since 1908 many modifications of Reynolds numbers have been proposed. For instance, when mechanical mixing is considered, the Reynolds number was defined (Hixson et al. 1937) as a function of rotational speed and diameter of an agitator, as presented by Eq. (2).

$$Re_{mix} = \frac{n d_{ag}^2 \rho}{\eta} \tag{2}$$

On the other hand, when oscillated or pulsated mixing processes, for example pulsed tube, are considered, the oscillatory Reynolds number (3) was presented and used by Fabiyi and Skelton (2000).

$$\text{Re}_o = \frac{2\pi f x_0 \rho D}{\eta} \tag{3}$$

where D is the column diameter (m), x_o is the oscillation amplitude (m) and f is the oscillation frequency (Hz). The product of $2\pi f x_o$ is the maximum oscillatory velocity (m s^{-1}). However, Ni and co-authors (2003) stated that in an analysis of pulsed columns along with oscillatory Reynolds number, the Strouhal number (Eq. 4) is needed. Strouhal number was also used in an analysis of forward-reverse mixing in vessel (Woziwodzki 2011).

$$St = \frac{D}{4\pi x_0} \tag{4}$$

However, considering the hydraulic mixing specification it is possible to calculate the local Reynolds number for specific time steps and oscillatory Reynolds number for the specific experimental configuration and cycle. The local Reynolds numbers for three characteristic dimensions $d_{e,1}$, $d_{e,2}$ and $d_{e,4}$ [which corresponds to Eqs. (5–7)] at four specific times i.e. t_1, t_2, t_3 and t_4 (presented in Fig. 2).

$$d_{e,1} = \frac{D_{in}^2 - d_{out}^2}{D_{in} + d_{out}} \rightarrow \text{Re}_{1,i} = \frac{\rho}{\eta} \frac{H_T}{t_{s,i}} \frac{D_{in}^2 - d_{out}^2}{D_{in} + d_{out}} \tag{5}$$

$$d_{e,2} = d_{in} \rightarrow \text{Re}_{2,i} = \frac{\rho}{\eta} \frac{h_T}{t_{s,i}} d_{in} \tag{6}$$

$$d_{e,4} = \frac{2\pi \overline{d_{in}} h_{cl}}{\pi \overline{d_{in}} + h_{cl}} \rightarrow \text{Re}_{4,i} = \frac{\rho}{\eta} \frac{h_T}{2 t_{s,i}} \frac{\pi d_{in}^2}{\pi \overline{d_{in}} + h_{cl}} \tag{7}$$

On other hand, the oscillatory Reynolds number can be calculated for two characteristic dimensions $d_{e,1}$ and $d_{e,2}$ under assumptions that the period of oscillation is equal to the time of a cycle and the amplitude is equal to the half of the distance which liquid-gas interface moved in a cycle in outer and inner compartments, which corresponds to Eqs. (8) and (9).

$$\text{Re}_{o,1} = \frac{\pi t_{cycle} (H_{max,i} - H_{min,i}) \rho d_{e,1}}{\eta}; \quad St_{o,1} = \frac{d_{e,1}}{4\pi t_{cycle}} \tag{8}$$

$$\text{Re}_{o,2} = \frac{\pi t_{cycle} (h_{max,i} - h_{min,i}) \rho d_{e,2}}{\eta}; \quad St_{o,2} = \frac{d_{e,2}}{4\pi t_{cycle}} \tag{9}$$

where: H_{max}—maximum level of liquid in outer compartment, H_{min}—minimum level of liquid in outer compartment, h_{max}—maximum level of liquid in inner compartment, h_{min}—minimum level of liquid in inner compartment,

Despite the original definition of Reynolds number or any of the above modifications, these Reynolds numbers are considering continuous or periodical but instantly changing processes. However, when considering the hydraulic mixing as time dependent, non-continuous and lacking single characteristic dimension process either of Reynolds number defined by Eqs. (1) and (3) can be used, therefore Mitkowski and co-authors (2016) proposed the modified Reynolds number for hydraulic mixing $(\overline{Re_{HM}})$ represented by Eq. (10) with Eq. (11).

$$\overline{Re_{HM}} = \frac{\rho}{\eta} \frac{4\overline{d_{in}}h_{cl}}{D_w + d_{out} + d_{in}} \sum_{\substack{i=1 \\ j=1}}^{n} \left(\Delta h_{s,i,j} \frac{t_{cycle,j}}{t_{s,i}t_p} \right) \tag{10}$$

$$\Delta h_{s,i,j} = \left| H_{in,s,i,j} - H_{f,s,i,j} \right| + \left| h_{in,s,i,j} - h_{f,s,i,j} \right| \tag{11}$$

where: d_{in}—inner diameter of inner compartment (m), $\overline{d_{in}}$—mean diameter of inner compartment (m), D_{in}—inner diameter of outer compartment (m), d_{out}—outer diameter of inner compartment (m), h_{cl}—height of clearance (m), $t_{cycle,i}$—time of i-th cycle (s), $t_{s,i}$—characteristic time of a step during an i-th stage (s), t_p—hydraulic mixing process time (s), h—height of liquid in inner compartment (m), H—height of liquid in outer compartment (m), $h_{f,s,i}$—final height of liquid in inner compartment in a step i (m), $H_{f,s,i}$—final height of liquid in outer compartment in a step i (m), $h_{in,s,i}$—initial height of liquid in inner compartment in a step i (m), $H_{in,s,i}$—initial height of liquid in outer compartment in a step i (m).

The details of all sets of twelve experiments discussed here are listed in Table 1 along with the values of local Reynolds number calculated for each characteristic step in hydraulic mixing process together with local oscillatory Reynolds number and modified Reynolds number for hydraulic mixing. At first, it has to be pointed out that the time of mixing (t_p) depends on the place where dye has been injected. The shortest time to obtain well mixing was reached for experiment No. 1 (177.88 s) while the longest one for experiment No. 5 (2423.76 s). For experiments in which clearance height has been set up to 20 and 40 mm the times of mixing are shorter when dye is injected centrically (CoM) than to outer compartment (HoM). Different behaviour has been observed for experiments with the clearance height equal to 60 mm, i.e. experiments 5, 6, 11 and 12. The shortest process time for those experiments was obtained when the dye solution has been injected to outer comportment and $H_{c,o}/D$ was equal to 1 (experiment No. 11). The longest time, equal to 2423.76 s, was obtained for experiment No. 5, which was mainly due to the occurrence of not well coloured region in outer compartment close to the gas-liquid interface. Similar observation has been noted for experiment No. 6 (Fig. 7a–d) and No. 5 (Fig. 7e–h).

The characteristic local Reynolds numbers, which are presented in Table 1, show that the highest values of Reynolds numbers are achieved for $Re_{2,t3}$. Most of the time, flows during performed experiments have laminar character, except $Re_{2,t1}$, and in most of $Re_{1,t3}$, $Re_{4,t3}$ and $Re_{2,t4}$. The minimum Reynolds number equal to 280 ($Re_{1,t2}$, exp. No. 3&9) is achieved during equalization to atmospheric pressure in outer compartment. The maximum value of 10,494 ($Re_{2,t3}$, exp. No. 5&11) is

Table 1 Details of performed experiments and comparison of local and oscillatory Reynolds number with modified Reynolds number for hydraulic mixing

No.[a]	$H_{c,o}/D$	h_{cl} (m)	P (hPa)	No. of cycles CoM	No. of cycles HoM	t_p (s) CoM	t_p (s) HoM	Local Reynolds numbers $Re_{1,t1}$	$Re_{2,t1}$	$Re_{4,t1}$	$Re_{1,t2}$	$Re_{2,t2}$
1&7	1	0.02	135	5	9	177.88	329.96	1191	**3587**	1650	535	1607
2&8	2	0.02	670	10	19	494.88	991.96	925	**2579**	1186	643	1911
3&9	1	0.04	110	12	29	394.00	941.14	1519	**4098**	1777	280	776
4&10	2	0.04	570	15	22	694.24	1062.06	1085	**3163**	1371	765	2252
5&11	1	0.06	100	66	15	2423.76	549.088	1724	**4413**	1809	291	745
6&12	2	0.06	560	29	15	1139.44	590.36	1114	**3025**	1240	1088	**2598**

No.[a]	Local Reynolds numbers $Re_{4,t2}$	$Re_{1,t3}$	$Re_{2,t3}$	$Re_{4,t3}$	$Re_{1,t4}$	$Re_{2,t4}$	$Re_{4,t4}$	$Re_{0,1}$	$Re_{0,2}$	$St_{0,1} \times 10^5$	$St_{0,2} \times 10^5$	$\overline{Re_{HM}} \times 10^{-4}$ CoM	HoM
1&7	739	**2782**	**8317**	**3825**	576	1832	843	523	1568	3.34	3.38	1.17	1.71
2&8	879	1529	**4577**	**2105**	1294	**3920**	1803	948	2800	1.51	1.61	2.18	3.77
3&9	336	**3216**	**10,275**	**4454**	630	2011	872	493	1496	3.86	3.83	3.08	4.18
4&10	976	2049	**6551**	**2840**	1593	**5122**	2220	918	2846	1.63	1.60	6.34	9.58
5&11	305	**3280**	**10,494**	**4302**	600	1830	750	370	1104	3.26	3.38	6.12	4.48
6&12	1065	**2661**	**8366**	**3429**	**2425**	**6862**	**2813**	991	2981	1.51	1.55	3.12	2.74

[a]Numbers of experiments are ordered that the first number represent experiment with dye injection in CoM while the second number with dye injection in HoM; $H_{c,o}$ initial height of liquid in HM, D inside diameter of HM, *CoM* injection of dye solution in the center of HM, *HoM* injection of dye solution in the half distance between inner wall of outer compartment and outside wall of inner compartment, h_{cl} height of clearance (m), P maximum applied pressure during hydraulic mixing (hPa), t_p process time (s). *Bold font* turbulent flow, *italic font* transient flow, *normal font* laminar flow
Experimental data for experiments 1, 2, 3, 4, 7, 8, 9 have been published in Mitkowski et al. (2016)

obtained in inner compartment, while the high pressure is applied to outer compartment. The turbulent flow is dominant in inner compartment when high pressure is applied, regardless of which compartment pressure is applied to; therefore uniform mixing in inner compartment is achieved faster than in outer compartment (Mitkowski et al. 2016) It is noticeable that local Reynolds number calculated for the outer compartment (i.e. $Re_{1,t1}$, $Re_{1,t2}$, $Re_{1,t4}$) represent rather laminar flow except when high pressure is applied to inner compartment (i.e. $Re_{1,t3}$). Although, the calculated local Reynolds numbers in clearance usually present laminar flow, in a few crossing 2500, the liquid flowing through the clearance changes the flow direction and provides turbulence to the outer compartment. In general, during levelling (t_2 and t_4) the local Reynolds numbers achieved values characteristic for laminar flow except the $Re_{2,t4}$ for experiments 2&8, 4&10 and 8&12, and for $Re_{2,t2}$ in experiments 6&12, which is related to faster levelling due to higher hydraulic pressure caused by the higher ratio of $H_{c,0}/D$ and the highest value of clearance. Although laminar flows present in outer compartment, with lowest Reynolds numbers, are obtained for experiments with $H_{c,0}/D$ equal to 1, the uniform mixing is achieved faster than for $H_{c,0}/D$ equal to 2, which could be explained by the high value of transferred volume from one compartment to another. Therefore, the use of the local values of Reynolds number do not provide information about the total performance of hydraulic mixing (Mitkowski et al. 2016).

When hydraulic mixing is considered as the oscillatory mixing, calculated values of oscillatory Reynolds number do not reach 3000 in inner compartment and 1000

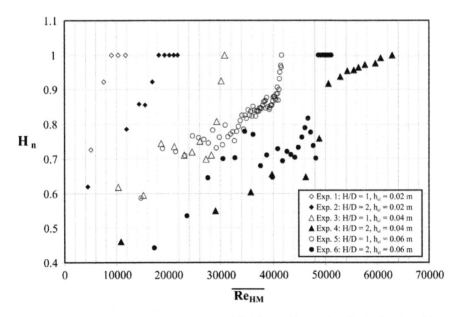

Fig. 3 Normalized height in function of modified Reynolds number for hydraulic mixing. Experimental data for experiments 1, 2, 3, 4, 7, 8, 9 have been published in (Mitkowski et al. 2016)

in outer compartment (see Table 1). From that point of view, it is clear that mixing in inner compartment is better than in outer one. On the other hand, the Strouhal number is very low, because it remains within the range from 1.51×10^{-5} to 3.86×10^{-5}. Such low values could suggest that the effective eddy propagation is limited.

In order to compare, experimental results of hydraulic mixing differing in initial volume of mixed liquid and height of clearance which are reported in Table 1. The results are presented in Fig. 3 as the function of normalized height of dye (H_n, Eq. 12) and the modified Reynolds number for hydraulic mixing ($\overline{Re_{HM}}$).

$$H_n = \frac{H_{d,n}}{H_{c,0}} \qquad\qquad (12)$$

where: $H_{d,n}$—height of dye (m), $H_{c,0}$—initial liquid height (m).

It has to be noted that the tracking of dye for experiments in which dye was injected to outer compartment is difficult because in both compartments the unmixed regions are observed. Therefore, the course of normalized heights of dye in experiments 1–6 is presented in Fig. 3. It is important that for all experiments, except No. 1, the plateaus of H_n between 0.6 and 0.85 were observed. It is significant that for $H_{c,0}/D = 2$ and height of clearance equal to 20 mm only one plateau is observed which is at around $H_n = 0.86$. With the increase of clearance to 40 mm at $H_{c,0}/D = 1$, the plateau at around $H_n = 0.6$ and $H_n = 0.75$ are observed. For $H_{c,0}/D = 2$ only one plateau is found (exp. No. 2 and 3, in Fig. 3) although for higher values of $\overline{Re_{HM}}$ the step-like increase is observed (exp. 4 and 5, Fig. 3) Very interesting results have been found for clearance height equal to 60 mm which caused for $H_{c,0}/D = 1$ a significant increase in $\overline{Re_{HM}}$ required to reach well mixing but, on the other hand, allowed to decrease $\overline{Re_{HM}}$ for $H_{c,0}/D = 2$, although the process time increased.

6   Observed Phenomena

During performed experiments various physical phenomena, such as formation of concave-convex structures at liquid-gas interface in inner compartment during air delivering, chaotic strokes of coloured liquid and separated coloured zones, were observed. While delivering air into the inner compartment (t_3 in Fig. 2), the formation of concave-convex structures at liquid interface was observed, which is presented in frame-by-frame mode with 0.02 s time step in Fig. 4. When the liquid level in inner compartment approaches its lower limit, the concave and convex structures are formed and they are diminishing one after the other (Fig. 4b–e), maximum of which was manifested by the formation of centrically located peak (Fig. 4g). That peak, at a first glance, seems to disintegrate from the liquid surface and form a droplet. The phenomena occur sequentially as shown by Fig. 4a–l.

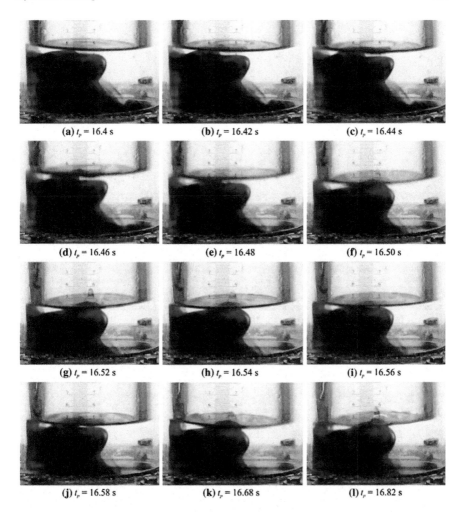

(a) t_p = 16.4 s   **(b)** t_p = 16.42 s   **(c)** t_p = 16.44 s

(d) t_p = 16.46 s   **(e)** t_p = 16.48   **(f)** t_p = 16.50 s

(g) t_p = 16.52 s   **(h)** t_p = 16.54 s   **(i)** t_p = 16.56 s

(j) t_p = 16.58 s   **(k)** t_p = 16.68 s   **(l)** t_p = 16.82 s

Fig. 4 Formation of a convex–concave liquid interface and waving in inner compartment (Exp. 3, cycle 1, step 10). Taken with permission from (Mitkowski et al. 2016)

According to (Mitkowski et al. 2016), the phenomenon of forming the liquid peak could be due to two processes occurring in the system. The first one is the compression of air in both compartments and the second one is the liquid flow behaviour in the clearance. The initially observed waving of air-liquid interface in compartments could be related to a sudden acceleration of liquid, which is than transferred to the concave-convex structures due to more steady flow and finally diminishing of the concave-convex structure due to slowing down acceleration to the final stop. It is important to notice that waving of interfacial surface continues up to five seconds after the air inlet is closed. That observation suggests that there is

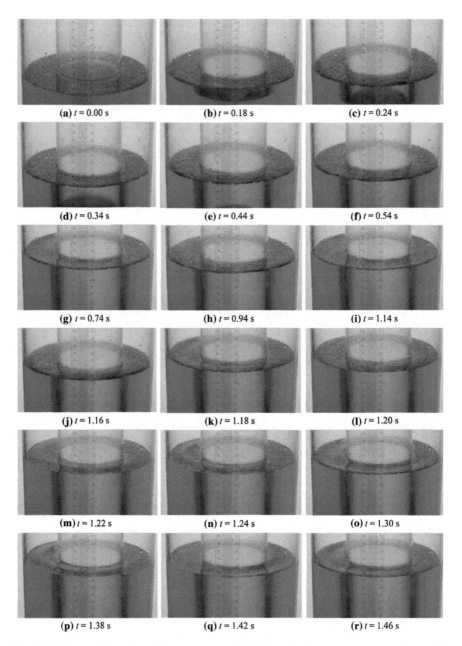

(a) $t = 0.00$ s   (b) $t = 0.18$ s   (c) $t = 0.24$ s

(d) $t = 0.34$ s   (e) $t = 0.44$ s   (f) $t = 0.54$ s

(g) $t = 0.74$ s   (h) $t = 0.94$ s   (i) $t = 1.14$ s

(j) $t = 1.16$ s   (k) $t = 1.18$ s   (l) $t = 1.20$ s

(m) $t = 1.22$ s   (n) $t = 1.24$ s   (o) $t = 1.30$ s

(p) $t = 1.38$ s   (q) $t = 1.42$ s   (r) $t = 1.46$ s

Fig. 5 Waving and formation of convex–concave liquid interface in outer compartment in step 10 (time counted from starting a step). Taken with permission from (Mitkowski et al. 2016)

an axial downward flow in inner compartment which is assisted with radial interfacial dispersion. The waving is more intensive in inner compartment than in outer one (Fig. 5), therefore these observations justify that better and faster mixing occurs in inner volume, which is also supported by the observed differences in mixing times between experiments with delivering the dye solution to inner or outer compartments (Table 1).

The explanation of occurring the long-time unmixed zones, as those observed in Fig. 6a–d, can be related to the phenomenon of creating chaotic streaks which form pathways of liquid containing dye into upper parts of hydraulic mixer. Figure 6a, b deliver important observation that streaks are more intense in close neighbourhood of outer wall of outer comportment which stays in line with the observed more intensive dye colour close to outer wall than to inner one (Fig. 7a–d).

Although, when closer look is taken to Fig. 7e, f it seems that close to the gas-liquid interface, i.e. 2–3 cm from the interface, the coloured liquid travel close to the outer wall of inner compartment.

In Fig. 7 another interesting phenomenon can be observed, namely, the separated zone which consists of at least two coloured zones separated by one zone which does not contain dye (Fig. 7e–h). It is important to note, that for experiments with $H_{c,0}/D = 1$ there is only one zone which does not contain dye, while for those with $H_{c,0}/D = 2$ two zones were observed. The second zone is very thin and close to gas-liquid interface as it is observed in Fig. 7d. The creation of uncoloured zones can be related to the occurring streaks phenomenon which delivers dye to the upper parts of HM, which than could be transferred to uncoloured zones through diffusion. However, it has to be noticed that uncoloured zone travels downwards at $t_p = 348$ s (Fig. 7c) and is around 3 cm from the interface while at $t_p = 694$ s is around 5 cm (Fig. 7d).

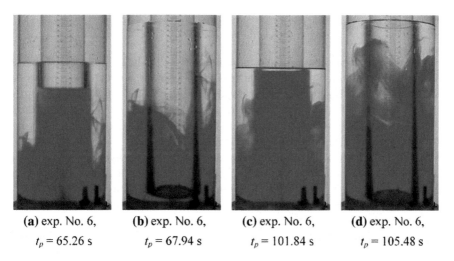

(a) exp. No. 6,   (b) exp. No. 6,   (c) exp. No. 6,   (d) exp. No. 6,
$t_p = 65.26$ s $t_p = 67.94$ s $t_p = 101.84$ s $t_p = 105.48$ s

Fig. 6 Streaks of dye solutions in outer compartment

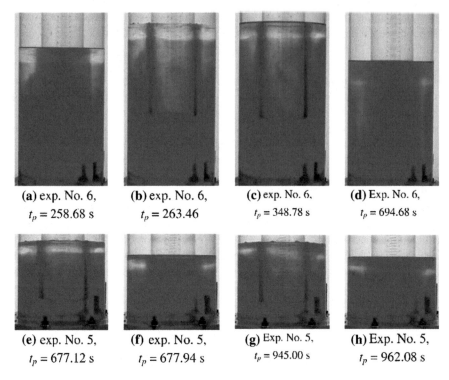

(a) exp. No. 6,
$t_p = 258.68$ s

(b) exp. No. 6,
$t_p = 263.46$

(c) exp. No. 6,
$t_p = 348.78$ s

(d) Exp. No. 6,
$t_p = 694.68$ s

(e) exp. No. 5,
$t_p = 677.12$ s

(f) exp. No. 5,
$t_p = 677.94$ s

(g) Exp. No. 5,
$t_p = 945.00$ s

(h) Exp. No. 5,
$t_p = 962.08$ s

Fig. 7 Long-time unmixed and separated coloured zones observed during hydraulic mixing

7 Conclusions

The studies performed have proved the concept of hydraulic mixing and deliver the analysis of the hydraulic mixing process. Although the authors delivered some explanations with regards to the observed phenomena, it should be kept in mind that these explanations are hypothetical and have forms of hypotheses. In general, it is evident that all these behaviours govern the process of hydraulic mixing and that hydraulic mixing has, to some extent, a chaotic behaviour.

The performance of hydraulic mixing can be measured by hydraulic mixing time and modified Reynolds number for hydraulic mixing because classic definition of Reynolds number could provide only information about local hydrodynamics while oscillation Reynolds number highly depends on initial liquid height. The hydraulic mixing highly depends on the set of various parameters such as height of clearance and initial height of liquid. In general, the increase of heights of clearance causes extension of hydraulic mixing time, although there was an exception for the clearance height equal to 60 mm. For that clearance and delivering of the dye to outer compartment a significant smaller mixing time was observed. These observations are also expressed through adequate modified Reynolds number.

The observed phenomena: long-time unmixed regions, formation of convex-concave interface structures, streaks of dye solutions and separated coloured zones, present the complex behaviour of hydraulic mixing process which needs further experimental and theoretical studies. Note, that complex behaviour provides also possibilities for future improvements especially in design and incorporating various internals and process parameter optimization.

It has to be highlighted that descriptions and explanations of results obtained pose many difficulties and give even more challenges for future work. Two main directions for future studies are considered. The first one is going to focus on mathematical description and modelling of hydraulic mixing process. The second direction is going to centre around further experimental studies which would include influence of process parameters, i.e. applied pressures and characteristic process time steps, and design of internals focused on process intensification. Moreover, the application to real process case study which would utilize its mild mixing conditions is looked for.

Acknowledgements This work was supported by PUT research grant no. 03/32/DSPB/0702.

References

Bałdyga J, Bourne J, Dubuis B et al (1995) Jet reactor scale-up for mixing-controlled reactions. Chem Eng Res Des 73:497–502

Bellard F (2015) FFmpeg. http://ffmpeg.org/. Accessed 7 Apr 2015

Espinosa-Solares T, Morales-Contreras M, Robles-Martínez F et al (2008) Hydrodynamic characterization of a column-type prototype bioreactor. Appl Biochem Biotechnol 147: 133–142

Fabiyi ME, Skelton RL (2000) On the effect of pulsing conditions on reaction rates in a pulsed baffled tube photoreactor (PBTPR). Trans IChemE 78:399–404

Forney LJ, Nafia N (1998) Turbulent jet reactors. Chem Eng Res Des 76:728–736

Ghanem A, Lemenand T, Della Valle D, Peerhossaini H (2014) Static mixers: mechanisms, applications, and characterization methods—a review. Chem Eng Res Des 92:205–228

Hardt S, Schönfeld F (2003) Laminar mixing in different interdigital micromixers: II. Numerical simulations. AIChE J 49:578–584

Hixson AW, Luedekel VD, York N (1937) Wall friction systems in liquid agitation systems. Ind Eng Chem 29:927–933

Karcz J (2013) CFD modelling of the fluid flow characteristics in an external-loop air-lift reactor. Chem Eng Trans 32:1435–1440

Katoh S, Yoshida F (2010) Bioreactors. Biochemical engineering: a textbook for engineers chemists and biologists. Wiley-VCH Verlag GmbH & Co. KGaA, Weinheim, Germany, pp 97–132

Lee C-Y, Wang W-T, Liu C-C, Fu L-M (2016) Passive mixers in microfluidic systems: a review. Chem Eng J 288:146–160

Mitkowski PT, Adamski M, Szaferski W (2016) Experimental set-up of motionless hydraulic mixer and analysis of hydraulic mixing. Chem Eng J 288:618–637

National Institutes of Health (2015) ImageJ 1.48. http://imagej.nih.gov/ij/. Accessed 7 Apr 2015

Ni X, Mackley MR, Harvey AP et al (2003) Mixing through oscillations and pulsations—a guide to achieving process enhancements in the chemical and process industries. Chem Eng Res Des 81:373–383

Open Microscopy Environment (2015) Bio-Formats Documentation, release 5.1.0, https://www-legacy.openmicroscopy.org/site/support/bio-formats5.1/, Retrieved 10 December 2016

Patwardhan AW, Gaikwad SG (2003) Mixing in tanks agitated by jets. Chem Eng Res Des 81:211–220

Reynolds O (1883) An experimental investigation of the circumstances which determine whether the motion of water shall be direct or sinuous, and of the law of resistance in parallel channels. Philos Trans R Soc London 174:935–982

Rott N (1990) Note on the history of the Reynolds number. Annu Rev Fluid Mech 22:1–12

Schönfeld F, Hessel V, Hofmann C (2004) An optimised split-and-recombine micro-mixer with uniform chaotic mixing. Lab Chip 4:65–69

Smith R (2005) Chemical Process Design and Integration. Wiley, Chichester

Stokes GG (1851) On the effect of the internal friction of fluids on the motion of pendulums. Trans Cambridge Philos Soc IX:8

Thakur RK, Vial C, Nigam KDP et al (2003) Static mixers in the process industries—a review. Chem Eng Res Des 81:787–826

Woziwodzki S (2011) Unsteady mixing characteristics in a vessel with forward-reverse rotating impeller. Chem Eng Technol 34:767–774

Woziwodzki S (2013) Mixing of viscous Newtonian fluids in a vessel equipped with steady and unsteady rotating dual-turbine impellers. Chem Eng Res Des 92:435–446

Xu C, Huang Y (2012) Maintenance-free pulse jet mixer. Powder Technol 230:158–168

Chemical Engineering in Biomedical Problems—Selected Applications

Arkadiusz Moskal and Tomasz R. Sosnowski

1 Introduction

The term aerosol describes two-phase systems consisting of fine particles or dro-plets dispersed in a continuous air phase. The air surrounding us contains suspended particles that are of natural or anthropogenic (man-made) origin. In each breath the thousands of airborne particles enter into our breathing system, and many of them are captured (deposited) there. After deposition they interact with the organism either locally in the lungs or in other organs after penetration to the systemic circulation. The effect of those interactions can often be hazardous, however aerosols can be also used as carriers of medicines delivered to/via the lungs. In both cases the knowledge of aerosol particles mechanics is needed to predict the dose of deposited matter, and place of the deposition. In this work we demonstrate that problems of aerosol inhalation and their interaction with the organism can be effectively analyzed with scientific tools of chemical and process engineering. In order to prove that, we will discuss two examples: the process of aerosol penetration to the lungs via the upper airways (oro-pharyngeal region) and the phenomena related to direct physicochemical interactions of deposited particles with the liquids which cover the lung surface.

A. Moskal (✉) · T. R. Sosnowski
Faculty of Chemical and Process Engineering,
Warsaw University of Technology, Warsaw, Poland
e-mail: arkadiusz.moskal@pw.edu.pl

© Springer International Publishing AG, part of Springer Nature 2018
M. Ochowiak et al. (eds.), *Practical Aspects of Chemical Engineering*,
Lecture Notes on Multidisciplinary Industrial Engineering,
https://doi.org/10.1007/978-3-319-73978-6_21

2   Aerosol Mechanics and Inhalation

As compared to aerosol flow and separation in technological systems (e.g. air filters, cyclones, electrostatic precipitators), inhaled aerosol particles experience very special conditions. First of all, not all particles can be inhaled—they have to be small enough to be effectively carried into the organism by the inspired airflow. Another factor—usually absent in technological systems—is the oscillatory flow pattern of inhaled aerosol in the respiratory tract, which sets certain limits on the residence time of particles inside the body but also significantly influences the dynamics of particles' motion with the flowing air. This issue will be discussed in this work in more detail. Finally, the geometry of air ducts (bronchial tree) through which particles travel inside the organism is very complex, so the analysis of flow and particle dynamics is usually much more complicated than in the common technical systems. This is why behavior of inhaled aerosol is typically analyzed only in smaller regions of the respiratory system. The aerosol particles trajectories in human breathing system depend on the particle morphology and forces acting on the particles during flow. The flow regimes in the human breathing system are restricted to dilute two-phase flows in a Newtonian fluid with significant influence of gravity and Brownian forces with negligible influence of heat and mass transfer effects. The flow is "diluted" if the effect of particle—particle interactions are not significant and they are not taken into account (Loth 2000). The second property of diluted flows is that the particles do not influence one another with respect to fluid dynamic forces. As the average inter-particle spacing is equal to $\left(\frac{\pi}{6\alpha}\right)^{\frac{1}{3}}$ the criterion for negligible particle-particle fluid dynamic interaction is:

$$\alpha^{\frac{1}{3}} \ll 1 \tag{1}$$

where α is the volume fraction of particles per mixed-fluid volume. Aerosol particles dynamics can be analyzed using Eulerian or Lagrangian approach. The Eulerian representation defines the particle velocity and concentration at the points of the system that generally coincide with those used for continuous-phase grid. The Lagrangian approach is based on the finding of the particles trajectories through the system and the particles characteristics are defined along the particles trajectories. This approach is more useful in tracing the dynamics of aerosol particles characterized by different properties. The Lagrangian approach to the analysis of particle dynamics is based on analysis of trajectory in the physical space of each individual particle (so this method is commonly called *particle trajectory* method). To determine the history of the particle position and velocity in the system, the most appropriate method is to apply the Newton`s second law. For unsteady rectilinear, non-uniform creeping motion of a solid sphere in a motionless, incompressible, viscous fluid the fundamental equation called Basset-Boussinesq-Oseen (BBO), modification of earlier works of Basset, (1888), Boussinesq, (1903) and Oseen, (1927), ws proposed in the form [for details see (Loth 2000)]:

$$m_p \frac{d\vec{V}_p}{dt} = \vec{F}_{DR} + \vec{F}_{VAM} + \vec{F}_{PG} + \vec{F}_{ext} \tag{2}$$

The left hand side of the equation relates to the acceleration of the particle mass. The first term on the right hand side of the Eq. (2) is resultant viscous *drag and resistance force*. For solid spherical particle the drag force can be written as:

$$F_D = -\frac{1}{8}\pi d^2 \rho_f C_D V_{rel}^2 \tag{3}$$

where C_D is the drag coefficient, which is primarily dependent on the particle Reynolds number. The second term on the RHS of the Eq. (2) is called *virtual added mass force* and denotes the force necessary to accelerate half of the fluid mass m_f displaced by the spherical particle and can be calculated from:

$$F_{VAM} = \frac{1}{2}m_f \frac{dV_{rel}}{dt} \tag{4}$$

The third term on the RHS of the Eq. (2) represents the forces caused by the pressure gradient created in particle surroundings by the accelerating fluids and can be globally called the *lift forces*. These forces are significant to the particles which characteristic diameter is relatively large. The general formulation for particle lift force (perpendicular to the drag force) is complicated by many factors, but for low Reynolds number, one may distinguish two main lift effects connected with local stress acting on various parts of the particle and with effect of the torque acting on the particle. These two effects are referred to as Saffman lift (shear induced lift) and Magnus lift (spin induced lift) (Brandon and Aggarwal 2001). In the cases of small aerosol particles traveling through the air the effect of the Saffman and Magnus forces may be neglected. The last term in the Eq. (2) includes all other external interactions affecting the particle trajectory e.g. gravity, electrostatic interactions, magnetic interactions, etc. For aerosol systems, when the density of air is usually three orders of magnitude smaller than density of the solid particles, the equation of motion reduces to the form that included only drag and external interactions:

$$\frac{dV_p}{dt} = -\frac{18\mu_f}{\rho_p d_p^2}(V_p - V_f) + \frac{6}{\rho_p \pi d_p^3}F_{ext} \tag{5}$$

The aerosol particle suspended in the fluid experiences other interactions, which were not taken into account in above considerations. Under atmospheric conditions, an aerosol particle suffers ca 10^{21} collisions with molecules of air, per second. Thus, the motion of aerosol particle may be either deterministic or stochastic, depending on the particle size and airflow specification. For sufficiently large particles ($\gg 1$ μm in diameter in air at standard temperature and pressure conditions), thermal fluctuations of the particle momentum are negligible and the particle trajectory is a deterministic, so can be found from the solution of the Eq. (5). For very small particles ($\ll 1$ μm) a momentum transfer during the collisions with air molecules should not be omitted and Eq. (5) is no longer valid. Such small particles move randomly through the air (*Brownian motion*). One of the methods to account for this phenomenon is to add the additional fluctuating force term ('Brownian force') on the right hand side of the Eq. (5), which then becomes the stochastic differential equation in the form:

$$\frac{dV_p}{dt} = -\frac{18\mu_f}{\rho_p d_p^2}\left(V_p - V_f\right) + \frac{6}{\rho_p \pi d_p^3} F_{ext} + \frac{6}{\rho_p \pi d_p^3} F_B \tag{6}$$

The Eq. (6) is called Langevin equation and is simply Newton`s second law of motion for a particle moving in a viscous fluid subject to the effective Brownian force which is modeled as randomly fluctuating body force. Abuzeid et al. (1991), and Chen et al. (2002), used the effective random body force approach to study process involving the motion of aerosol particles with random excitation. Adopting the aforementioned cost–effective approach, the Brownian force F_B that is exerted on the aerosol particle may be modeled as a white noise with spectral intensity, S_0, given by:

$$S_0 = \frac{216\mu_f k_B T}{\pi^2 d_p^5 \rho_p^2 C_s} \tag{7}$$

where T is the absolute temperature, k_B is Boltzmann`s constant, μ_f is the fluid dynamic viscosity, d_p is the particle diameter, and C_s is the Cunningham slip correction factor. For single spherical aerosol particle suspended in the air, the Brownian force is given by:

$$F_B = m_p \alpha_B Z \tag{8}$$

where α_B is defined as the characteristic magnitude of the acceleration of the Brownian excitation and is expressed as follows:

$$\alpha_B = \sqrt{\frac{2\pi S_0}{\Delta t}} \tag{9}$$

where Δt is the time step which is used in the numerical integration of the trajectory equation. Z in Eq. (8) is the dimensionless vector with random direction, and is calculated as:

$$Z = Z_x \widehat{e}_x + Z_y \widehat{e}_y + Z_z \widehat{e}_z \tag{10}$$

where $\{\widehat{e}_x, \widehat{e}_y, \widehat{e}_z\}$ are the unit vectors in the inertial frame of references, and $\{Z_x, Z_y, Z_z\}$ are identical (because of isotropy) normally distributed random variables with zero mean and unit variance. The Brownian force on aerosol particle is exerted at all time instants, t, but its magnitude and direction are changed randomly (through the random updating of value of $\mathbf{Z}$) at the end of every time step Δt. A simulator based on Eq. (8) can be used to model the Brownian motion of aerosol particles in a human breathing system.

3 Modeling of Aerosol Particles Deposition in the Upper Part of Human Breathing System

As mentioned earlier, aerosol particles inhaled from the atmospheric environment can create a hazard for human health, (Gehr and Heyder 2000), or can be precursor of the therapeutic action in the case of aerosol-therapy. Assessment of the potential health effects and validation of the devices used during aerosol-therapy should begin with the estimation of the deposition efficiency of such particles in the different parts of the respiratory system. The process of aerosol transport and deposition in the lungs can be analyzed with methods similar to the ones commonly used in the process-engineering problems related to dispersed solid-fluid systems. Accordingly, mathematical modeling coupled with suitable in vitro experiments should help in the prediction of the amount of deposited matter and the "hot spots" of the deposition in the system. As an example of using numerical predictions for deposition of aerosol particles, the oro-pharyngeal region will be chosen. The mouth and throat are the second way after nose, through which the ambient air is introduced into the human respiratory system, however it is a common route for drug delivery by inhalation. Different patterns of particle deposition in this region explain the variability of particle bioavailability in the lungs (Borgstrom et al. 2006). For many inhaled pharmaceutical aerosols, a high fraction of the delivered dose is lost due to deposition in the oral airway resulting in ineffective delivery to deep lung regions, (Martonen et al. 2000; Finlay 2002). During investigation of the aerosol particles deposition in oro-pharyngeal parts of the breathing system two issues need to be stressed as the important features of the discussed problem:

- Particle deposition is governed by the local dynamics of aerosol particles that are carried through the complex geometry of the airways by inspiratory/ expiratory airflow;
- The dose of deposited particles is not necessarily the most appropriate measure of potential health effects.

The second issue concerns the submicron and nano-size particles, which are widely present in the occupational environment, e.g. they are generated during welding, grinding, combustion processes and by diesel engines, (Maynard and Kuempel 2005). Their toxic potential is attributed to the high surface area per unit mass, (Kreyling et al. 2004). Such particles are capable of inducing apparent health effects, although their weight fraction in the total mass of deposited material is almost negligible. In such cases, the number rather than mass of deposited particles should be taken into account. Deposition of inhaled aerosol particles in the mouth and throat was analyzed by Sosnowski et al. with the geometrical model of the oro-pharyngeal part of the breathing system which is presented in the Fig. 1a (Sosnowski et al. 2006, 2007). The problem was solved numerically, but simultaneous experimental studies using silicone-rubber reconstruction of this geometry were used to verify the results (Sosnowski et al. 2005). Computations of the airflow field and aerosol particles dynamics, during realistic inhalations, have been performed with the commercial CFD package FLUENT.

Two breathing curves have been characterized by the volume of inhaled air, V_{inh} [dm^3] given by the function:

$$V_{inh}(t) = a_i \left[1 - \cos\left(\frac{\pi}{T_i}t\right) \right] \qquad (11)$$

where for curve I: $a_I = 1.383$ dm^3, $T_I = 3.41$ s and for curve II: $a_{II} = 1.386$ dm^3, $T_{II} = 2.75$ s. The air flow turbulence was solved using a modified $k - \varepsilon$ model (mean flow tracking algorithm) with a low Reynolds number solver and the enhanced wall treatment. Aerosol dynamics were calculated by analyzing all forces (including Brownian diffusion) acting on each spherical particle introduced into the system, which allowed for calculation of the particle trajectory using Eq. (6). The deposition was defined by "trap" condition at the airway walls, which meant that particle was always lost from further calculation at the point of touching the wall. Computational analysis was focused on particles in the size range 0.3–10 μm, for two material densities: 1100 and 2200 kg/m^3. The results allow examining of the spatial and temporal distribution of particles with the certain size deposited in the oro-pharynx. The total particle deposition was calculated by integration of the local mass fluxes along the whole computational domain for the assumed total mass

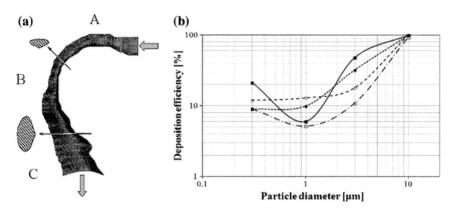

Fig. 1 **a** Numerical model of oropharynx with marked segments: A—the inlet region (mouth), B —oro-pharyngeal bend, C—throat; **b** total deposition of aerosol particles in the model of oropharynx (circles—breathing curve I, squares—breathing curve II; empty symbols—$\rho_p = 1100$ kg/m^3, filled symbols—$\rho_p = 2200$ kg/m^3)

introduced into the object. These results are presented in the Fig. 1b. Deposition of particles larger than 1 μm was found to be higher for particles with the higher material density at stronger flow intensities (curve "II"), which can be easily explained by the prevalence of the impaction mechanism of deposition for such particles. Low deposition efficiency is observed for 1 μm particles, for which impaction becomes less important and other deposition mechanisms are also ineffective. Their higher deposition efficiency corresponds to slower inhalation (curve "I"), that is, to the longer particle residence time in the oro-pharyngeal region. It suggests that the sedimentation combined with diffusional effects are prevalent for particles of this size. Increased deposition of 0.3 μm particles suggests the influence of Brownian effects on the dynamics of aerosol in the sub-micrometer size range. More information can be obtained from the calculated distribution of particle deposition in time and space. Figure 2a shows the temporal relative deposition in the whole oro-pharynx, and Fig. 2b—the spatial distribution of the deposition during the inspiratory part of the breathing cycle. The temporal relative deposition

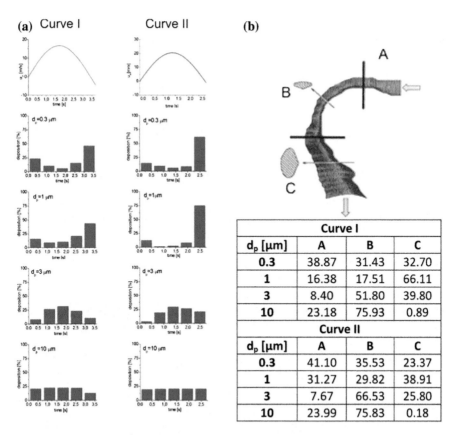

Curve I			
d_p [μm]	A	B	C
0.3	38.87	31.43	32.70
1	16.38	17.51	66.11
3	8.40	51.80	39.80
10	23.18	75.93	0.89
Curve II			
d_p [μm]	A	B	C
0.3	41.10	35.53	23.37
1	31.27	29.82	38.91
3	7.67	66.53	25.80
10	23.99	75.83	0.18

Fig. 2 Calculated temporal **a** and spatial **b** distribution of deposition of particles with different diameter in the model of the oro-pharynx during non-steady flow (particle density: 2200 kg/m³)

is defined as a ratio of mass deposited at the Δt_i, i.e. the flux $m(t_i)$ over the surface area of the object, to the total mass deposited in the object:

$$TD(t) = \frac{m_i(\Delta t_i)}{\sum_{i=1}^{N} m_i(\Delta t_i)} \tag{12}$$

The effect of transition between inspiration and expiration with incorporation of mixing effect at the transition moment was also considered. Each bar in the Fig. 2b represents temporal deposition efficiency (averaged over the indicated time-interval), which is normalized with respect to the total collection efficiency, (see Fig. 1b) for particles with a given size. Deposition of 10 µm particles is almost independent of the actual airflow rate, which can be explained by the particle mass being high enough for the effective impaction at the walls of the airway; even at relatively low airflow rates. As indicated in Fig. 2b, the region of the most efficient deposition of 10 µm particles is located in segment B (oro-pharyngeal bend characterized by the narrowing and strong curvature of the duct), and virtually no such particles are deposited beyond it, as they simply do not penetrate to segment C. The situation is different for smaller particles. For 3 µm particles, the efficiency is increased during a middle part of the inspiration, i.e. for the highest flows. This suggests that impaction still remains an important deposition mechanism for such particles; however, higher velocities are needed to make this mechanism predominant. For a faster inhalation (curve 'II'), the deposition remains effective also during a decline of the airflow, which resembles the results obtained for larger particles.

Again, the bend area (Segment B) is the main region of deposition, however, some 3 µm particles are able to penetrate and deposit in the posterior part of the oro-pharynx (Segment C). Particles with diameter 1 µm are collected with a low overall efficiency (see Fig. 1b) and, as seen in the Fig. 2a, their deposition takes place mainly during the flow decreasing and the transition to expiration. The preferred region of deposition is located in the segment C, but particles are appreciably deposited also in two other parts of the oro-pharynx. Similar results have been found for particles with diameter equal to 0.3 µm, although their deposition is more uniform along the oro-pharyngeal model. Deposition of sub-micron particles is governed by the Brownian diffusion phenomenon; however, the observed enhancement during flow reversal suggests also the strong influence of air mixing effect, and, as a result, reduction of the thickness of diffusional boundary layer. This can be attributed to aerodynamic effects caused by the interaction between inspired and expired air, which are stronger for the faster transition. Such effects seem unimportant for particles of high inertia (e.g. 10 µm), which are deposited with almost equal efficiency in all stages of inspiration. The most important finding from the presented analysis is that the assumption of constant, i.e. mean flow in the respiratory system (which is a common simplification in CFD modeling) leads to the unrealistic physical picture of the aerosol behavior in the human oro-pharynx. Even if the particle residence time in any part of this system is in the sub-second range, particle dynamics is determined by the local aerodynamic conditions, which are continuously changing during breathing cycle. As shown above this effect is significant to sub-micron ones, but less important for large particles. It partly explains why predictions of particles deposition at mean airflow rate can be useful for coarse aerosol particles. However, the correct analysis of nano- and sub-micron particles requires a more realistic, dynamic description.

4   Experimental Methods for Analysis of Particle-Lung Surfactant Interactions at the Physicochemical Level

Particles, which are deposited in the respiratory system immediately get in contact with fluids which cover the lung epithelium. In the bronchial tree particles land on the mucus, but in the respiratory zone (alveoli) they interact with the lung surfactant (LS).

The lung surfactant is composed mainly of phospholipids and special proteins, and—according to its name—it has strong surface-active properties. Presence of the LS in the lungs minimizes the work of breathing (i.e. expansion of a very large surface ~ 100 m^2), but it is also important for mass transfer processes in the pulmonary region (Sosnowski 2016). The LS is very surface-active at dynamic conditions of breathing, which are related to oscillatory variations of air/liquid interfacial area. In such situation, temporal and spatial variations of the surface tension along the interface take place due to surfactant mass exchange between the bulk and the free interface, and due to the flow of the interfacial region. The ensuing surface tension gradients generate the Marangoni flows, which may form the mechanism of removal of particulate deposits from the pulmonary region (Gradoń and Podgórski 1989; Gradoń et al. 1996), Fig. 3.

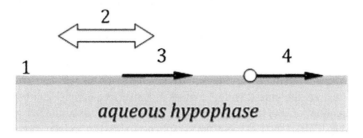

Fig. 3 Schematic explanation of particle transfer in pulmonary region: lung surfactant film (**1**) on the gas/liquid interface which oscillates in time (**2**), is responsible for generation of surface tension gradients. These gradients induce the flows of the interface (**3**), eventually leading to displacement of deposited particles (**4**)

Such behavior of gas-liquid interface can be reproduced in the laboratory (in vitro) conditions using specialized devices, for instance Langmuir-Wilhelmy balance, oscillating bubble or oscillating drop tensiometers e.g. Wustneck et al. (2002), Kondej and Sosnowski (2013), Sosnowski et al. (2017). By tracing the dynamic changes of the surface tension, it is possible to detect physicochemical changes in the LS system caused by deposited particles (Kondej and Sosnowski 2016). For harmonic variations of interfacial area:

$$\Delta A = \Delta A_{max} \cos \omega t \tag{13}$$

the response in form of harmonic surface tension variations:

$$\Delta \gamma = \Delta \gamma_{max} \cos(\omega t + \varphi) \tag{14}$$

may be analyzed using rheological approach. We deal with surface dilation, and the response of the system to this type of deformation is visco-elastic, with ε_D [N m^{-1}] being the surface dilatational elasticity, and μ_D—surface dilatational viscosity [N s m^{-1}], where:

$$tg\varphi = \frac{\omega \mu_D}{\varepsilon_D} \tag{15}$$

All these values should be considered only as the apparent (effective) ones, since the change of the surface tension is related to the intrinsic mechanical properties of the interface, but also to the surfactant mass exchange between the air/liquid surface and the liquid. As an example of the analysis of the LS properties in the presence of nanoparticles which can penetrate to the lungs during inhalation, we will present the impact of nanometric graphene oxide (GOx) flakes on the model LS (Curosurf–Chiesi, Italy) studied with the oscillating drop method with PAT-1 M device (Sinterface, Germany) at physiological conditions: $36.6 \pm 0.2°C$, oscillation frequency: 15 min^{-1}. Figure 4 shows the measured variation of ε_D and μ_D as a function of dose of GOx in the model LS (Curosurf). Interfacial dynamics of the LS is changed by GOx already at concentrations 0.25 mg/ml and the further change is observed when nanoparticles content increases up to 0.5 mg/ml. As a result, variations of the dynamic surface tension during successive compression and expansion of the air/liquid interface are altered when compared to the situation without nanoparticles (i.e. the pure LS). Differences in LS surface activity can be explained by modification of the mass transfer in the system contaminated with GOx nanoparticles. The results suggest that graphene oxide can induce direct physico-chemical changes in the lung surfactant in vivo, therefore the inhalation of air contaminated with GOx nano-particles cannot be considered as completely safe.

Fig. 4 Changes of the apparent surface dilatational elasticity ε_D and surface dilatational viscosity μ_D in the air/water system with the model lung surfactant (Curosurf) and graphene oxide flakes (GOx)

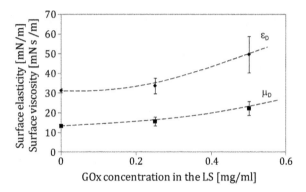

5 Conclusions

In this work we have demonstrated that scientific tools of chemical engineering are useful in the analysis of aerosol inhalation and interactions with the organism, both being important issues from both toxicological and therapeutic perspectives. Aerosol flow and deposition in the complicated structure of human respiratory system during oscillatory flow of breathing is a non-trivial scientific problem, which requires a comprehensive treatment of fluid and particle mechanics.

In such cases all intuitive predictions may be misleading, therefore only extensive CFD analysis can provide the quantitative data of practical significance for the medicine. Another biomedical issue presented here deals with the physicochemical interactions of deposited particles with lung fluids. Such analysis is the important step in the assessment of harmful effects of inhaled aerosols. Scientific methods of chemical and process engineering also in this case allow highlighting the importance of certain hydrodynamic and mass transfer phenomena that occur in the human organism.

Acknowledgements The research related to the lung surfactant was financially supported by NCN Project No. 2014/13/B/ST8/00808. The research related to the aerosol mechanics was financially supported by NCN Project No. 2015/19/B/ST8/00599.

References

Abuzeid S, Busnaina AA, Ahmadi G (1991) Wall deposition of aerosol particles in a turbulent channel flow. J Aerosol Sci 22(1):43–62
Basset AB (1888) A treatise on hydrodynamics, vol 2. Deighton, Bell and Co, Cambridge
Boussinesq J (1903) Theorie Analytique de la Chaleur, vol. 2. L`Ecole Polytechnique, Paris
Borgström L, Olsson B, Thorsson L (2006) Degree of throat deposition can explain the variability in lung deposition of inhaled drugs. J Aerosol Med 19(4):473–483
Brandon DJ, Aggarwal SK (2001) A numerical investigation of particle deposition on a square cylinder placed in a channel flow. Aerosol Sci Technol 34:340–352

Chen S, Cheung CS, Chan CK et al (2002) Numerical simulation of aerosol collection in filters with staggered parallel rectangular fibers. Comput Mech 28:152–161

Finlay WH (2002) The mechanics of inhaled pharmaceutical aerosols. Academic Press, San Diego

Gehr P, Heyder J (2000) Particle-lung interaction. Marcel Dekker, New York

Gradoń L, Podgórski A (1989) Hydrodynamical model of pulmonary clearance. Chem Eng Sci 44:741–749

Gradoń L, Podgórski A, Sosnowski TR (1996) Experimental and theoretical investigations of transport properties of DPPC monolayer. J Aerosol Med 9:357–367

Kondej D, Sosnowski TR (2013) Alteration of biophysical activity of pulmonary surfactant by aluminosilicate nanoparticles. Inhal Toxicol 25:77–83

Kondej D, Sosnowski TR (2016) Effect of clay nanoparticles on model lung surfactant: a potential marker of hazard from nanoaerosol inhalation. Env Sci Pollut Res 23:4660–4669

Kreyling WG, Semmler M, Moller W (2004) Dosimetry and toxicology of ultrafine particles. J Aerosol Med 17:140–152

Loth E (2000) Numerical approaches for motion of dispersed particles, droplets and bubbles. Prog Energ Combust 26:161–223

Martonen TB, Musante CJ, Segal RA et al (2000) Lung models: strengths and limitations. Resp Care 45(6):712–736

Maynard AD, Kuempel ED (2005) Airborne nanostructured particles and occupational health. J Nanoparticle Res 7:587–614

Oseen CW (1927) Neuere Methoden und Ergebnisse in der Hydrodynamik. Akademische Verlagsgesellschaft, Leipzig

Sosnowski TR, Moskal A, Gradoń L (2005) Deposition of pharmaceutical powders in the oro-pharyngeal cast during inspiration. 2. Experiments. In: Abstracts of European Aerosol Conference, Ghent, Belgium, p 726

Sosnowski TR, Moskal A, Gradoń L (2006) Dynamics of oropharyngeal aerosol transport and deposition with the realistic flow pattern. Inahal Toxicol 18:1–9

Sosnowski TR, Moskal A, Gradoń L (2007) Mechanisms of aerosol particle deposition in the oro-pharynx under non-steady airflow. Ann Occup Hyg 51(1):19–25

Sosnowski TR, Kubski P, Wojciechowski K (2017) New experimental model of pulmonary surfactant for biophysical studies. Coll Surfaces A Physicochem Eng Aspects 519:27–33

Sosnowski TR (2016) Selected engineering and physicochemical aspects of systemic drug delivery by inhalation. Current Pharm Design 22(17):2453–2462

Wüstneck R, Wüstneck N, Moser B et al (2002) Surface dilatational behavior of pulmonary surfactant components spread on the surface of a pendant drop. 1. Dipalmitoylphosphatidylcholine and surfactant protein C. Langmuir 18:1119–1124

Hybrid and Non-stationary Drying— Process Effectiveness and Products Quality

Grzegorz Musielak, Dominik Mierzwa, Andrzej Pawłowski,
Kinga Rajewska and Justyna Szadzińska

1 Introduction

Drying is the technological process of thermally removing (through evaporation or sublimation) volatile substances (moisture: water or other solvent) to yield a solid product. It is one of the most commonly used unit operation in almost all industrial sectors e.g. food, wood, ceramic, chemicals, pharmaceuticals, construction materials etc. The main goals of drying are: improvement of product durability (e.g. a dry food is less susceptible to spoilage caused by bacteria, molds and insects), facilitation of handling (packaging, handling, storage, and transportation of a dry material is easier and cheaper because of the mass and the volume reduction), improvement/possibility of further processing (e.g. reducing of stickiness improves milling, mixing or segregation, sintering of ceramics), quality enhancement (e.g. drying influences color, flavor, appearance, nutritional values of food product, internal structure of various materials) etc. (Itaya et al. 2014; Sokhansanj and Jayas 2014).

Drying is a thermal process and the energy required to evaporate moisture could be transferred to the material in various ways. The most popular is convection—heat is supplied by heated air or gas flowing over the surface (direct or convective drying). The second possibility is conduction—heat is supplied through heated surfaces placed within the dryer (indirect or conduction drying). Another possibility is absorption of the electromagnetic wave. Depending on the wave frequency, drying is classified as solar (75×10^6–3×10^9 MHz), infrared (50×10^6–300×10^6 MHz), microwave (915 or 2450 MHz) and dielectric drying (10–100 MHz). The last opportunity is acoustic wave absorption (sonic and ultrasonic ranges). Drying has been identified as the one of the most energy intensive unit operation. Several

G. Musielak (✉) · D. Mierzwa · A. Pawłowski · K. Rajewska · J. Szadzińska
Department of Process Engineering, Institute of Technology and Chemical Engineering,
Poznan University of Technology, Poznań, Poland
e-mail: grzegorz.musielak@put.poznan.pl

© Springer International Publishing AG, part of Springer Nature 2018
M. Ochowiak et al. (eds.), *Practical Aspects of Chemical Engineering*,
Lecture Notes on Multidisciplinary Industrial Engineering,
https://doi.org/10.1007/978-3-319-73978-6_22

authors reported that this unit operation utilizes up to 25% of all industrial energy usage (Strumiłło et al. 2014). Because of recent environmental and power engineering problems, i.e. abatement of greenhouse gasses emission, depletion of fossil fuels, etc. (Mujumdar 2004), it has become extremely important to reduce consumption of energy in all industrial sectors.

On the other hand, drying processes, especially the most popular convective drying, could worsen the products quality in terms of mechanical properties, shape, dimensions, nutritional value, color, texture, etc. (Leeratanarak et al. 2006; Kowalski and Pawłowski 2010a; Kowalski and Szadzińska 2012). Excessive shrinkage, permanent deformations, loss of coherency, destruction of the internal structure, caramelisation, enzymatic and non-enzymatic reactions (e.g. Maillard reaction), pigments degradation, ascorbic acid or lipids oxidation constitute only some examples of many deteriorative processes which proceed during drying and affect final products quality (Itaya and Hasatani 1996; Mujumdar 2007; He and Wang 2015).

For these reasons, new sustainable drying technologies are sought to reduce the time of drying and energy consumption, and to minimize its negative influence on the final products quality (Kudra 2004; Kudra and Mujumdar 2009; Strumiłło et al. 2014) and to produce high-quality products at the possibly minimum capital and operating costs (Mujumdar 2004; Fushimi and Dewi 2015). Many innovative drying techniques have been reported in the literature, such as impinging stream drying, superheated steam drying, pulsed drying (Kudra and Mujumdar 2009), vacuum drying (Parikh 2015), atmospheric freeze drying (Claussen et al. 2007), microwave–freeze drying (Wang et al. 2010), vacuum–microwave drying (Song et al. 2009), heat pump drying (Jangam and Mujumdar 2011).

Subsequent newly-developed drying techniques are focused on breaking the limits of convective drying. There are two main trends in looking for an appropriate solution i.e. application of intermittent conditions or utilization of few drying techniques in one process (hybrid drying) (Chou and Chua 2001; Kumar et al. 2014).

2 Convective Non-stationary Drying

It is known that time-dependent drying schemes could improve the process. Variable drying conditions were investigated in various drying techniques (Kudra and Mujumdar 2009). These studies confirmed that the use of variable drying conditions can accelerate the process, increase its energy efficiency and improve product quality. Many materials shrink during drying. Heterogeneous distribution of moisture content and temperature during drying results in non-uniform shrinking and, consequently, in drying stresses generation (Musielak 2004; Musielak and Śliwa 2015). During convective drying, the outer layers are dried first. These layers shrink compressing the core of the sample which contains more moisture. Unwanted cracking stresses are generated in this time. Especially in the case of wood, mostly in roundwood, differences in shrinkage in tangential and radial direction significantly affect cracking during drying. This can be prevented by introducing cyclic changes in the drying medium parameters (temperature, humidity). In this way, the temperature and moisture content gradients in dried material can be reduced (Zhang and Mujumdar 1992; Herrithsch et al. 2008; Kowalski and Pawłowski 2010b). The easiest way to modify and improve convective processes is the application of periodical changes of process parameters like, e.g. air temperature and humidity. This operation allows to reduce the product fracture by decrease in drying stresses and strains level during processing. However, those processes have to be investigated carefully as any parameter changes involve modification in drying kinetics and energetically effectiveness. To illustrate the advantages of such processes, two examples presenting drying of kaolin clay cylinders and wood samples are shown. Nonetheless, each of them has to be considered separately in reference to pure convective drying as in all cases processes were carried out in different dryers.

The first example concerns the drying of a kaolin cylinder (Pawłowski 2011). The process was done in a convective oven that allows simultaneous control of the temperature and humidity of the drying medium (air). The temperature alterations were realized by the periodic reduction of the air temperature, while the control of the humidity by the periodic increase of this parameter. The introduced changes were intended to temporarily slow down the process to reduce drying stresses. Figure 1 presents the average drying rate and energy consumption for reference process (pure convective drying—CV) and processes with temperature (INT1) or humidity (INT2) variations. Periodical changes improve significantly product quality, and not affect visibly the drying rate. However, in the second case (control of humidity) they require additional amount of energy, even few times more, to compensate the variability of process parameters. Nevertheless, appropriate designing of intermittent drying allow to achieve processes with even similar energy consumption as in case of pure convective processes.

In the second example, the convective non-stationary drying was carried out for walnut wood (*Juglans regia* L.). The branches were transversely cut into 35 mm high and 50 mm diameter samples, then barked and dried. The convective stationary drying was carried out at a drying air temperature of 100 °C to achieve equilibrium moisture content and lasted 11 h. Non-stationary drying was conducted

Fig. 1 Average drying rate (DR) and energy consumption (EC) per kg of evaporated water for kaolin drying in constant and intermittent conditions

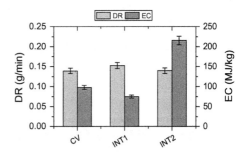

(a)   **(b)**   **(c)**

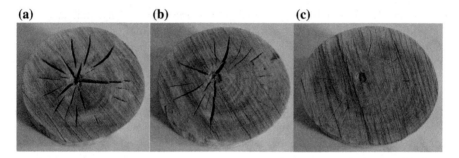

Fig. 2 Pictures of wood dried in: **a** stationary conditions, **b** non-stationary conditions with temperature changes **c** non-stationary conditions with humidity changes (Pawłowski 2011)

in two ways. First, with periodically variable temperature of the drying medium. The amplitude of the changes was 50 °C (max 100 °C, min 50 °C) and the frequency of the changes was introduced in three ways, adapted to process kinetics. In the second case, the humidity of the drying medium was periodically changed. The amplitude of the changes was about 70% relative humidity (max 75%, min 5%), and the frequency of changes was entered in the sequence: humidification time 15 min every 45 min or humidification time 20 min every 30 min.

The main advantage of such non-stationary drying programs was to maintain the drying time to the same level as for stationary drying. This is not obvious because lowering the temperature or increasing the humidity of the drying medium generally increases the drying time. Figure 2 shows photos of wood which was dried as described above. The way of cutting samples and the species of wood favor the formation of heart shakes. They are visible on the transverse cross section of the trunk as slits, widest at the core, and tapering towards the perimeter. These are visible in the photos that are part of the starry cracks, which greatly reduces the quality of the wood. It is possible to conclude on the basis of visual analysis that in the case described the most effective way of carrying out the process is drying under periodically changing conditions with changes in the moisture content of the drying medium if the quality criterion assumes no cracks in the material. Non-stationary convection drying, despite the higher energy input, could produce good results, especially for materials with high shrinkage and structural anisotropy, such as wood.

3 Hybrid Drying

Another possibility for the development of drying is combining different drying techniques into one process. Terms *hybrid drying* refers to the processes which connect two or more drying technique. The main aim of the different drying techniques coupling is the utilization of the advantages and elimination of the drawbacks of the particular techniques. Solely convective drying is long lasting and energy consuming. It results from the low efficiency of the drying apparatus and specific kinetics of the process. When the vast majority of moisture has been evaporated and dry areas have appeared on the material surface, the evaporation front shifts to the interior of the material and the falling drying rate period begins. At this stage of the drying, the rate of the process is fully controlled by the diffusion of the moisture from the interior to the surface of the material being dried (or evaporation front). Thus, the overall drying rate depends little on temperature and volumetric flow of the drying medium (air). Raising or maintaining of the same process parameters in this period is not reasonable and leads to the higher energy consumption. To overcome this problem energy could be delivered to dried material in different then convection way.

3.1   Convective–Microwave Drying

During the microwave drying, the material is heated in the whole volume and water evaporation (moisture) takes place inside the body (not only at the evaporation front —surface). Moisture in the form of vapor is easily evacuated from the interior and collected by the hot drying medium. In this way, the predominant role of the diffusion is the depressed and significant growth of the drying rate is observed. Moreover, due to the increase of the (vapor) pressure inside the material the liquid moisture may be pushed out the material by the flowing vapor (the so-called pumping phenomenon) (Wang et al. 2014).

Drying of kale is shown as an example of the convective–microwave hybrid process. Purely convective processes (CV; $T = 30/50$ °C, $v = 2$ m/s) was compared with constant (CV − MW) and intermittent (INT) application of the microwaves (MW; $f = 2.45$ GHz, $I = 0.1$–1 kW). Drying at intermittent conditions was in this case realized by the periodical application of the microwaves during convective drying of kale. The length of the microwave enhancement was controlled by the temperature of the sample surface. When the temperature rose above 60 °C, the microwaves were switched off and process undergoes to 15 min 'relaxation period' realized at 30 °C. The main purpose of relaxation periods was to equalize the temperature and humidity gradients occurring in the material during intensive microwave drying. The difference between INT1 and INT3 processes lies in the way of microwave power regulation. The kinetic and quality parameters were presented in Fig. 3.

The results obtained clearly indicate that hybrid processes conducted at intermittent conditions may lead to the increase of the drying rate (DR) and drop of the energy consumption (Fig. 3a). The differences between solely convective and hybrid processes are in this case particularly visible. Of course, the qualitative and quantitative effects depend strictly on the construction of the drying schedule. The best results, in terms of process kinetics, were observed during microwave-assisted

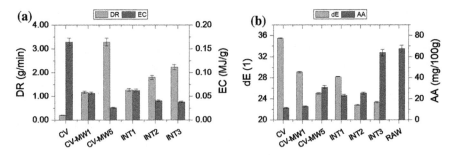

Fig. 3 Average values of **a** drying rate (DR) and energy efficiency (EC), **b** relative color change (dE) and ascorbic acid content (AA) obtained during hybrid drying of kale

convective drying (CV − MW5; I = 500 W) and intermittent drying (INT3; $I = 1000$ W).

As for the quality of products, it can be stated that all drying procedures influenced significantly the color of the material. The relative color change (dE) was significantly higher than 20 apart from the variant of drying. The worst color was observed for the samples dried with the convective process (CV), whereas products obtained from hybrid schedules (CV − MW, INT) were characterized by smaller values of dE (Fig. 3b). The ascorbic acid content per 100 g of fresh material (AA) was also higher for samples dried with combined techniques (Fig. 3b). Ascorbic acid is usually considered as a "quality marker" for many food products. If the retention of this thermal-resistant compound is high, it may be assumed that retention of other nutrients is also at a satisfactory rate. The highest amount of AA was observed after drying with pulsed microwave enhancement (I = 1000 W) during the intermittent program (INT3, Fig. 3b).

3.2 Convective–Microwave–Infrared Drying

Usually during the first period of drying, moisture-rich material is covered by a film of water. At this stage of drying, the rate of the process is controlled by the speed of evaporation of water from the free surface (water film). Because of the constant latent heat of evaporation, the overall speed of drying depends on the amount of energy delivered to the surface of the body. The more energy (higher temperature of the air) is delivered, the faster drying is conducted. Thus, it is reasonable to utilize more efficient, than convection, heat transport mechanism e.g. thermal radiation.

One of the most efficient methods of heat supply is infrared radiation (Ratti and Mujumdar 2014). This technique is quite easy to realize and relatively cheap in purchase and operation. Moreover, because of simple construction, the IR heaters can be successfully assembled to the existing dryers (e.g. tunnel dryers), without the necessity of the meaningful reconstruction in the device structure. In the case of the thin layer products (herbs, tea, leaves, seasonings, painting/coating layers, paper, pharmaceuticals, chemicals etc.) the infrared radiation is used for drying as a solely technique. Mostly, IR is applied rather as the enhancement during essential drying realized by convection, with microwaves, under the vacuum or during freeze drying. Two examples of convective–microwave–infrared hybrid drying are shown below: drying of potatoes and kaolin cylinder. During this research, the potatoes were dried with the convective (CV; T = 55 °C, v = 1.2 m/s), microwave (MW; f = 2.45 GHz, I = 100 W) and infrared radiation (IR; I = 250 W) techniques. The difference between the particular process of drying lays on the applied technique (CV, MV or IR) and the way of application (continuous'ct', periodical 'it').

In Fig. 4 the results obtained for potatoes dried with the hybrid processes are shown. One can see the average drying rate (DR) in the hybrid processes is significantly higher in comparison to solely convective (CV) or microwave (MW) processes. In effect, the overall drying time decreased meaningfully which leads to the lower energy consumption. Energy consumption was denoted as an energy consumed per kg of evaporated moisture. It can be easily noticed that hybrid programs of drying require less energy per kg of moisture to be evaporated than pure CV or MW ones (Fig. 4a). All these results follow from the comprehensive action of the particular drying techniques. Microwaves heated the interior of the material, the moisture evaporates and accelerates its transport to the surface. Infrared radiation caused faster evaporation of the pushed with microwaves moisture and finally dry hot air collect the vapor from the surface and took out the dryer.

The quality of products was in this case judged on the basis of several factors such as tactile feelings, odor, shape, water activity and color change. In Fig. 4b the relative color change (dE) and water activity (aw) of samples are presented. Relative color change characterizes the difference in color between fresh and processed material. The higher value of the dE is, the more difference between these two colors. The positive influence of the hybrid techniques on the color of the products obtained may be easily distinguished (Fig. 4b). However, the color changes were in all of the conducted processes meaningful (dE > 12), but the combined processes were characterized by smaller values of this parameter. Finally, water activity (aw) was also small enough to assure the safety and stability of the

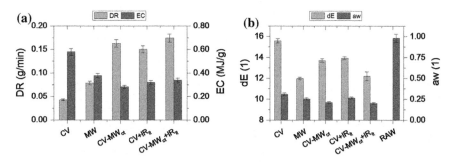

Fig. 4 Average values of: **a** drying rate (DR) and energy efficiency (EC), **b** relative color change (dE) and water activity (aw) obtained during hybrid drying of potatoes

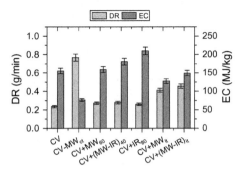

Fig. 5 Drying of kaolin cylinders—average drying rate (DR) and energy consumption per kg of evaporated water (EC) achieved in convective-microwave-infrared drier

product during longer storage or transportation. In all of drying programs, this parameter attained the value below 0.4 which is assumed as a critical one. According to Labuza et al. (1972), if water activity of a given material is below 0.6 then the growth of the pathogenic microorganisms (mold, yeast, bacteria) and the rate of deteriorative processes (lipid oxidation, enzymatic and non-enzymatic browning, hydrolysis) is inhibited.

The second example is drying of kaolin cylinders. Figure 5 presents the comparison of processes carried out with utilization of hot air (CV; T = 85 °C, v = 2 m/s), microwave (MW; f = 2.45 GHz, I = 100 W) and infrared (IR; I = 250 W) energy where some of them were performed with periodical energy application ('it'). It is visible, that microwaves on a par with infrared affected the significant increase in drying rate, and what's more important for some of the processes (CV − MW$_{ct}$ and periodical ones) allow to reduce the overall energy consumption. The reduction of energy consumption can be explained by higher summary efficiency of energy sources as also significant reduction of drying time (even over 60%), when additional energy is applied. From the analysis of the aforementioned parameters it can be stated that only appropriately designed hybrid drying processes can significantly influence the effectiveness of ceramic drying. In other cases the increase in drying rate will be redeemed by the increase in energy consumption and sometimes even decreased product quality.

3.3　Convective–Microwave–Ultrasound Drying

Another strategy is to apply the enhancement factor which may reduce both internal and external resistances during drying. One example is acoustic drying realized with the use of mechanical waves from the ultrasonic range. In this case, essential drying (e.g. hot air drying) is enhanced with the high power ultrasound during the whole process, periodically or at a given part of the process. The positive influence of the ultrasound on the drying kinetics and the products quality is doubtless and adequately discussed in the literature (Musielak et al. 2016; Chemat et al. 2017). Drying of apples, red pepper, raspberries and kaolin is shown below to imaging convective–microwave–ultrasound hybrid drying. The first example is the hybrid drying of the apple with convective (CV; T = 70 °C, v = 2 m/s), microwave (MW; f = 2.45 GHz, I = 100 and 250 W) and the ultrasound (US; f = 25 kHz, I = 200 W) techniques.

The results obtained during this research are presented in Fig. 6. In this case, the hybrid processes were faster in comparison to pure convective drying. Average drying rate (DR) was two- or three-fold higher compared to CV process (Fig. 6a). Moreover, the microwave was found as a more efficient enhancement factor than ultrasound. Whereas application of the ultrasound during convective drying influenced slightly both the rate of the process (DR) and the energy consumption (EC) but microwaves caused significant changes in these parameters (Fig. 6a). Nevertheless, the energy consumption of all hybrid schedules (especially microwave-assisted ones) was smaller in comparison to convective process.

Unfortunately, in this particular example neither positive nor negative influence of the hybrid processes on the quality of products was observed. Relative color change parameter (dE) of the samples dried with hybrid processes was at the same level or even higher in comparison to the solely convective one. It probably results from the different type of the experimental material and more rigid drying conditions (T = 70 °C, v = 2 m/s). Apple tissue is much more sensitive than potatoes and contains a lot of carbohydrates which may be easily converted into unfavorable compounds of the brown color (due to browning and Maillard reactions). Obtained results prove that proper application of the different drying techniques may result in the reduction of the drying time and energy consumption.

Subsequently, examples of hybrid drying of red pepper and raspberries were chosen. Red pepper samples in the form of strips and raspberry fruits were dried in a laboratory-scale hybrid dryer using four different drying programs such as convective drying (CV) as a reference test, hybrid drying, i.e. convective–microwave (CV − MW) and convective-ultrasound drying (CV − US), as well as

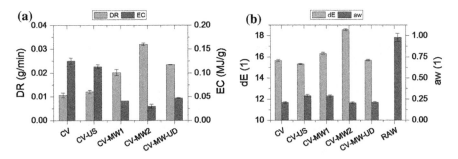

Fig. 6 Average values of: **a** drying rate (DR) and energy efficiency (EC), **b** relative color change (dE) and water activity (aw) obtained during hybrid drying of apples

hybrid-intermittent drying (INT). Each hybrid drying program was based on convection (CV) supported by microwaves (MW) and/or airborne ultrasound (US). The first three drying procedures including convective drying were carried out at stationary conditions (continuous application of MW/US), but the last procedure was conducted at non-stationary conditions (intermittent application of MW and US). In order to evaluate the effect of hybrid-intermittent drying on product quality, several quality factors such as total colour change (dE), water activity (aw) as well as retention of vitamin C (AA), carotenoids (CA) and anthocyanins (AC) were assessed.

Figure 7 shows the average drying rate (DR) and the average drying time (DT) for four different drying programs used in this study.

The results of drying rate and total drying time have shown that both microwaves and ultrasound have a positive effect on the drying kinetics of food materials. In case of drying rate the CV program proved to be the slowest and the longest drying process, as it revealed the lowest value of drying rate and the highest value of total drying time. On the other hand, the highest value of DR and the smallest value of DT were observed for CV − MW. In turn, for CV − US the drying time reduction was much smaller than for CV − MW, but its overall drying time was 55 and 70% shorter in comparison to CV, for raspberries and red pepper respectively. Furthermore, the influence of acoustic waves on drying rate of convective drying process was still meaningful, as DR was approximately two/three times higher than for CV. In case of the INT processes the drying rate was significantly higher than for CV. Moreover, intermittent MW and US in convective drying reduced the total drying time of raspberries by 87% and pepper by 60% in comparison with the CV process.

The energy efficiency of drying process is presented in Fig. 8. The CV schedule due to long-lasting processing time as well as a rather high drying temperature consumed the highest amount of energy and resulted in the lowest energy efficiency. In case of CV − MW drying the lowest value of EC was recorded. The CV − US procedure was characterized by a 18 and 40% reduction in EC as compared to CV, for red pepper and raspberries respectively. The INT drying

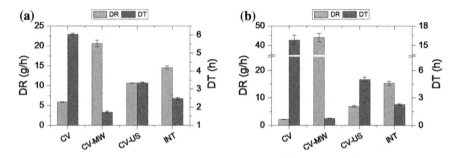

Fig. 7 Average drying rate (DR) and drying time (DT): **a** red pepper, **b** raspberries

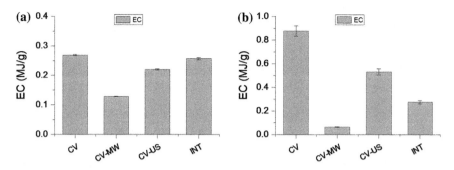

Fig. 8 Average energy consumption (EC): **a** red pepper, **b** raspberries

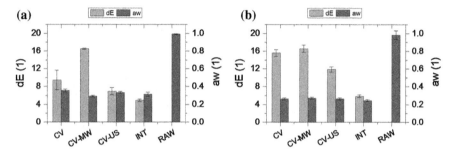

Fig. 9 Average colour change (dE) and water activity (aw): **a** red pepper, **b** raspberries

program demonstrates much lower energy efficiency as compared to CV − MW, but better than for CV, especially in case of raspberry samples.

The effect of different drying methods on the total colour change and water activity is shown in Fig. 9. In case of total colour change, it was found that raspberries dried in intermittent mode (INT) better retained their natural colour (smaller value of dE) in comparison to CV, CV − MW and CV − US. Moreover, each raspberry as well as pepper sample after drying was characterized by water activity below 0.4, the so-called critical water activity value, above which the speed of many reactions such as lipid oxidation, Maillard, hydrolysis, enzymatic and microbial growth increases significantly.

The influence of various drying schedules on the retention of vitamin C (AA), carotenoids (CA) and anthocyanins (AC) is shown in Fig. 10. CV − MW contributed to the lowest retention of vitamin C of all the analysed drying programs. In contrast, after CV drying a 68% of AA was observed. For the CV − US and INT program, a similar value of vitamin C retention was obtained, i.e. about 80%. As follows from Fig. 10, the retention of natural dyes, i.e. carotenoids (CA) and anthocyanins (AC) in the studied wavetables and fruits, correspond well with the total colour change (Fig. 9). The lowest CA and AC, i.e. 41 and 47%, were observed for the samples dried by CV − MW, where the highest difference in dE

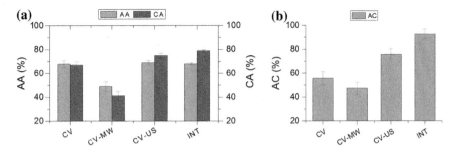

Fig. 10 Average retention of vitamin C (AA) and carotenoids (CA) in: **a** red pepper, **b** anthocyanins (AC) in raspberries

Fig. 11 Convective–microwave–ultrasounds hybrid drying of kaolin cylinder—average drying rate (DR) and energy consumption per kg of evaporated water (EC)

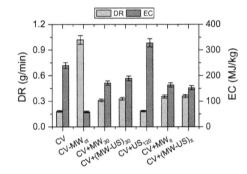

was observed. In case of CV, 67% of CA and 56% of AC were retained. In turn, the samples dried by CV − US and INT were characterized by a significantly higher retention of carotenoids and anthocyanins.

The next example of convective–microwave–ultrasound drying is drying of kaolin cylinder. Figure 11 presents the average drying rate (DR) and energy consumption (EC). Each combined drying process (apart CV + US$_{120}$ program) reaches higher drying rates than CV. Due to the improvement of water removing, aside from CV + US$_{120}$ program all combined processes allow to reduce also the energy consumption per kg of removed water in the range from about 20% (CV + (MW − US)$_{30}$) up to almost 80% (CV − MW$_{ct}$). This fact is very promising as any reduction of energy consumption in the industry is very much appreciated. These results could be compared with convective–microwave–infrared hybrid drying. Comparing similar processes presented in Figs. 5 and 11, it can be stated that construction of the drier have a significant impact on process effectiveness.

With regards to constructive materials like different types of ceramic, except for drying kinetics or process energy efficiency, one more parameter is very important from the usable point of view, namely product quality. It can be assessed in different ways, depending on investigated parameter, like e.g. appearance, structure deformations, product strength, insulating properties, thermal resistance etc.

Fig. 12 An exemplary strength of kaolin samples dried in hybrid processes

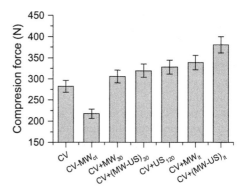

Figure 12 presents the strength of kaolin cylinders processed by using hybrid drying including hot air (CV), microwave (MW) and ultrasound energy (US) utilization. The strength was evaluated during compression test in which maximum compression force was collected. It can be seen, that very fast and energy efficient processes like e.g. $CV - MW_{ct}$ or $CV + MW_{30}$ lead to material strength decrease in comparison to CV process. Nevertheless application of additional energy source in the form of high power ultrasounds e.g. in $CV + (MW - US)_{30}$ process changes this dependency. Processes in which ultrasounds were applied give stronger products and their strength is even over 30% higher than in reference CV drying. What is very interesting and have to be kept in mind, not always all quality parameters are coherent with other ones. Such effect was observed e.g. for $CV + (MW - US)_{it}$ dried samples where the surface had more cracks than samples dried according $CV + MW_{30}$ but the product strength was higher for the samples dried with ultrasounds. This can be an effect of acoustic wave vibration what affects compression of material particles inside dried body.

3.4 Microwave–Vacuum Drying

Microwave–vacuum drying is a method of dewatering that has become widespread in the last two decades. This process combined the advantages of a pressure decrease (low process temperature and fast mass transfer) and microwave drying (rapid energy transfer). As a result, a process can be obtained which is faster than the convective process, has a lower temperature and, both, a faster mass transfer and rapid energy transfer. The quality parameters of products dried using the microwave–vacuum technique are usually slightly lower than those obtained during freeze drying, but they are much higher than those obtained using direct hot air drying (Hu et al. 2006). Therefore, microwave–vacuum drying is tested for drying thermo-sensitive materials such as plants, especially fruits and vegetables (Cui et al.

2004; Figiel 2009), medicines (McLoughlin et al. 2003), or wood (Oloyede and Groombridge 2000).

In this chapter drying of carrots, oak and pine wood will be presented as examples. In the first series of experiments carrot samples were dried using four levels of fixed microwave powers: 150 W ("A0"), 200 W ("A1"), 250 W ("A2"), and 300 W ("A3"). The abbreviated designations for each series of measurements are in parentheses.

It is obvious that the higher the power, the higher the drying rate (Fig. 13a). But this drying acceleration caused damage to both of the dried materials. Visual assessment of carrot samples showed that only samples dried using the lowest microwave power (150 W) did not have any visible damage, thus the greater the drying power the worse the appearance of the samples. This is confirmed by the measurements of color changes by drying. The value of the color difference parameter dE was the lowest in the case of program A0 (150 W) (Fig. 13b). However, the use of such low microwave power caused drying time extension (Fig. 13).

Additionally, in case of carrots the results of β-carotene loss (Fig. 14) showed that, both, a long drying time and a higher microwave power had a negative influence on the β-carotene content in the product. The A1 program had the best drying results (95.43%). But, as it has been mentioned before, this caused visible damage to the material. For this reason, when using a constant microwave power during drying it was difficult to choose a program which would provide good results. For that reason, non-stationary microwave–vacuum drying was studied.

The slowest constant drying program (A0) was accelerated by using a higher power microwave at short intervals (the obtained programs were marked with the

Fig. 13 Microwave–vacuum drying of carrots: **a** drying rates (DR), **b** average color change (dE)

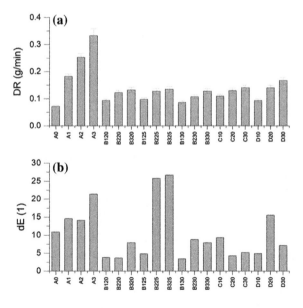

letter B). It was assumed that a higher microwave power be used for 5 min. Between these intervals a microwave power of 150 W at intervals of 20, 25 and 30 min was used. This created nine series of experiments. In the abbreviation of the processes' names the first numeral indicates the microwave's power level (1–200, 2–250, and 3–300 W) and the last two numerals correspond to the duration time of the 150 W microwave-power intervals. Average drying rate of these experiments are shown in Fig. 13a. As expected, the use of higher power for short periods of time resulted in an acceleration of the process (compare B125, B225, B325 with A0). The experiments showed that the higher the microwave power applied, the shorter the drying time was.

Both visual validation and the values of the color difference parameter dE (Fig. 13b) showed that the best results were obtained when the difference between the two microwave powers used was the lowest (B1 programs). It was also observed that increasing the microwave power at the end of the process (the last interval of the higher power applied) resulted in damages to the materials (B225 and B325 programs). The color difference parameter is between 3.5 and 25. This means that the dried samples' color change varied to a great extent. Therefore, these processes allowed to improve the parameter, although there was also the risk of a great change in color. In the case of the β-carotene content the best results were obtained when programs with 20-min intervals of 150 W microwave power were used (see Fig. 14). But the loss of β-carotene is higher than in the A programs, (especially A1), and there were no benefits in the drying time. Thus, it can be stated that drying at periodically changed microwave power is not a good solution.

It was the reason for the subsequent series of experiments. In the course of these series a higher microwave power was initially used, and when dimensionless moisture contents crossed the set value the microwave power was reduced to 150 W. The cross values were fixed at 0.54 and 0.43 of moisture ratio MR (ratio of actual to initial moisture content). This created six series of experiments. The abbreviated designations of individual drying programs were C (microwave power lower when MR = 0.54) and D (microwave power lower when MR = 0.43). The number in the abbreviation indicates the initial microwave power (10–200, 20–250, 30–300 W). The average drying rates under two-stage microwave–power drying

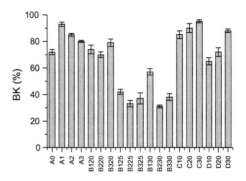

Fig. 14 Microwave–vacuum drying of carrots—β-carotene retention

conditions are shown in Fig. 13a. It is obvious that the higher the initial microwave power, the faster the drying. Therefore, in these experiments significant acceleration of drying were obtained as compared to the slowest program (A0).

Visual observations and the colorimetric measurements showed that the color change of the samples decreased in comparison to constant power drying (except program D20). It has to be pointed out that the results were better for the C than the D programs. Also, the β-carotene loss in carrot was higher for the D programs (Fig. 14). However, the obtained results showed that β-carotene preservation was higher when the initial microwave power was the highest (Fig. 14), and that the results for the C30 (97.34%) and D30 (90.60%) programs were similar to those obtained in A1 (95.43%). For this reason it can be concluded that the two-stage program with the greatest differences between the two used microwave powers allowed to obtain the best results.

During the third series of experiments the best results (no damage of samples, little color change, short time and high β-carotene preservation) were obtained. The results of the C30 drying program were the best. Therefore, the two-stage program, with the greatest difference between the two microwave powers that was used, was recognized as the best method of microwave–vacuum drying. The study confirmed that using variable power microwave drying allows to, both, shorten the time and improve the quality of the product.

Drying of wood is chosen as the second example of microwave–vacuum drying. In the case of timber the high strength requirements (for construction wood) and the color of the wood is important. Species containing larger amounts of tannins, including oak and pine, darken relatively quickly in contact with iron salts and air and light, which may limit their use in the furniture industry, musical instruments or artwork. Therefore, it was decided to present the influence of microwave–vacuum drying (MVD) on process kinetics and quality (mechanical strength) and visual color change.

Samples of oak wood (*Quercus* L.) measuring $15 \times 20 \times 150$ mm with 65% initial moisture content and pine wood (*Pinus silvestris* L.) measuring of $15 \times 25 \times 150$ mm at initial moisture content of 95% were used in the studies. Criterion of the end of the process was to achieve a 10% moisture content. After the drying process, static bending tests were performed. The design of the dryer allowed the continuous measurement of sample mass and pressure in the vacuum chamber. Measurements were made for three levels of microwave input power: 200, 300 and 400 W. The air pressure in the chamber ranged between 15–40 mmHg depending on the stage of the process. Comparative convective drying for pine wood was carried out at a drying air temperature of 120 °C, the duration of the process was 140 min. The use of microwave as a heat source shortens the drying time according to the principle, the higher the power, the shorter the drying time (15–40 min for both species), which improves the efficiency of the process. Strength tests did not show the relationship between drying parameters and mechanical properties, which can be seen as an advantage. As far as their impact on the mechanical properties of the samples is concerned, no deformation, cracks or mechanical deterioration in relation to convection drying are observed. Volumetric heating of the material and compatibility of the diffusion and thermodiffusion moisture flux have reduced drying stresses. Unfortunately, there is no compatibility in this matter in the literature. According to literature (Oloyede and Groombridge 2000) microwave drying reduces the strength of wood by up to 60% compared to the strength of wood traditionally dried. Hansson and Antii (2003) claim that the drying method does not, in principle, affect the mechanical strength of the wood. The use of low pressure air as a drying medium has reduced the boiling point of the water and the oxygen content of the chamber, which significantly reduced the darkening of the wood. This is in good agreement with literature (Seyfarth et al. 2003; Sandoval-Torres et al. 2010).

4 Conclusions

Non-stationary drying and hybrid drying provide an essential improvement to simple and stationary techniques. Regardless of the techniques utilized in a particular combined drying process, the advantages of the hybrid and non-stationary drying are clear. The rate of drying is usually significantly higher leading to the reduction of the operational time. Shorter drying and milder drying conditions usually give the high-quality products. Finally, the energy efficiency of such processes is greater compared to solely convective processes. They are good ways of enhancement efficiency of existing drying systems as well as modernization of these systems. The key meaning factor is a proper construction of the hybrid schedules to avoid deterioration of the products quality or ineffective energy consumption. New drying programs have to be designed properly to take into account as many parameters as possible to improve the overall effectiveness of drying and not only one of them.

Acknowledgements This work was carried out as a part of the research project No 03/32/DSPB/ 0705 founded by Poznan University of Technology. The studies on drying of red pepper and raspberries were conducted as a part of the research project No 2014/15/D/ST8/02777 sponsored by the National Science Centre in Poland.

References

Chemat F, Rombaut N, Sicaire A-G et al (2017) Ultrasound assisted extraction of food and natural products. Mechanisms, techniques, combinations, protocols and applications. A review. Ultrason Sonochem 34:540–560
Chou SK, Chua KJ (2001) New hybrid drying technologies for heat sensitive foodstuffs. Trends Food Sci Technol 12(10):359–369
Claussen IC, Yang TS, Strommen I et al (2007) Atmospheric freeze drying—a review. Dry Technol 25(4–6):947–957

Cui Z-W, Xu S-Y, Sun D-W (2004) Effect of microwave-vacuum drying on the carotenoids retention of carrot slices and chlorophyll retention of Chinese chive leaves. Dry Technol 22 (3):563–575

Figiel A (2009) Drying kinetics and quality of vacuum-microwave dehydrated garlic cloves and slices. J Food Eng 94(1):98–104

Fushimi C, Dewi WN (2015) Energy efficiency and capital cost estimation of superheated steam drying processes combined with integrated coal gasification combined cycle. J Chem Eng Jpn 48(10):872–880

Hansson L, Antti AL (2003) The effect of microwave drying on Norway spruce woods strength: a comparison with conventional drying. J Mater Process Technol 141(1):41–50

He Q, Wang X (2015) Drying stress relaxation of wood subjected to microwave radiation. BioResources 10(3):4441–4452

Herrithsch A, Dronfield J, Nijdam J (2008) Intermittent and continuous drying of red-beech timber from the green conditions. In: Proceedings of the 16th international drying symposium, Hyderabad, India, pp 1114–1121

Hu Q-G, Zhang M, Mujumdar AS et al (2006) Effects of different drying methods on the quality changes of granular edamame. Dry Technol 24(8):1025–1032

Itaya Y, Hasatani M (1996) R&D needs—drying of ceramics. Dry Technol 14(6):1301–1313

Itaya Y, Mori S, Hasatani M (2014) Drying of ceramics. In: Handbook of industrial drying, 4th Edition. CRC Press, pp 717–728

Jangam SV, Mujumdar AS (2011) Heat pump assisted drying technology—overview with focus on energy, environment and product quality. In: Tsotsas E, Mujumdar AS (eds) Modern drying technology. Wiley-VCH Verlag GmbH & Co. KGaA, Weinheim, pp 121–162

Kowalski SJ, Pawłowski A (2010a) Drying of wood with air of variable parameters. Chem Process Eng 31(1):135–147

Kowalski SJ, Pawłowski A (2010b) Drying of wet materials in intermittent conditions. Dry Technol 28(5):636–643

Kowalski SJ, Szadzińska J (2012) Non-stationary drying of ceramic-like materials controlled through acoustic emission method. Heat Mass Transf 48(22):2023–2032

Kudra T (2004) Energy aspects in drying. Dry Technol 22(5):917–932

Kudra T, Mujumdar AS (eds) (2009) Advanced drying technologies, 2nd edn. CRC Press, Boca Raton (USA)

Kumar C, Karim MA, Joardder MUH (2014) Intermittent drying of food products: a critical review. J Food Eng 121(1):48–57

Labuza TP, McNally L, Gallagher D et al (1972) Stability of intermediate moisture foods. 1. Lipid oxidation. J Food Sci 37(1):154–159

Leeratanarak N, Devahastin S, Chiewchan N (2006) Drying kinetics and quality of potato chips undergoing different drying techniques. J Food Eng 77(3):635–643

McLoughlin CM, McMinn WAM, Magee TRA (2003) Microwave-vacuum drying of pharmaceutical powders. Dry Technol 21(9):1719–1733

Mujumdar AS (2004) Research and development in drying: recent trends and future prospects. Dry Technol 22(1–2):1–26

Mujumdar AS (2007) An overview of innovation in industrial drying: current status and R&D needs. Transp Porous Media 66(1–2):3–18

Musielak G (2004) Modeling and numerical simulation of transport phenomena and drying stresses in capillary-porous materials. Poznań University of Technology, Poznań

Musielak G, Mierzwa D, Kroehnke J (2016) Food drying enhancement by ultrasound—a review. Trends Food Sci Technol 56:126–141. https://doi.org/10.1016/j.tifs.2016.08.003

Musielak G, Śliwa T (2015) Modeling and numerical simulation of clays cracking during drying. Dry Technol 33(14):1758–1767

Oloyede A, Groombridge P (2000) The influence of microwave heating on the mechanical properties of wood. J Mater Process Technol 100(1–3):67–73

Parikh DM (2015) Vacuum drying: basics and application. Chem Eng 122(4):48–54

Pawłowski A (2011) Efficiency of drying of saturated porous materials in non-stationary conditions. Poznań University of Technology, Poznań

Ratti C, Mujumdar AS (2014) Infrared drying. In: Handbook of industrial drying, 4th Edition. CRC Press, pp 405–420

Sandoval-Torres S, Jomaa W, Marc F et al (2010) Causes of color changes in wood during drying. For Stud China 12(4):167–175

Seyfarth R, Leiker M, Mollekopf N (2003) Continuous drying of lumber in a microwave vacuum kiln. In: Proceedings of 8th international IUFRO wood drying conference, pp 159–163

Sokhansanj S, Jayas DS (2014) Drying of foodstuffs. In: Handbook of industrial drying, 4th Edition. CRC Press, pp 521–544

Song XJ, Zhang M, Mujumdar AS, Fan L (2009) Drying characteristics and kinetics of vacuum microwave–dried potato slices. Dry Technol 27(9):969–974

Strumiłło C, Jones PL, Żyłła R (2014) Energy aspects in drying. In: Mujumdar AS (ed) Handbook of industrial drying, 4th Edition, CRC Press, pp 1077–1100

Wang R, Zhang M, Mujumdar AS (2010) Effect of osmotic dehydration on microwave freeze-drying characteristics and quality of potato chips. Dry Technol 28(6):798–806

Wang Y, Zhang M, Mujumdar AS (2014) Microwave-assisted drying of foods—equipment, process and product quality. In: Tsotsas E, Mujumdar AS (eds) Modern drying technology. Wiley-VCH Verlag GmbH & Co. KGaA, Weinheim, pp 279–315

Zhang W, Mujumdar AS (1992) Deformation and stress analysis of porous capillary bodies during intermittent volumetric thermal drying. Dry Technol 10(2):421–443

The Use of Pressure Membrane Separation for Heavy Metal Removal or Recovery

Arkadiusz Nędzarek

1 Introduction

Pressure membrane separations using a sieve effect are effective separation/ concentration techniques for solutions of various components (elements, ions, organic compounds, microorganisms) (Drost et al. 2014, 2016; Nędzarek et al. 2015a, b). The driving force of the process is the differential pressure on the feed side and the permeate. Due to the size of particles separated in the process, except for reverse osmosis (RO), microfiltration (MF), ultrafiltration (UF) and nanofiltration (NF) are distinguished. The volume of separated substances is 0.0001, 0.1, 0.01, and 0.001 μm, respectively. Plastics (polymeric membranes) as well as ceramic and metal sinter (e.g. Staszak et al. 2013) are used for the production of membranes. Due to the diversity of polymer materials and their interaction with filtered solutions, polymeric membranes are among the best tested and most commonly used. Over the last two decades the popularity of inorganic membranes has increased, including ceramic membranes. These membranes, in contrast to polymeric membranes, are resistant to high temperatures, chemically resistant, can be used over a wide pH range and are hydrophilic. They are also characterized by long life and high hydraulic permeability, which is why they are used for separation of compounds with high fouling potential, such as proteins (Kuca and Szaniawska 2009; Barredo-Damas et al. 2010; Kujawski et al. 2016).

Membrane filtration (RO, UF and NF) is a promising technique for the removal and/or concentration of heavy metals. The effect is high process efficiency, easy handling and space saving (Fu and Wang 2011). In membrane technology a relatively new technique is nanofiltration. Compared to reverse osmosis, NF allows for higher flow rates, and with respect to ultrafiltration, much better retention of such

A. Nędzarek (✉)
Department of Aquatic Sozology, West Pomeranian University
of Technology in Szczecin, Szczecin, Poland
e-mail: anedzarek@zut.edu.pl

© Springer International Publishing AG, part of Springer Nature 2018
M. Ochowiak et al. (eds.), *Practical Aspects of Chemical Engineering*,
Lecture Notes on Multidisciplinary Industrial Engineering,
https://doi.org/10.1007/978-3-319-73978-6_23

Fig. 1 Stream flow configuration: **a** dead end mode, **b** cross flow filtration

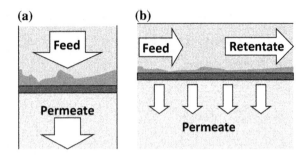

molecules as sugars, amino acids, peptides and ions. The distribution of NF feeder components is the result of a complex spatial mechanism, Donnan effect and dielectric effects. Modification of the selective layer of NF membrane with polymer chains offers an opportunity of creating a three-dimensional network, so that filtration can be regulated by manipulating its pore size e.g. by changing pH or ionic strength (Lou and Wan 2013).

The possible retention of low molecular weight substances has led to the recent increase in the use of nanofiltration in water treatment, reuse of waste water, desalination of seawater, fractionation and selective removal of dissolved substances from complex process solutions (Barakat 2011; Lou and Wan 2013).

Membrane filtration is carried out in two configurations: dead end mode (feeder stream is perpendicular to the membrane) and cross flow filtration (feeder and permeate streams are crossed) (Fig. 1).

Due to the sieving characteristics and the formation of fouling and/or scaling, the membrane separation is accompanied by a decrease in the volumetric permeate stream over time (Das et al. 2006; Bes-Piá et al. 2010). These phenomena are caused by depositing organic and inorganic substances on the outer surface of the membrane and in its pores (Amy 2008; Bellona et al. 2010; Arkhangelsky et al. 2012). The so-called filter cake thus formed acts as an additional filter capable of retaining feeds with a lower molecular weight than the membrane cut-off (Oliveira et al. 2013).

2 Removal/Concentration of Heavy Metals

Industrial wastewater with heavy metals is mainly subjected to the purification process, but metal recovery offers greater benefits. In this case, membrane separations may be an alternative technique for such methods as chemical precipitation, extraction, adsorption, ion exchange, and electrochemical removal (Kurniawan et al. 2006; Barakat 2008, 2011; Fu and Wang 2011; Keng et al. 2014).

The effectiveness of these processes varies. Very high efficiency is shown by ion exchange and chemical precipitation. For example, Al-Enezi et al. (2004) showed a 99% reduction in heavy metals by extraction from wastewater and sediment.

Chen et al. (2009) obtained precipitation of Cu, Zn, Cr and Pb by CaO at 99.37–99.6%. The efficiency of ion exchange and adsorption of metals depend on the type of metal and on the ion-exchange resins and adsorbents used (Alyüz and Veli 2009; Fu and Wang 2011; Barakat 2011). For example, hydroxyapatite microparticles give higher copper adsorption (>84%) than nickel (79%), iron (74%), zinc (72%) and cadmium (49%) (Moayyeri et al. 2013). Aziz et al. (2008) showed more than 57% removal of Zn using the lignite fly ash. In ion exchange, synthetic resins are mainly used, and the use of zeolites and montmorillonites is currently limited to laboratory experiments (Fu and Wang 2011; Keng et al. 2014). It should be emphasized that toxic sludge can develop in these processes, and the technologies used may have high-energy requirements (Eccles 1999; Keng et al. 2014). Still new precipitants, ion exchangers or absorbing materials are being tested. For example, low-cost absorbents from natural organic materials (e.g. cocoa shell, sugar beet pulp) converted to activated carbon are tested in the adsorption process (Kurniawan et al. 2006).

Removal of metals, by each of the aforementioned methods, strictly depends on the concentration of H+ ions. It should be remembered that the form in which a metal is present depends on pH: the acidic environment is dominated by metal cations, while with increasing pH the hardly soluble carbonates and hydroxides are formed. Also removing metals by means of precipitating, absorbing or ionizing agents requires appropriate reaction. For example, the highest removal efficiency of lead in the form of hydroxide was obtained at pH 10. However, further reduction of H+ concentration resulted in the reduction of the lead removal rate, which was due to the formation of plumbates. Similarly, in modified processes, the concentration of H+ ions should be at the optimum level thus providing the maximum removal of metals. For example, in coagulation with aluminum or iron salts, on $Al(OH)_3$ and $Fe(OH)_3$ fluffs, hardly soluble heavy metal compounds are absorbed and H+ ions concentration should provide on one hand the minimum solubility of aluminum (III) or iron (III) hydroxides and, on the other hand, the formation of hardly soluble bonds of removed heavy metal. Optimal value of the reaction is individual for a particular metal and for the coagulant used. For example, the maximum removal efficiency of Pb, Cu, Zn or Ni is achieved at pH 10–11 while for as it is reached at pH 7 (Sigworth and Smith 1972; Culedge and O'Connor 1973).

Also, the efficiency of membrane separation in heavy metal removal is varied (Fu and Wang 2011; Pellegrin et al. 2016). As shown in Table 1, removal efficiency (in %) of heavy metal varies from 40 to almost 100. The choice of filtration type depends on the type of particles to be retained. High removal efficiency of heavy metals is achieved by reverse osmosis (Moshen-Nia et al. 2007). For example, Abu Qdaisa and Mous-sab (2004), while treating effluents containing copper and cadmium ions with RO, achieved 98 and 99% removal efficiency respectively. However, the disadvantage of this method is high energy consumption required to generate high pressure and restoration of the membranes (Fu and Wang 2011). Significantly lower trans membrane pressure is used in ultrafiltration and nanofiltration. These techniques use membranes whose size is larger than the size of heavy metal ions. Therefore, the lower rates of rejection of metals by these membranes are

Table 1 Comparison of heavy metal removal using reverse osmosis (RO), nanofiltration (NF), enhanced micellar (MEUF) and polyelectrolytes (PEUF)

Membrane	Metal	Removal efficiency (%)	References
RO	Cd	>97	Kheriji et al. (2015)
RO	Cu, Ni	99.5	Mohsen-Nia et al. (2007)
RO	Cu	70–95	Zhang et al. (2009)
NF	Cu	47–66	Chaabane et al. (2006)
NF	Cd	40–70	Kheriji et al. (2015)
NF	Cr(VI)	99.5	Muthukrishnan and Guha (2008)
RO + NF	Cu	>95	Csefalvay et al. (2009)
RO + NF	Cu	95–99	Sudilovskiy et al. (2008)
NF + BSA	Cu, Cd, Pb, Zn, Fe	88–98	Nędzarek et al. (2015a)
MEUF (SDS)	Cd	83–89	Staszak et al. (2013)
	Ni	>85.52	Khalid et al. (2015)
MEUF (DBSA)	Pb	>99	Ferella et al. (2007)
MEUF (CTAB)	Cr(VI)	>95	Nguyen et al. (2015)
PEUF (PVAm)	Hg	50–99	Huang et al. (2015)
PEUF (CMC)	Cu, Cr(III), Ni	97.6, 99.5, 99.1	Barakat and Schmidt (2010)
PEUF (PEI)	Cu, Zn, Ni, Cr, Co, Cd	>90	Almutairi et al. (2012)

MEFU Micellar enhanced ultrafiltration; *PEUF* Polyelectrolyte enhanced ultrafiltration; *BSA* Bovine serum albumin; *CMC* Carboxy methyl cellulose; *CTAB* Cetyl trimethylammonium bromide; *DBSA* Dodecylbenzenesulfonic acid; *PEI* Polyethylenimine; *PVAm* Polyvinylamine; *SDS* Sodium dodecyl sulfate

achieved. For example, Chaabane et al. (2006) in the NF process they achieved a removal efficiency of copper in the range of 47–66%. However, Nędzarek et al. (2015a) in the NF process (at pH 2.0–4.6) obtained the following retention: 40% for Cd, 47% for Fe, 50% for Pb, 36% for Zn, and 72% for Cu. At such a low pH, the metals are present as free ions and the lack of soluble charged hydroxides of these metals prevents the formation of an active layer on the surface of the membrane. This results in high permeability. Increased metal rejection can be achieved by modifying the pH of the feed. In the alkaline environment, the proportion of hardly soluble metal hydroxides rises, which, by electrostatic interaction, form an active filter layer on the surface of the membrane leading to an increase in the retention degree (Luo and Wan 2013). This increase in metal retention was demonstrated by, for example, Nędzarek et al. (2015a), who by increasing the pH of the feed to 9.0 in the tested NF recorded the following retention: 89% for Cd, 80% for Fe, 85% for Pb, 69% for Zn, and 98% for Cu.

The effect of concentration of H+ ions on separation processes is complex. This is due to the multi-directional interaction between the membrane and the solution depending on the chemical composition of the membrane itself. For example, the polymeric polyamide membrane has hydrophilic properties and dissociable carboxylic and amine groups and may have a positive or negative surface charge depending on pH (Manttari et al. 2006). Similarly, ceramic membranes also show a change in surface charge depending on pH. At pH below PZC the membrane surface charge is positive and negative above this PZC (Nędzarek et al. 2015a). Positive surface charge causes electrostatic repulsion of metal ions and higher permeability of the membrane. With the pH increase, the negatively charged membrane, as a result of electrostatic attraction, can absorb the counterions on the membrane surface and in its pores. This leads to the formation of electric double layer, thereby increasing the electro-viscous effect with simultaneous reduction of the permeate stream and greater retention of the feed components (Bowen and Yousef 2003).

Increased metal retention in NF can be achieved, for example, by complexing metals such as bovine serum albumin (BSA). The retention growth can be achieved already in the acid environment. For example, at pH 2.0–4.6 the metal retention in the Metal + BSA configuration was 81% for Cd, 65% for Fe, 55% for Pb, 62% for Zn, and 75% for Cu, while at pH 9.0 the retention of these metals was 95% % for Cd, 88% for Fe, 89% for Pb, 90% for Zn and 98% for Cu (Nędzarek et al. 2015a). The increase in metal retention is due to BSA metal complexing and the formation of fouling due to BSA-membrane interaction. For organic/inorganic mixtures, the size of the components and the size of their charges are important—with larger sizes the organic compounds also have greater charges so that their interaction with the membrane leads to the formation of a fouling layer throughout the pH range of the solution and retention growth (Freger et al. 2000).

In ultrafiltration, the pore sizes are also larger than the hydrated complexes of heavy metals so they can easily pass through the UF membranes. To achieve a high efficiency of metal removal, modified systems such as micellar enhanced ultrafiltration (MEUF) and polyelectrolyte enhanced ultrafiltration (PEUF) are tested. MEUFs use surfactants whose concentration in aqueous solution must exceed the critical micelle concentration to aggregate into micelle-binding metals. This results in large structures that can be trapped by the UF membrane. In order to obtain high retention, surfactants with electrical charge opposite to metal ions should be used. As shown in Table 1, MEUF metal rejection rates are in the range of 88–99%. Similarly, high rejection of metals is recorded in PEUF. This technique uses water-soluble polymers which with metal ions form macromolecules with a molecular weight greater than the cut-off of the UF membrane. MEUF and PEUF efficiency depends on the nature of the metal, its concentration, surfactant and complexing agent, pH of the solution, ionic strength and membrane interaction (Fu and Wang 2011).

3 Membrane Wash

The decreased permeate flow efficiency observed during filtration (e.g. Acero et al. 2010; Bellona et al. 2010) is a disadvantage for both economic and operational reasons (including that filtrate quality deteriorates). Therefore, it is necessary to periodically wash the membrane. The simplest method of washing is periodic hydraulic flushing of membranes, which results in increased productivity (e.g. Katsoufidou et al. 2005; Bonisławska et al. 2016). This process removes the outer layer of pollution (so called reversible fouling). Forward flushing can be used (membrane is cleaned in the flow direction) and the backwash (washing from the center to the outside). These processes are schematically illustrated in Fig. 2.

However, in order to recover the initial capacity of the stream it is necessary to carry out the chemical cleaning that will remove fouling and/or scaling of membranes. These impurities also contribute to the change in the type and density of surface charge (Religa et al. 2011). As a consequence, the change in membrane selectivity results in a change in the retention of feed components and may lead to an increase in the concentration polarization of the membrane and an additional decrease in its permeability. Therefore, chemical cleaning should simultaneously remove impurities from the membrane, restore its original efficiency and its original charge. The most beneficial effects are obtained when using integrated acid wash systems (e.g. HNO_3, HCl) and sodium hydroxide. For example, Kowalik-Klimczak et al. (2014) analyzing the effect of various washing baths on the permeability and selective properties of the NF membrane has shown that the use of individual washing liquors in the form of hydrochloric acid or sodium hydroxide prevents recovery of the negative membrane surface charge and changes its selective properties (reduces retention of Cr^{+3} and Cl^-). The zeta potential (at pH 4) of the new membrane was −4.0 mV, whereas after washing with HCl it was 0.8 mV and 5.8 mV NaOH. Only the use of sodium hydroxide bath after hydrochloric acid treatment led to the recovery of the negative surface charge of the tested membrane.

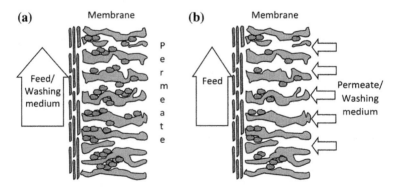

Fig. 2 Membrane wash diagram: **a** forward wash, **b** return wash

4 Summary

Membrane filtration is an innovative technique for industrial waste treatment containing heavy metals and can be an alternative way to recover them. Comparatively high levels of heavy metals elimination compared to conventional methods (i.e. chemical precipitation, adsorption, ion exchange) are obtained in reverse osmosis as well as for MEUF and PEUF. However, the accumulation of solid and colloidal particles on the feed side causes membrane fouling and permeate flux decline. To improve the efficiency of the process, it is necessary to wash the membranes periodically.

References

Abu Qdaisa H, Moussab H (2004) Removal of heavy metals from wastewater by membrane process: a comparative study. Desalination 164:105–110

Acero JL, Benitez FJ, Leal AI, Real J, Teva F (2010) Membrane filtration technologies applied to municipal secondary effluents for potential reuse. J Hazard Mater 177:390–398

Al-Enezi G, Hamoda MF, Fawzi N (2004) Ion exchange extraction of heavy metals from wastewater sludges. J Environ Sci Health A Tox Hazard Subst Environ Eng 39(2):455–564

Almutairi FM, Williams PM, Lovitt RW (2012) Effect of membrane surface charge on filtration of heavy metal ions in the presence and absence of polyethylenimine. Desalin Water Treat 42:131–137

Alyüz B, Veli S (2009) Kinetics and equilibrium studies for the removal of nickel and zinc from aqueous solutions by ion exchange resins. J Hazard Mater 167:482–488

Amy G (2008) Fundamental understanding of organic matter fouling of membranes. Desalination 231:44–51

Arkhangelsky E, Wicaksana F, Cou S, Al-Rabiah AA, Al-Zahrani SM, Wang R (2012) Effect of scaling and cleaning on the performance of forward osmosis hollow fiber membranes. J Membr Sci 415–416:101–108

Aziz HA, Adlan MN, Ariffin KS (2008) Heavy metals (Cd, Pb, Zn, Ni, Cu and Cr(III)) removal from water in Malaysia: post treatment by high quality limestone. Bioresource Technol 99:1578–1583

Barakat MA (2008) Removal of Cu(II), Ni(II), and Cr(III) ions from wastewater using complexation-ultrafiltration technique. J Environ Sci Technol 1(3):151–156

Barakat MA (2011) New trends in removing heavy metals from industrial wastewater. Arab J Chem 4:361–377

Barakat MA, Schmidt E (2010) Polymer-enhanced ultrafiltration process for heavy metals removal from industrial wastewater. Desalination 256:90–93

Barredo-Damas S, Alcaina-Miranda MI, Bes-Piá A et al (2010) Ceramic membrane behavior in textile wastewater ultrafiltration. Desalination 250:623–628

Bellona C, Marts M, Drewes JE (2010) The effect of organic membrane fouling on properties and rejection characteristics of nanofiltration membranes. Sep Purif Technol 74:44–54

Bes-Piá A, Cuartas-Uribe B, Mendoza-Roca JA et al (2010) Study of the behavior of different NF membranes for the reclamation of a secondary textile effluent in rinsing processes. J Hazard Mater 178:341–348

Bonisławska M, Nędzarek A, Drost A et al (2016) The application of ceramic membranes for treating effluent water from closed-circuit fish farming. Arch Environ Prot 42(2):59–66

Bowen WR, Yousef HNS (2003) Effect of salts on water viscosity in narrow membrane pores. J Colloid Interf Sci 264:452–457

Chaabane T, Taha S, Taleb Ahmed M et al (2006) Removal of copper from industrial effluent using a spiral wound module—film theory and hydrodynamic approach. Desalination 200:403–405

Chen QY, Luo Z, Hills C et al (2009) Precipitation of heavy metals from wastewater using simulated flue gas: sequent additions of fly ash, lime and carbon dioxide. Water Res 43:2605–2614

Csefalvay E, Pauer V, Mizsey P (2009) Recovery of copper from process Walters by nanofiltration and reverse osmosis. Desalination 240:132–142

Culedge JH, O'Connor JT (1973) Removal of arsenic V from water by adsorption on aluminium and ferric hydroxides. J Am Water Works Ass 65(8):548–552

Das C, Patel P, De S et al (2006) Treatment of tanning effluent using nanofiltration followed by reverse osmosis. Sep Purif Technol 50:291–299

Drost A, Nędzarek A, Bogusławska-Wąs E et al (2014) UF application for innovative reuse of fish brine: product quality, CCP management and the HACCP system. J Food Process Eng 37:396–401

Drost A, Nędzarek A, Tórz A (2016) Reduction of proteins and products of their hydrolysis in process of cleaning post-production herring (*Clupea harengus*) marinating brines by using membranes. Membr Water Treat 7(5):451–462

Eccles H (1999) Treatment of metal-contaminated waste: why select a biological process? Trends Biotechnol 17:462–465

Ferella F, Prisciandaro M, Michelis ID, Veglio F (2007) Removal of heavy metals by surfaktant-enhanced ultrafiltration from wastewaters. Desalination 207:125–133

Freger V, Arnot TC, Howell JA (2000) Separation of concentrated organic/inorganic salt mixtures by nanofiltration. J Membr Sci 178:185–193

Fu F, Wang Q (2011) Removal of heavy metal ions from wastewater: a review. J Environ Manage 92:407–418

Huang Y, Du JJR, Zhang Y et al (2015) Removal of mercury (II) from wastewater by polyvinylamine-enhanced ultrafiltration. Sep Purif Technol 154:1–10

Katsoufidou K, Yiantsios SG, Karabelas AJ (2005) A study of ultrafiltration membrane fouling by humic acid and flux recovery by backwashing: experiments and modeling. J Membr Sci 266:40–50

Keng PS, Lee SL, Ha ST et al (2014) Removal of hazardous heavy metals from aqueous environment by low-cost adsorption materials. Environ Chem Lett 12:15–25

Khalid M, Usman M, Siddiq M, Rasool N, Saif MJ, Imran M, Rana UA (2015) Removal of Ni(II) from aqueous solution by using micellar enhanced ultrafiltration. Water Sci Technol 72(6):946–951

Kheriji J, Tabassi D, Hamrouni B (2015) Removal of Cd(II) ions from aqueous solution and industrial effluent using reverse osmosis and nanofiltration membranes. Water Sci Technol 72(7):1206–1216

Kowalik-Klimczak A, Religa P, Gierycz P (2014) Chemiczne czyszczenie membran nanofiltracyjnych stosowanych do regeneracji chromowych ścieków garbarskich. Inż Aparat Chem 53:261–262

Kuca M, Szaniawska D (2009) Application of microfiltration and ceramic membranes for treatment of salted aqueous effluents from fish processing. Desalination 241:227–235

Kujawski W, Kujawa J, Wierzbowska E et al (2016) Influence of hydrophobization conditions and ceramic membranes pore size on their properties in vacuum membrane distillation of water-organic solvent mixtures. J Membr Sci 499:442–451

Kurniawan TA, Chan GYS, Lo WH et al (2006) Comparisons of low-cost adsorbents for treating wastewaters laden with heavy metals. Sci Tot Environ 366:409–426

Lou J, Wan Y (2013) Effect of pH and salt on nanofiltration—a critical review. J Membr Sci 438:18–28

Manttari M, Pihlajamaki A, Nystrom M (2006) Effect of pH on hydrophilicity and charge and their effect on the filtration efficiency on NF membranes at different pH. J Membr Sci 280:311–320

Moayyeri N, Saeb K, Biazar E (2013) Removal of heavy metals (lead, cadmium, zinc, nickel and iron) from water by bio-ceramic absorbers of hydroxy-apatite microparticles. Int J Marine Sci Eng 3(1):13–16

Mohsen-Nia M, Montazeri P, Modarress H (2007) Removal of Cu^{2+} and Ni^{2+} from wastewater with a chelating agent and reverse osmosis processes. Desalination 217:276–281

Muthukrishnan M, Guha BK (2008) Effect of pH on rejection of hexavalent chromium by nanofiltration. Desalination 219:171–178

Nędzarek A, Drost A, Harasimiuk F et al (2015a) The influence of pH and BSA on the retention of selected heavy metals in the nanofiltration process using ceramic membrane. Desalination 369:62–67

Nędzarek A, Drost A, Harasimiuk F et al (2015b) Application of ceramic membranes for microalgal biomass accumulation and recovery of the permeate to be reused in algae cultivation. J Photoch Photobio B 153:367–372

Nguyen HT, Chang W, Nguyen NC et al (2015) Influence of micelle properties on micellar-enhanced ultrafiltration for chromium recovery. Water Sci Technol 72(11):2045–2051

Oliveira RC, Docé RC, Barros ST (2013) Clarification of passion fruit juice by microfiltration: analyses of operating parameters, study of membrane fouling and juice quality. J Food Eng 111:432–439

Pellegrin ML, Burbano MS, Sadler ME et al (2016) Membrane processes. Water Environ Res 88 (10):1050–1124

Religa P, Kowalik A, Gierycz P (2011) A new approach to chromium concentration from salt mixture solution using nanosiltration. Separ Purif Technol 82:114–120

Staszak K, Karaś Z, Jaworska K (2013) Comparison of polymeric and ceramic membranes performance in the process of micellar enhanced ultrafiltration of cadmium(II) ions from aqueous solutions. Chem Pap 67(4):380–388

Sudilovskiy PS, Kagramanov GG, Kolesnikov VA (2008) Use of RO and NF for treatment of copper containing wastewaters in combination with flotation. Desalination 221:192–201

Sigworth EA, Smith B (1972) Adsorption of inorganic compounds by activated carbon. J Am Water Works Ass 64(6):386–391

Zhang LN, Wu YJ, Qu XY, Li ZS, Ni JR (2009) Mechanism of combination membrane and electro-winning process on treatment and remediation of Cu^{2+} polluted water body. J Environ Sci 21:764–769

Prospective Application of High Energy Mixing for Powder Flow Enhancement and Better Performance of Hydrogen and Energy Storage Systems

Ireneusz Opaliński, Karolina Leś, Sylwia Kozdra, Mateusz Przywara, Jerome Chauveau and Anthony Bonnet

1 Introduction

Processing of solids with mechanochemical methods has been routinely used in the past, mainly as milling technique in powder metallurgy involving cold welding, fracturing and rewelding of powder particles. Recently, due to dynamic development in the fields of electronics, information technology, new materials and growing environmental and health concerns, mechanochemistry gained more scientific attention and developed from an art to science in many areas of applications.

The main advantage of mechanochemical methods is their solvent-free and energy saving approach to many existing and new technologies because they not require solvents or even water and that is no need for subsequent product drying. Other advantages are their ability to remarkably modify and also to create engineered particulates with improved or even new physical/chemical properties (Baláž et al. 2013).

The idea of mechanochemistry, i.e. mechanically induced physical or chemical conversions of solids finds its practical implementation in relatively simple technique of high-energy mixing/milling (HEM) using planetary or vibrating ball mills. The outstanding feature of this technique having unique significance for powder technology is flowability improvement of cohesive powders commonly used in

I. Opaliński (✉) · K. Leś · S. Kozdra · M. Przywara
Department of Chemical and Process Engineering, Faculty of Chemistry,
Rzeszow University of Technology, Rzeszow, Poland
e-mail: ichio@prz.edu.pl

J. Chauveau
KYNAR Europe Development Department, ARKEMA CERDATO,
Serquigny, France

A. Bonnet
Global Fluoropolymers R&D, Director Development Technical Polymers,
ARKEMA, Colombes, France

© Springer International Publishing AG, part of Springer Nature 2018
M. Ochowiak et al. (eds.), *Practical Aspects of Chemical Engineering*,
Lecture Notes on Multidisciplinary Industrial Engineering,
https://doi.org/10.1007/978-3-319-73978-6_24

chemical, pharmaceutical and food process industries. The mechanism of HEM, mixing parameters optimization and some new applications of this technique are given in the first part of this work.

Of available alternative energy sources such as solar, wind, ocean and geothermal, hydrogen as an energy of future has gained much research interest. This is due to its safety, convenience to transportation and simplicity of conversion to other usable form of energy with no environmental pollution. Despite the advantages of hydrogen-based energy systems, the greatest challenge now is the development of technology for efficient hydrogen storage. The advantages and some distinctive features of HEM makes it a promising synthesis route for invention of hydrogen storage materials having the desired properties obtained in just a single step at low temperature and solvent-free conditions. The applicability of HEM for enhancement of hydrogenation properties of lithium nitride in a mixture with polymer powder was the aim of the second part of the Chapter.

Development of 3C technology (computers, communications and consumer electronics) strongly reinforces research in the field of new energy storage systems, particularly highly efficient, rechargeable batteries for both stationary and mobile applications. Devices which are fitting such demands are lithium batteries utilizing lithium metal as the anode material. Wide application of the batteries is however limited due to safety concerns resulting from formation of lithium dendrite on the anode surface, the need to utilize of volatile and flammable carbon compounds and the requirement to work under safety hazards since they contain a flammable electrolyte which may be kept pressurized.

Lithium-ion-polymer (LIP) batteries are new generation of rechargeable, chemical energy storage devices. In contrast to lithium- or lithium-ion-batteries loaded with a liquid electrolyte, they are operating with a safe, solid polymer electrolyte. LIP batteries provide a noticeable advantage over earlier lithium batteries in terms of electrical capacity, utility and durability. The features that make the LIP batteries particularly useful are their flexibility allowing them to be fitted in any array of devices and the rate of self-discharge which is much lower than for others.

The solid polymer electrolytes (SPE) can be typically classified as one of three types: dry-, gelled- and porous SPE. To examine applicability of high energy mixing for electrochemical activation of solid dry-type polymer electrolyte, in this work homopolymer PVDF was used. This is the part three of the Chapter.

2 Powder Flow Enhancement

The quality and reliability of processes involving powders depends mostly on powder flowability, which deteriorates increasingly with decreasing particle diameter. An effective way to manage the cohesive powder is high energy mixing (HEM). The concept of HEM consists in attaching of fine (G-guest) particles onto surface of larger (H-host) particles by firstly, bringing them mechanically close together and then allowing them to be held by van der Waals forces in firm contact. This mechanically induced process of particle coating results in de-agglomeration of originally cohesive powder bed making it flowing freely.

As it has been proven in our experiments, such mechanism of powder flow enhancement is an effective tool for many cohesive powders making them usable in chemical and pharmaceutical industries. Examples include calcium carbonate, Disulphiram, potato starch, Apyral and others. Some details concerning experimental technique and process optimum estimation are given below.

The main parameters affecting efficiency of HEM process are the speed and the time of mixing, the amount of fine admixture (fumed silica Aerosil® 200 with particle diameter around 10 nm) and the amount of process control agent (PCA), which is added to the powder mixture during milling to avoid powder caking and sticking. In order to find optimal values of the parameters it is assumed that relationship between input variables (mixing parameters and the amount of Aerosil and PCA admixtures) and response variables (flowability indices: angle of repose and compressibility index) is non-linear and that the process variables are not entirely independent of each other. For this reason the response function is usually given in the form of second-order function with interactions according to the equation:

$$y = (b_0 + \varepsilon) + \sum_{i-1}^{n} b_i x_i + \sum_{i-1}^{n} b_{ii} x_{ii}^2 + \sum_{i-1}^{n} \sum_{j-i \mid 1}^{n} b_{ij} x_i x_j \tag{1}$$

where x_i—process variables, ε—residual associated with the experiment, b_0—the constant term, b_i—equation coefficients, y—response value (angle of repose or compressibility index).

The response function is used to determine process conditions as needed to obtain expected powder flowability. The efficient method to fit Eq. 1 with experimental data and to evaluate the influence of main process variables on the powder flowability is Response Surface Methodology (RSM) coupled with Design of Experiments (DOE). The first step of RSM, before running the experiments, is preparing a proper experiment design and usually Central Composite Rotatable Design (CCRD) is recommended for this purpose. RSM and CCRD require assuming some values of coded levels for the process variables as shown in Table 1.

Five coded variable levels are needed to generate linear, quadratic and interaction terms. Codification is also essential to evaluate the fit of the response function to experimental data, to assess significance of the coefficients and to estimate the influence of process variables on the powder flowability factors. After performing the statistical analysis, the function with given coded levels may be recalculated for

Table 1 Codification of process variables

Coded variable	$-\alpha$	-1	0	$+1$	$+\alpha$
Actual variable	minimum	low	center	high	maximum

actual levels of the variables. The resulting function can eventually be represented on 3D surface plots.

For the examined calcium carbonate powder (Leś et al. 2015), the response function created according to Eq. 1 for angle of repose (Eq. 2) and compressibility index (Eq. 3) for coded levels of the variables, after removing not significant coefficients, are given below:

$$y_1 = 42.0 - 0.637x_1 - 0.504x_2 - 1.02x_3 + 0.587x_4 + 0.427x_1^2 + 0.264x_2^2$$
$$+ 0.539x_3^2 + 0.256x_1x_2 + 0.306x_1x_3 + 0.286x_2x_3 + 0.206x_2x_4 - 0.244x_3x_4$$
$$(2)$$

$$y_2 = 45.6 - 0.962x_1 - 1.08x_2 - 1.04x_3 + 0.562x_4 + 0223x_2^2 \qquad (3)$$

Typical response surface plot based on a response function for calcium carbonate is presented in Fig. 1. The graphs in the Figure show the evident minimum of flowability factors as a function of process variables (mixing speed, mixing time, amount of Aerosil and of isopropanol).

Having performed the process of HEM supported with Response Surface Methodology, it is then possible to improve considerably the flowing properties of calcium carbonate: as seen from Fig. 1, the angle of repose was decreased from initial value of 55 to about 40° and compressibility index was diminished from 48 to 39%.

Similar results were obtained for other materials. For example, for potato starch (Leś et al. 2017) it is also possible to obtain flowability improvement but due to some particle size reduction resulting from ball-milling, it is better to perform HEM with lower speed and longer time or higher speed and shorter time. As a result the angle of repose can be considerably reduced (from 50 to about 30°) and to some extent this also concerns the compressibility index (reduced from 22 to 21.8%)—Fig. 2.

We succeeded in achieving some interesting results also for Apyral powder (aluminium trihydroxide), and Disulphiram (1-(diethylthiocarbamoyldisulfanyl)-N,

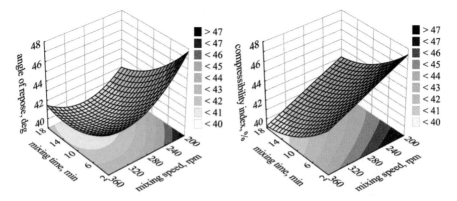

Fig. 1 Improvement of flowability factors obtained in the process of HEM for calcium carbonate

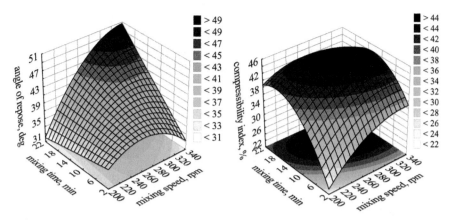

Fig. 2 Improvement of flowability factors obtained in the process of HEM for potato starch

Table 2 Changes of flowability indices when using HEM process for Apyral and Disulphiram

Material	Process variables				Reduction of flowability indices from initial to final values	
	Mixing speed (rpm)	Mixing time (min)	Amount of Aerosil (wt%)	Amount of isopropanol (mL)	Angle of repose (°)	Compressibility index (%)
Apyral	300	14	3.5	1.1	13	10
Disulphiram	140	8	2	0.4	20	5

N-diethyl-methanethioamide), both mixed with Aerosil (to be published). Examples of the process variable values required for flowability improvement are presented in Table 2.

Rapid development of high-energy mixing devices opened new areas of physical treatment and surface modifications of particulate solids. The idea of HEM was taken one step further to alter chemical properties of solid particles and to make them energy carriers. The important present-day examples are hydrogen storage in solid metal hydrides and energy storage in batteries with solid electrolytes.

3   Hydrogen Solid-State Storage Systems

Approaches that have been developed so far include hydrogen stored as a compressed gas, as a cryogenic liquid, in the form of physically adsorbed molecules of H_2 in nanostructured materials (carbon nanotubes) and as chemical compounds—metal hydrides. High costs needed to pressurize the gas (70 MPa), to liquefy it (20 K) and low hydrogen storage density (2 wt%) of nanostructured materials, makes the storage of hydrogen in metal hydrides a very interesting option.

Adsorption of hydrogen by metals is critically affected by surface condition of a metal. As hydrogenation progresses, the hydride layer grows and free surface active sites promoting H_2 molecules for dissociation and adsorption are no longer available. Because diffusion of H_2 through the hydride layer is rate-limiting step of hydrogenation process, a kind of surface activation is needed to solid-phase reaction enhancement.

The ball-mixing/milling method renews and increases the surface due to formation of micro- and nanostructural defects like cracks, dislocations, vacancies and other providing many sites with low activation energy in a crystalline solid trapped in a close-fitting space between colliding balls. New surfaces which are continuously created under ball-milling and constant mass transfer when mixing are regarded as two major factors responsible for the enhanced reactivity of metals toward hydrogen (Balema 2011).

Lightweight elements have recently received attention as a hydrogen storage material (Rusman and Dahari 2016). These include lithium nitride Li_3N as an important compound because of its high theoretical hydrogen storage (10.4 wt%) and good reaction reversibility (Chen et al. 2002).

With the intention of improving the hydrogen storage properties, an attempt of using Li_3N in a composition with poly(vinylidene fluoride-co-hexafluoropropylene): (PVDF-HFP) polymer was taken up. The composition of the mixture samples examined and the mixing parameters applied in the experiments are shown in Table 3.

To monitor structural changes occurring in the composites under HEM conditions, X-ray diffraction (*Bruker Nanostar-U*) was used. The XRD pattern of the pure, unprocessed Li_3N sample is shown in Fig. 3. The Figure displays that the structure of this sample was clearly polycrystalline Li_3N phase with large grains as suggested by the sharp Bragg's peaks corresponding to the α- and β-Li_3N polymorphic phases. Sensitivity of such structures to hydrogen storage is regarded as rather low and the way the reactivity of Li_3N towards hydrogen can be enhanced is to use it in a composition with PVDF-HFP co-polymer under HEM conditions.

Structural changes of the ball-milled samples are shown in Fig. 4 where XRD patterns obtained for the modified samples were compared to a reference sample (PVDF-HFP). Two main features of the ball-mill treatment should be considered. The first one is an obvious lowering in the intensity and broadening in the Bragg's lines for all tested samples. This is a result of a continuous decreasing in the grain

Table 3 Sample compositions and HEM parameters

Samples composition	PVDF-HFP (wt%)	Li_3N (wt%)	Aerosil® 200 (wt%)	Time (h)	Rotational speed (rpm)
PVDF-HFP-250-Li_3N	93	5	2	1	250
PVDF-HFP-350-Li_3N	93	5	2	1	350
PVDF-HFP-250	98	0	2	1	250
PVDF-HFP	100	0	0	–	–

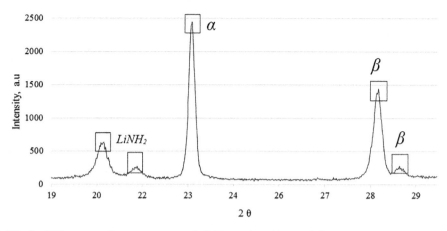

Fig. 3 XRD pattern of pure, unprocessed Li_3N sample (with α and β Li_3N phases shown). The peaks attributed to the $LiNH_2$ are supposed to be the result of reaction Li_3N with hydrogen from surrounding air (Furukawa et al. 2016)

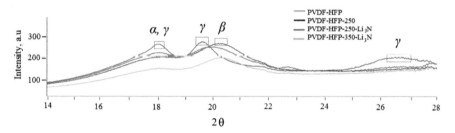

Fig. 4 XRD patterns of examined samples (with α and β crystalline phases shown)

size of ball-milled powder leading to progressive solid-phase amorphization, which is a favorable attribute for the hydrogenation process.

The second important feature of HEM as concerns hydrogen storage are polymorphic transformations reflected mainly by the Bragg's signals in the range of 2θ from 16 to 28°. The signals suggest appearing changes directed towards reducing α- and γ-phases and forming β-phase as a dominant structure of the ball-milled powder. The observed phase transformations are complex factors resulting from energy input provided by HEM and Li_3N addition and they are more apparent for higher values of milling speed.

Using dedicated software *DIFFRAC. EVA*, the increase in amorphous phase content resulting from HEM treatment and the presence of Li_3N in the sample was then estimated. This is shown in Fig. 5 where an advantageous effect of the increase in milling speed can also be identified for the samples milled at 250 and 350 rpm respectively.

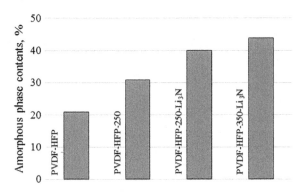

Fig. 5 Increse in amorphous phase contents for ball-milled samples

Moreover, it becomes evident that regardless of crystalline nature of Li_3N there is no crystalline contents increase in the milled samples. This is evident that presence of Li_3N results in an additional enhancement of the crystalline- to amorphous-phase transformation, which can be attributed to a mechanical-strain energy input causing further destruction of long-range periodic structure and the introduction of more lattice defects to the milled particulates.

In conclusion, the results obtained so far give a sound suggestion that desired improvement of hydrogen storage properties can be attained using composite materials constituted of light metal compounds together with some active and inert additives under HEM conditions.

4   Energy Solid-State Storage Systems

At present many polymers such as poly(vinyl chloride) (PVC), poly(acrylonitrile) (PAN), poly(methyl metacrylane) (PMMA) and poly(vinylidene fluoride) (PVDF) are used as solid polymer electrolyte (SPE) matrices (Xiao et al. 2009). Among them, PVDF has been considered to be very interesting due to its good resistance to chemical and physical factors and good thermal stability up to 150 °C. Moreover, high dielectric constant of PVDF ($\varepsilon = 8.4$) results in a good anodic stability of the polymer, which is important to obtain better ionization of lithium salt in SPE structure (Schaefer et al. 2012).

As most polymers, PVDF reveals insulating properties and this is the reason for using certain active additives, like Li_3N, Li_2O_3 or $LiNH_2$, which are giving the polymer some conductive properties needed in a final energy storage device. Inert additives (Al_2O_3, SiO_2, MgO) are also used to prevent formation and growth of lithium dendrites on the anode surface in battery charging periods. The inert additives in polymer matrix additionally increase its electrical conductivity and improve contact on the border of the electrolyte-electrode (Schaefer et al. 2012). This is explained by an increase in the content of amorphous phase and limited recrystallization which can occur during battery charging.

Successful design, assembling and subsequent SPE usage require deep modification of their components, both, in terms of physical and chemical properties needed for energy storage purposes. As the polymer and additives are usually applied in the form of powders, high-energy mixing (HEM) is a proper choice and useful technique for their processing. One more valuable feature provided by HEM is a possibility of making the SPE components as homogeneous as possible, which is an important factor for reliability and stability of battery at its final usage stage.

PVDF in both homo- and co-polymer forms (*Kynar* 761, *Kynar Flex*, ARKEMA) were used in this work as basic components of SPE and they were modified using HEM in various combinations with active ($LiNH_2$ and Li_3N) as well inert (Aerosil$^®$ 200) solid additives. Below, the relevant variants of mixed solids are shown (Table 4) and appropriate results for these mixtures are given.

The mechanochemical activation of PVDF using HEM method was carried out in a planetary ball mill with a tightly sealed chamber allowing for loading and unloading under Argon protective atmosphere in a glove box. A typical procedure was applied: solids were placed in a mixing chamber and ball-mixed for a specified period of time at given mixing speed. Then the solids were removed from the chamber. The parameters of HEM were as follows: mixing time and speed: 15 min and 300 rpm respectively. They were determined experimentally to obtain the best mixture homogeneity and possibly reliable and clear analysis results.

SPE in the final form of regular thin layers (films) were prepared from the solutions of ball-mixed powder mixtures in acetone that were then spread on a plane glass surface. All operations for samples with lithium amide were performed in Ar protective atmosphere. The obtained polymer films were dried and analyzed.

Figure 6 shows the FT IR (*Thermo Scientific, Nicolet* 8700) spectra of the tested samples (ATR technique). All specific spectra-bands for the polymer matrix can be noted. Most of the vibration characteristics for the polymer structure are located in the 600–1500 cm^{-1} region. The observed pattern originates from oscillations of large parts of the polymer skeleton. Some regions have been identified and characterized with regards to the crystalline structure. The signals shown in the spectrum indicate the presence of α and β crystalline phase conformations. Signals at 1400, 1070 and 840 cm^{-1} were attributed to β conformation whereas those at 976, 795 and 766 cm^{-1} were assigned to α conformation. Other signals displayed in the spectrum correspond to the vibration of the polymer amorphous phase (Martins et al. 2014).

Table 4 Methods of the preparation of the SPE sample (PVDF: 93 wt%, Aerosil: 2 wt%, $LiNH_2$: 5 wt%)

SPE sample	Methods of sample preparation
PVDF-LiNH$_2$-Aerosil-1	All the sample components mixed in planetary ball mill under Argon atm.
PVDF-LiNH$_2$-Aerosil-2	Polymer and inert additive mixed in planetary ball mill and active additive added separately
PVDF-LiNH$_2$-Aerosil-3	Manually mixed the sample components

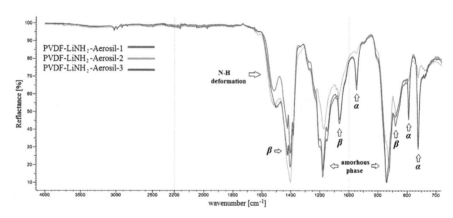

Fig. 6 FT IR spectra of the SPE samples with labelled α- and β-phase characteristic bands

HEM increases the intensity of 1400 cm^{-1} signal which can be attributed to β conformation of PVDF and this is an advantageous feature of mechanochemical solid processing as the electrical properties of the polymer get better with increasing of β conformation content. As concerns other β signals (at 1070 and 840 cm^{-1}), their intensity can be regarded as unchanged due to almost the same relevant peak areas. Furthermore, for this sample, intensity of signals (976, 795, 766 cm^{-1}) assigned for α conformation decreases, which is a clear indication that polarization properties of SPE are getting better. This is a further indication of HEM effectiveness towards improvement of solid electrolyte properties.

XRD analysis was performed to determine crystallographic structure of PVDF and to verify the FT IR structural results obtained earlier. The comparison of the XRD patterns obtained for the examined samples is shown in Fig. 7. Using dedicated software *DIFFRAC.EVA*, the content of the PVDF amorphous phase in the samples was calculated and it is shown in Fig. 8. Increasing intensity of mechanochemical solids treatment increases the content of amorphous phase and it is a strong indication for improvement of ionic conductivity and better performance of SPE.

Electrochemical parameters of SPE can be measured with Electrochemical Impedance Spectroscopy (EIS). The necessary requirement for obtaining correct results with this technique is a tight contact of SPE sample and the surface of the instrument electrodes. For this reason it was not possible to perform the EIS measurements for LiNH$_2$-3 sample as it was fragile, nonflexible and of irregular, uneven surface.

The EIS instrument employed was *PASRTAT* 2273 operating in the frequency window from 0.1 Hz to 100 kHz at AC amplitude of 10 mV at room temperature. The EIS measurement results are shown in Fig. 9 in the form of Nyquist plots.

The bulk resistance of the films was calculated as a value obtained from the intersection of regression straight line fitted to the first ten measured points at the side of high frequency with the axis of real impedance on the Nyquist diagram.

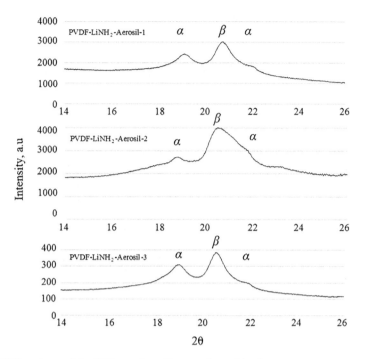

Fig. 7 XRD patterns of the SPE samples with crystallographic phases

Fig. 8 Amorphous phase contents for the SPE

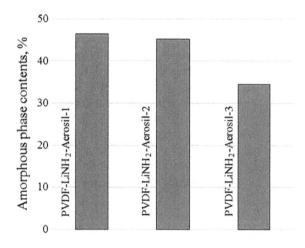

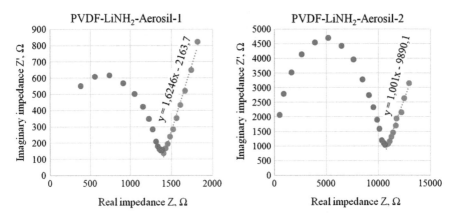

Fig. 9 Nyquist plots for films made of the SPE samples at temp. 25 °C

Having the values of the bulk resistance, the ionic conductivity can be calculated as follows:

$$\sigma = \frac{l}{R \cdot A} \qquad (4)$$

where R—the electrolyte bulk resistance, l—film thickness and A is the area of sample.

The calculated ionic conductivities of the prepared SPE films were as follows: 1.22×10^{-6} (S cm^{-1}) for PVDF-LiNH$_2$-Aerosil-1 and 0.177×10^{-6} (S cm^{-1}) for PVDF-LiNH$_2$-Aerosil-2. The achieved conductivity values can be regarded as an evidence of suitability of HEM for preparation and electrochemical activation of some polymers used as prospective SPE.

Further research employing the same HEM technique for processing homopolymer PVDF as a matrix and lithium nitride Li$_3$N (Kozdra et al. 2016) or LiClO$_4$ as activators were performed and obtained values of conductivity ranged from 0.63×10^{-5} to 2.0×10^{-5} (S cm^{-1}). Moreover, a strong influence of mixing time on conductivity of ball-milled samples was found and it is an important factor revealing suitability of mechanochemical methods for investigating and preparing energy storage materials (to be published).

References

Baláž P, Achimovičová M, Baláž M et al (2013) Hallmarks of mechanochemistry: from nanoparticles to technology. Chem Soc Rev 42:7571–7637

Balema P (2011) Mechanical processing in hydrogen storage research and development. Mater Matters 2:16

Chen P, Xiong Z, Luo J et al (2002) Interaction of hydrogen with metal nitrides and imides. Nature 420(6913):302–304

Furukawa T et al (2016) Dissolution behavior of lithium compounds in ethanol. Nucl Mater Energy 9:286–291

Kozdra S, Opaliński I, Leś K et al (2016) Modyfikacja mechanochemiczna kompozytów z poli (fluorku winylidenu) jako składników elektrolitów akumulatorów litowo-jonowych. Inż Ap Chem 55(6):233–236

Leś K, Kowalski K, Opaliński I (2015) Optimisation of process parameters in high energy mixing as a method of cohesive powder flowability improvement. Chem Process Eng 36(4):449–460

Leś K, Kozdraw S, Opalinski I (2017) Optimization of flow indices by response surface methodology in high-energy mixing of powders. Przem Chem 96(2):469–472

Martins P, Lopes AC, Lanceros-Mendez S (2014) Electroactive phases of poly(vinylidene fluoride): determination, processing and applications. Prog Polym Sci 39(4):683–706

Rusman NAA, Dahari M (2016) A review on the current progress of metal hydrides material for solid-state hydrogen storage. Int J Hydrogen Energy 41(28):12108–12126

Schaefer JL, Lu Y, Moganty SS et al (2012) Electrolytes for high-energy batteries. Appl Nanosci 2(2):91–109

Xiao Q, Wang X, Li W et al (2009) Macroporous polymer electrolytes based on PVDF/ PEO-b-PMMA block copolymer blends for rechargeable lithium ion battery. J Membr Sci 334(1–2):117–122

Superheated Steam Drying of Solid Fuels: Wood Biomass and Lignite

Zdzisław Pakowski and Robert Adamski

1 Introduction

Superheated steam drying offers several advantages (Pakowski 2011), which makes it attractive for drying of high-tonnage, high-moisture content raw materials, like solid fuels, either fossil as lignite, or renewable such as wood biomass. The following are the most important ones:

- The dryer exhaust gas is also steam although of a lower grade. Its heat of condensation can be recovered either outside or inside a dryer. Alternatively, steam can be upgraded by mechanical compression and send to other processes. The dryer can also work at high steam pressure so no recompression of the exhaust is necessary.
- The drying cycle can be closed i.e. any gaseous emission to atmosphere is eliminated.
- The atmosphere in the dryer is oxygen free, so a risk of oxidation or explosion is minimized.
- High product temperature eliminates germs, fungi and spores in the product therefore storage is safe and dust is not harmful to humans.

Disadvantages include:

- Drying cycle must be hermetically closed so feeding of the wet solid and discharge of dry solid require airlocks.
- All parts of equipment in contact with steam must be well insulated thermally to prevent condensation.

Z. Pakowski (✉) · R. Adamski
Faculty of Process and Environmental Engineering,
Lodz University of Technology, Łódź, Poland
e-mail: ZdzislawPakowski@p.lodz.pl

© Springer International Publishing AG, part of Springer Nature 2018
M. Ochowiak et al. (eds.), *Practical Aspects of Chemical Engineering*,
Lecture Notes on Multidisciplinary Industrial Engineering,
https://doi.org/10.1007/978-3-319-73978-6_25

Fig. 1 Experimental setup of TUL used for sorption isobar and drying kinetics determination. 1—Circulation fan, 2—superheater, 3—external steam generator, 4—sample chamber (one of two)

- High temperature steam with even low concentration of oxygen is highly corrosive and requires quality steel as material of construction.

The entering steam has higher temperature than its boiling point at the operational pressure. The difference between these two is called the degree of superheat. The degree of superheat, which is the driving force of the process, is usually 20–120 °C at atmospheric pressure and decreases with increasing pressure. During the drying cycle, the material initially heats up due to steam condensation until it reaches the boiling-point temperature. From now on the temperature remains constant until all unbound moisture is removed and then increases up to fresh steam temperature when bound moisture is being removed. This high material temperature makes it unsuitable for thermosensitive products. For organic fuels product temperature of 200 °C should not be exceeded, since it marks the onset of pyrolytic decomposition.

SSD experiments performed within this work were carried out in the lab-scale superheated steam dryer shown in Fig. 1.

Below, wood biomass and lignite as objects of SSD will be characterized on the basis of work performed in TUL Lodz within two research projects 2 T0C9 01028 in the years 2005–2008 and N N209 103537 in the years 2009–2012.

2 Wood Biomass

Wood biomass is fully renewable irrespectively of its origin i.e. natural forests or plantations. The yearly growth of wood biomass in world scale is ca. 12.5×10^9 m^3. For example, in Poland ca. 22 mln m^3 are harvested yearly, of which ca. 10% is waste. It is estimated that this amounts to ca. 244 PJ/annum of energy (Kotowski 2003). As harvested wood biomass contains ca. 60% water and its calorific value is low, equalling to 6.53 GJ/t. By drying it to 10% the calorific value can be increased to 17.80 GJ/t. It is accepted that combustion of biomass does not contribute net CO_2 to the atmosphere since it contains atmospheric CO_2 previously consumed by plants in the photosynthesis process.

Wood biomass in the form of chips can be SSD dried in rotary dryers or after milling in flash dryers. The resulting dry biomass is pelletized into a convenient commercial form and used as fuel. Also wood fibers used for MDF, LDF i HDF boards production can be dried using superheated steam.

2.1 Sorptional Equilibrium

Sorptional equilibrium between water in the gas phase (steam) and water in the solid phase at constant temperature T, in air drying is described as the relationship of relative pressure φ (also called water activity a_w) on solid moisture content X and called sorption isotherm

$$\varphi \equiv \frac{p}{p_s(T)} = f(T, X) \tag{1}$$

where p is partial pressure of water in the gas phase and $p_s(T)$ is saturated pressure of water at temperature T. In SSD the equilibrium takes place at constant partial pressure of water equal to the total pressure, therefore the relationship of these two takes the form of the sorption isobar at constant P_0

$$\varphi \equiv \frac{P_0}{p_s(T)} = f(T, X) \tag{2}$$

where P_0 is the total (process) pressure. Sorption isobars and isotherms are situated on the same surface in φ, t, X coordinates. In Fig. 2 grey lines of constant T are isotherms and thick lines are isobars plotted for three selected process pressures.

The knowledge of sorption isobars is essential for the formulation of the driving force in drying simulations. Sorption isobars for wood biomass were obtained for willow *Salix viminalis* coming from an energetic willow plantation. Two and three year old stems, 2 m long and 30 mm (3 year old) or 25 mm (2 year old) in diameter, were collected. The isobars were measured at 105, 110, 115, 120 and 125 °C. About 7 g samples in the form of chips were used.

The results were fitted with the Chen and Clayton equation (Pakowski et al. 2007) shown below:

$$a_w = \exp[a_1 T^{a_2} \exp(a_3 T^{a_4} X)] \tag{3}$$

where: a—equation coefficients of Eq. (3), a_w—water activity, T—(material) temperature (°C or K), X—moisture content (kg/kg); with $a_1 = -6.6412$, $a_2 = -0.2459$, $a_3 = -7.9150$, $a_4 = -0.1804$. The fitted function is shown in Fig. 3.

Fig. 2 Isotherms and three exemplary isobars of sorption for a given material

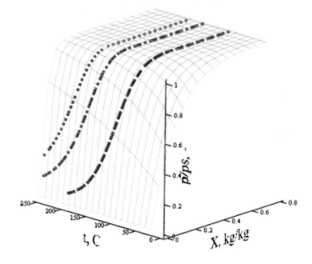

Fig. 3 Fitted sorption equilibrium equation of Chen and Clayton and experimental data points

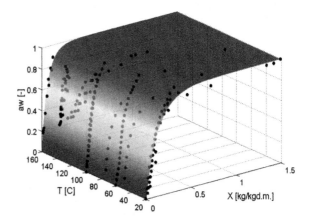

2.2 Drying Kinetics

Mass flux of water vapor on the particle surface w_D can be calculated as

$$w_D = \frac{\alpha(T_\infty - T^*)}{\Delta h_v} \qquad (4)$$

where: T^* is calculated from the sorption isobar at X of the surface (Pakowski and Adamski 2011), α—heat transfer coefficient (W m^{-1} K^{-1}), Δh_v—heat of vaporization (J kg^{-1}). This allows for the calculation of the duration for condensation and constant drying rate periods. However, for the falling rate period an adequate model of internal heat and mass transfer must be used. For drying of wood biomass samples in cylindrical form, the following model was formulated (Adamski and Pakowski 2008)

$$\frac{\partial}{\partial t}(\rho_S X) + \frac{1}{r}\frac{\partial}{\partial r}\left[r\left(j_{X,V+L} + j_{T,V+L} + j_{P,V}\right)\right] = 0 \tag{5}$$

$$\begin{aligned}\frac{1}{r}\frac{\partial}{\partial r}\left[r\left(\lambda_m \frac{\partial T}{\partial r}\right)\right] &+ \frac{\partial}{\partial t}[\rho_S c_S (1+X)T] \\ &= -\frac{1}{r}\frac{\partial}{\partial r}\left[r c_V \left(j_{X,V+L} + j_{T,V+L} + j_{P,V}\right)T\right] \\ &- \Delta h_V \frac{1}{r}\frac{\partial}{\partial r}\left[r\left(j_{X,V+L} + j_{T,V+L} + j_{P,V}\right)\right]\end{aligned} \tag{6}$$

where
diffusion flux

$$j_{X,V+L} = \rho_S D_{eff}\frac{\partial X}{\partial r} \tag{7}$$

thermodiffusion flux

$$j_{T,V+L} = \rho_S \delta D_{eff}\frac{\partial T}{\partial r} \tag{8}$$

Darcy flux

$$j_{P,V} = \rho_V \frac{k_V}{\eta_V}\frac{\partial p_V}{\partial r} \tag{9}$$

$$p_V = a_w(X,T)p_{sat}(T) \tag{10}$$

Boundary conditions in cylinder axis were of symmetry type while on the surface they were formulated as:

$$-D_{eff}\rho_S \frac{\partial X}{\partial r}\bigg|_{r=R} - \delta D_{eff}\rho_S \frac{\partial T}{\partial r}\bigg|_{r=R} - \rho_V \frac{k_V}{\eta_V}\frac{\partial p_V(X,T)}{\partial r}\bigg|_{r=R} = w_D \tag{11}$$

$$-\lambda_S \frac{\partial T}{\partial r}\bigg|_{r=R} = \alpha(T_\infty - T^*) - w_{Dr}(c_V T + \Delta h_V) \tag{12}$$

where:

D diffusion coefficient ($m^2\ s^{-1}$),
R total radius (m),
X moisture content (kg/kg),
c heat capacity ($J\ kg^{-1}\ C^{-1}$),
j mass flux ($kg\ m^{-2}\ s^{-1}$),
k permeability (m^2),
r radius (m),

t time (s),
w_D drying rate (kg m^{-2} s^{-1}),
α heat transfer coefficient (W m^{-1} K^{-1}),
δ thermodiffusion coefficient (K^{-1}),
λ thermal conductivity (W m^{-1} K^{-1}),
φ relative pressure (–),
μ viscosity (Pa s),
ρ density (kg m^{-3}),
L liquid,
V vapor,
S dry solid,
eff effective,
m wet solid,
∞ steam.

Evaporation rate at the surface was calculated from Eq. 4. SSD experiments were performed in the set-up shown in Fig. 1. Cylindrical samples 24.8 mm in diameter and 50 mm high were dried. On the basis of experimental results constants in the effective diffusion coefficient defined as

$$D_b = d_1 X^{d_2} \exp\left(\frac{-d_3}{RT}\right) \tag{13}$$

[where d are equation coefficients of Eq. (13)] were identified by solving an inverse problem as $d_1 = 8.083\mathrm{e}{-8}$ m^2/s, $d_2 = 5.173\mathrm{e}{-1}$, $d_3 = 3.857\mathrm{e}{+3}$ kJ/kmol. The permeability coefficient of water vapor was identified as (Pakowski et al. 2009):

$$k_V = \frac{\left(A + \frac{l}{\bar{p}}\right)\left(B + \frac{m}{\bar{p}}\right)}{\frac{l+m}{\bar{p}} + A + B} \tag{14}$$

[where: A, B, l, m are equation coefficients in Eq. (14), $\bar{P}$ is average pressure (Pa)] with $A = -1.181\mathrm{e}{-12}$, $B = 7.719\mathrm{e}{-13}$, $l = 4.676\mathrm{e}{-18}$ and $m = 2.199\mathrm{e}{-19}$. When mean pressure $\bar{P}$ is substituted into (14) in Pa the resulting k_V is in m^2. The resulting thermodiffusion coefficient was negligibly small and the thermodiffusion term was deleted before final identification of D_{eff} and k_V.

2.3   Drying Process Simulation

The results of simulation of SSD for wood biomass of *Salix viminalis* in the form of cylinders are shown in Figs. 4 and 5. Experiments were made at atmospheric pressure and at steam temperature of 170 °C. Four samples were dried. Samples were weighed continuously. Sample temperature was measured in two points,

Fig. 4 Change of volume averaged moisture content with time

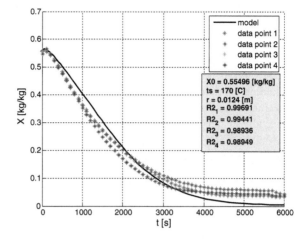

Fig. 5 Change of internal profiles of temperature along sample radius with time. – mesh–simulation, points– experimental data

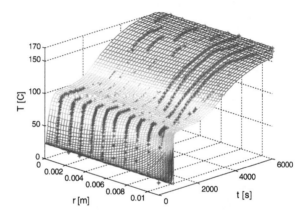

different in each sample in order to scan the entire radius, with thin thermocouples introduced lengthwise.

3 Lignite

Lignite or brown coal is the low-rank coal defined as coal of calorific value <16.5 MJ/kg. Its world-scale deposits are estimated as ca. 286 billion tons (World Energy Council 2016) while hard coal deposits are estimated as ca. 700 billion tons. Unfortunately, high moisture content of excavated lignite (typically 35–75% w.b.) decreases its calorific value and leads to higher index of CO_2 emission (104 kg/GJ for lignite vs. 95 kg/GJ for hard coal). Pre-drying of lignite can increase its calorific value, increasing the energy block efficiency by 4–6% and reducing the CO_2 index, besides it is demanded by some processes like oxyfuel. The interest in SSD of lignite was initiated in ca. 1969 in Australia, where it was applied on a commercial scale. The development of WTA technology (exhaust steam is condensed in heaters immersed in the bed) and the first pilot plant, which RWE opened in Niederaussem in 2007, spurred the interest again and despite of rather slow development of this technology in industry (only four plants based on WTA technology exist worldwide), research in this area continues (cf. Kiriyama et al. 2016; Levy 2005; Liu et al. 2017) especially in Japan, Indonesia and other lignite abundant countries.

Among SSD dryers for coal, three types suitable for large tonnage can be considered: rotary, fluid bed and flash. Flash dryers are used for pulverized coals (Ross 2005) but due to low HGI grindability index of lignite in wet state, this dryer may be available only for relatively low moisture content coals. Rotary dryers for SSD (Clayton et al. 2007) have intrinsic problem of sealing the rotating drum flanges and therefore are limited basically to atmospheric SSD. Fluid bed dryers remain the most flexible ones. Simulation of dryers used in their design requires a set of physical properties and kinetic data or a suitable model to calculate the drying time. The following is a summary of the work done by us in this field.

3.1 Sorptional Equilibrium

Sorptional equilibrium for lignite in air drying and isobaric conditions at atmospheric pressure in SSD were measured (Pakowski et al. 2011a). Both methods involved weighing of solid sample. This would be a hard task at elevated pressures. A novel isochoric autoclave method was proposed, which eliminates weighing (Kokocinska and Pakowski 2014). The method uses the recorded temperature and pressure in the autoclave and the virial equation of state to calculate mass of water in the gas phase. Thus, the balance of water in the solid can be closed without weighing the sample. According to this work the following equation of the BET type:

$$\varphi = \frac{(X - X_m) \cdot a_1 - 2X + \sqrt{\Delta}}{2(a_1 - 1)X} \tag{15}$$

with

$$\Delta = [(X - X_m)a_1 - 2X]^2 + 4(a_1 - 1)X^2 \tag{16}$$

and

$$X_m = -a_2 T + a_3 \tag{17}$$

with $a_1 = 2.0493e2$, $a_2 = -2.4517e-4$ and $a_3 = 1.5022e-2$ describes well ($R^2 = 0.960$) the equilibrium of Bełchatów lignite in the range of $T = 293-473$ K, $P_0 = 1-23.6$ bar and $X = 0-1.22$ kg/kg d.b.

3.2 Drying Kinetics

Diffusional model is a simple distributed parameter model to describe drying kinetics of coal particles. It uses only one fitted parameter—the effective diffusivity D_{eff}. D_{eff} is a function of solid temperature and moisture content and can be identified in an experiment where drying kinetics of a particle of simple geometry (cylinder or sphere) are experimentally measured. D_{eff} for Bełchatów lignite was identified at atmospheric pressure experiments (Pakowski et al. 2011b; Pakowski et al. 2013). The following model for drying of a finite cylinder was used:

$$\frac{\partial X}{\partial \tau} = D_{eff}\left(\frac{\partial^2 X}{\partial r^2} + \frac{1}{r}\frac{\partial X}{\partial r} + \frac{\partial^2 X}{\partial z^2}\right) \tag{18}$$

$$\frac{\partial T}{\partial \tau} = \frac{1}{\rho_S(c_L X + c_S)}\lambda_m\left(\frac{\partial^2 T}{\partial r^2} + \frac{1}{r}\frac{\partial T}{\partial r} + \frac{\partial^2 T}{\partial z^2}\right) \tag{19}$$

A symmetry condition was applied in the cylinder center and on the insulated cylinder bases while the following BC's were applied at the surface:

$$-\rho_S D_{eff}\frac{\partial X}{\partial r}\bigg|_{r=R} = w_D \tag{20}$$

$$-\lambda_m\frac{\partial T}{\partial r}\bigg|_{r=R} = \alpha(T_\infty - T) - w_D\Delta h_v \tag{21}$$

The 2D model was solved in COMSOL and the effective diffusivity was identified on the basis of SSD experiments performed on 22 mm in diameter and 35 mm tall lignite cylinders, at 200 °C, using the following equation:

$$D_{eff} = d_1 X^{d_2} \exp\left(-\frac{d_3}{RT}\right) \tag{22}$$

and the values of $d_1 = 1.82e-7$ m^2/s, $d_2 = 0.0773$ and $d_3 = 3549$ kJ/kmol H$_2$O were obtained.

Experiments were also carried out drying 25 mm spherical samples at elevated pressure in steam of 180 °C @ 2 bar and 200 °C @ 2 bar. The following model was used:

$$\frac{\partial X}{\partial \tau} = D_{eff}\left(\frac{\partial^2 X}{\partial r^2} + \frac{2}{r}\frac{\partial X}{\partial r}\right) \tag{23}$$

$$\frac{\partial T}{\partial \tau} = \frac{1}{\rho_S(c_L X + c_S)}\lambda_m\left(\frac{\partial^2 T}{\partial r^2} + \frac{2}{r}\frac{\partial T}{\partial r}\right) \tag{24}$$

A symmetry condition was applied in the sphere center while the following BC's were applied at the surface:

$$-\rho_S D_{eff}\frac{\partial X}{\partial r}\bigg|_{r=R} = w_{Dr} \tag{25}$$

$$-\lambda_m\frac{\partial T}{\partial r}\bigg|_{r=R} = \alpha(T_\infty - T) - w_D\Delta h_v \tag{26}$$

Identification of D_{eff} gave the following values: $d_1 = 1.26e{-}7$ m^2/s, $d_2 = 0.0892$ and $d_3 = 3025$ kJ/kmol H$_2$O. This results in values of $D_{eff} = 6.97e{-}8$ m^2/s at the beginning of the falling rate period to $6.87e{-}8$ m^2/s at the end of it when drying at 200 °C. These relatively high values indicate that lignite has large, open pores facilitating diffusion. It also allows for using alternative shrinking core model as presented by Chen et al. (2000). It can also be noticed that steam pressure has little influence on the value of effective diffusivity.

3.3   Drying Process Simulation

The results of simulation for SSD of a spherical particle at elevated pressures of 2 and 5 bar are shown in Figs. 6 and 7. These simulations indicate that the temperature has a dominant influence on the drying rate of lignite, while the influence of pressure is less pronounced. However, in theory the increase in the operational pressure increases steam density and heat capacity, thus increasing the heat transfer coefficient of steam phase. However, at the same time this increases the saturation temperature and thus decreases the temperature driving force. These counteracting factors suggest that at certain pressure there is a maximum of drying rate or a minimum in drying time.

Results obtained by simulations performed on SSD of a single particle 2 mm in size, using the diffusional model presented above and a shrinking-core model with identified permeability value of $0.4e{-}12$ m^2 @200 °C (Pakowski and Kwapisz 2014) are shown in Fig. 8. Drying times necessary to reduce moisture content from 1 to 0.1 kg/kg were computed.

The above results indicate that the optimum pressure is ca. 3 bar. This finding requires further experimental validation.

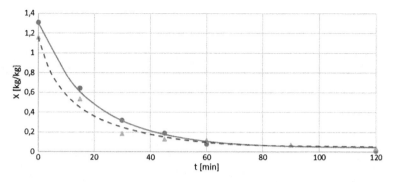

Fig. 6 SSD of a spherical lignite particle at 180 °C 2 bar solid line sim., filled circle exp. and 5 bar dashed line sim., filled triangle exp.

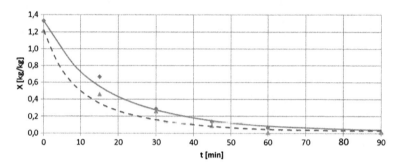

Fig. 7 SSD of a spherical lignite particle at 200 °C 2 bar solid line sim., filled circle exp. and 5 bar dashed line sim., filled triangle exp.

Fig. 8 Influence of operating pressure on drying time for a 2 mm spherical particle of lignite

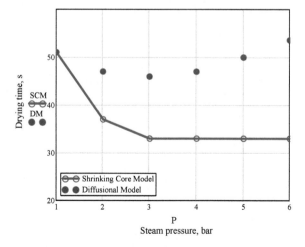

4 Conclusions

SSD is energetically much more efficient than air drying due to ease of heat recovery from exhaust steam. The steam can be used in situ and condensed inside internal heaters or externally by heat integration with other units. Important parameters of the investigated fuels were measured in order to enable process simulation. It was found that the sorption isotherms and isobars are lines located on the same equilibrium surface in X, T, φ coordinates. Moreover, isochores obtained by the novel isochoric method can also belong to the same surface. It was found that wood biomass equilibrium can be successfully described by the Chen and Clayton equation while for lignite a BET type equation provides better fit. To describe the drying kinetics of solid particles in the falling drying rate period a diffusional model was found to be satisfactory. In the case of wood biomass, a Darcy flow term caused by pressure gradient was added. In both cases the effective diffusivity can be described by the Arrhenius type equation corrected by the moisture content correction X^n. However, it was found that due to small values of n obtained by identification, it may be concluded that D_{eff} is almost independent on moisture content in the investigated moisture content range.

Simulation results for wood biomass indicate that the drying kinetics in simple geometry of a cylinder can be satisfactorily described by the developed model. The internal profiles of temperature are flat enough that for practical applications the internal thermal resistance can be neglected.

Simulation results for lignite indicate that the diffusional model can be used in parallel shrinking-core model to describe drying kinetics. The simulated influence of steam pressure at constant temperature over drying rate indicates that at ca. 3 bar a maximum of drying rate (minimum of drying time) may exist. This finding requires further experimental confirmation.

The aforementioned findings serve for the construction of the so called Single-Particle Models, which are the fundamental building blocks of the dryer model used for dryer design. The construction of a dryer model is, however, outside the scope of this work. Reader may refer to the work by Pakowski (2011).

Acknowledgements We gratefully acknowledge the help of the following students: S. Kwapisz, B. Krupińska, M. Kokocińska, K. Stolarek, M. Lasota and M. Gregor in experimental work.

References

Adamski R, Pakowski Z (2008) Superheated steam drying model for willow *Salix viminalis*, 16th International Drying Symposium (IDS 2008), Hyderabad, India 9–12 November 2008

Chen Z, Wu W, Agarwal PK (2000) Steam-drying of coal. Part 1. Modeling the behavior of a single particle. Fuel 79:961–973

Clayton S, Desai D, Hoadley A (2007) Drying of brown coal using superheated steam rotary dryer. In: Chen G (ed) 5th APDC, Hong Kong, pp 179–184

Kirayama T, Sasaki H, Hashimoto A et al (2016) Evolution of drying rates of lignite particles in superheated steam using single-particle model. Metall Mat Trans E 3(4):308–316

Kokocińska M, Pakowski Z (2014) High pressure desorption equilibrium of lignite obtained by the novel isochoric method. Fuel 109:626–634

Kotowski W (2003) Zgazowanie biomasy w małych elektrociepłowniach (biomass gasification in small power plants). Energetyka Cieplna i Zawodowa 1:34–38

Levy E (2005) Use of coal drying to reduce water consumed in pulverized coal power plants. Quarterly report for the period July 1, 2005 to September 30. DOE Award Number DE-FC26-03NT41729. Energy Research Center Lehigh University, 117 ATLSS Drive Bethlehem, PA 18015

Liu Z, Tamuer M, Aramaki H, Riechelmann D (2017) Simulation of lignite drying in a continuous fluidized bed. EuroDrying 2017, Liege, Belgium, 19–21 July, pp 54–55

Pakowski Z (2011) Projektowanie suszarek do suszenia parą przegrzaną (design of dryers for superheated steam drying). TUL Publishers. ISBN 978-83-7283-428-7

Pakowski Z, Adamski R (2011) On prediction of the drying rate in superheated steam drying process. Drying Techn 29:1492–1498

Pakowski Z, Krupińska B, Adamski R (2007) Prediction of sorption equilibrium both in air and superheated steam drying of energetic variety of willow *Salix viminalis* in a wide temperature range. Fuel 86(12–13):1749–1757

Pakowski Z, Adamski R, Kokocińska M (2009) Cross-fiber dry wood Darcy permeability of energetic willow *Salix viminalis v. Orm*. Drying Techn 27(12):1379–1383

Pakowski Z, Adamski R, Kokocińska M et al (2011a) Generalized desorption equilibrium equation of lignite in a wide temperature and moisture content range. Fuel 90:3330–3335

Pakowski Z, Adamski R, Kwapisz S (2011b) Effective diffusivity of moisture in low rank coal during superheated steam drying at atmospheric pressure. Chem Process Eng 32(4/2):43–51

Pakowski Z, Kwapisz S, Adamski R et al (2013) The kinetics of superheated steam drying of low rank coal at high pressure. In: XIII polish drying symposium Kołobrzeg, Poland, 5–6 Sept

Pakowski Z, Kwapisz S (2014) Comparison of two single particle models for superheated steam drying of low rank coal. In: IDS'2014, Lyon, France

Ross D (2005) Pressurised flash drying of Yallourn lignite. Fuel 84:47–52

World Energy Resources Coal (2016) World Energy Council

Extensional Flow of Polymer Solutions Through the Porous Media

Sylwia Różańska

1 Introduction

In the laminar flow of fluid through a channel with a permanent cross-sectional area, only simple shear occurs. In case of a fluid flow through a porous bed the cross-section surface of the channel is subject to constant change. In such flows, there appear additionally the extensional components associated with the converging-diverging flow paths (Sochi 2010). The strong extensional components during the flow of non-Newtonian fluids may lead to a noticeable increase in flow resistance, which must be accounted for in the calculations of the pressure drop.

The flow of non-Newtonian fluids through a porous bed occurs during rock fractures (crude oil and gas production), liquid filtration (Dehghanpour and Kuru 2011) and gel permeation chromatography. The extensional flows are also important of such processes as: spraying (Ochowiak et al. 2012), free settling of sphere (Arigo and McKinley 1998; Bot et al. 1998; Chen and Rothstein 2004), sedimentation (Arigo et al. 1995) and fluidization (Chhabra et al. 2001). This chapter shall describe basic information on the flow of polymer solutions through porous media, with special attention paid to the influence of the extensional components on the resistance flow.

S. Różańska (✉)
Institute of Chemical Technology and Engineering, Poznan University of Technology, Poznań, Poland
e-mail: sylwia.rozanska@put.poznan.pl

© Springer International Publishing AG, part of Springer Nature 2018   377
M. Ochowiak et al. (eds.), *Practical Aspects of Chemical Engineering*,
Lecture Notes on Multidisciplinary Industrial Engineering,
https://doi.org/10.1007/978-3-319-73978-6_26

2 Theoretical Considerations

2.1 Extensional Viscosity

Viscosity related to resistance that is demonstrated by the fluid during extensional flow is referred to as extensional viscosity (or elongational viscosity). In case of uniaxial stretching of the extensional viscosity it is defined by the equation (Petrie 2006):

$$\eta_E(\dot{\varepsilon}) = \lim_{t \to \infty} \left(\frac{\sigma_E(\dot{\varepsilon}, t)}{\dot{\varepsilon}} \right) \tag{1}$$

where

$$\dot{\varepsilon} = \frac{dw_x}{dx} = \frac{d\varepsilon}{dt} \tag{2}$$

is the strain rate referred to as stretch rate or extensional rate, σ_E is the "net tensile stress" (normal stress difference)

$$\sigma_E \equiv \sigma_{11} - \sigma_{22} \tag{3}$$

where σ_{11} and σ_{22} are normal stress components of stress tensor.

The term "net tensile stress" means that the measured stress is adjusted in accordance with the equation (Petrie 2006):

$$\sigma_E = \frac{Applied\ force}{Area} - \frac{Coefficient\ of\ surface}{Radius} \tag{4}$$

Equation (1) show that the "true" extensional viscosity shall remain an equilibrium value, not changing in time. Providing better equilibrium conditions during the determination of extensional viscosity for mobile fluids has turned out to be extremely difficult (Ferguson and Kembłowski 1991). Reaching the equilibrium conditions is possible only in case of the flow of fluids with very high viscosity (exceeding 1000 Pa s, mainly melted polymers). In case of fluids with low viscosity (the so called mobile fluids) maintaining the equilibrium conditions is not possible, (Ferguson et al. 1997), hence it is necessary to apply such methods in which transient extensional viscosity can be determinate only.

For the Newtonian fluids, Trouton (1906) demonstrated in a theoretical manner that the ratio between extensional viscosity η_E with uniaxial stretching and shear viscosity η equals 3, it is the so called Trouton ratio.

In case of non-Newtonian fluids, extensional viscosity and shear viscosity are the function of strain rate, a Trouton ratio may be calculated from the equation:

$$Tr = \frac{\eta_E(\dot{\varepsilon})}{\eta(\dot{\gamma})} \qquad (5)$$

where η_E and η are determined with the same strain rate ($\dot{\gamma} = \dot{\varepsilon}$).

It is not a single approach to the notion of Trouton ratio for non-Newtonian fluids. Jones et al. (1987) suggested calculating the Trouton ratio based on the following equation:

$$Tr_B = \frac{\eta_E(\dot{\varepsilon})}{\eta(\sqrt{3}\dot{\varepsilon})} \qquad (6)$$

where shear viscosity η is defined at the shear rate calculated as $\sqrt{3}\dot{\varepsilon}$. According to Jones et al. (1987), for clearly viscous non-Newtonian fluids the Tr_B ration value equals 3. Any deviations from the said value may be related to fluid elastic properties.

Figure 1 shows the dependence of extensional viscosity in the function of strain rate for high molecular dilute polymer solutions. In case of dilute solutions with high values of strain rate the extensional viscosity has a constant value and the ratio $Tr = 3$ (Fig. 1a). Over a certain value of $\dot{\varepsilon}$, extensional viscosity begins to increase rapidly (strain hardening region), which is related to the stretched polymer chains. When they are fully stretched, extensional viscosity reaches a constant high value. Such a course remains typical for polymer solutions with flexible chain construction e.g., poly(ethylene oxide) (PEO) or polyacrylamide (PAA). For the solutions of such polymers in strain hardening region, the values of Tr ratio may be very high, reaching even 1000 (Hudson et al. 1988; Dontula et al. 1997; Gauri and Koelling 1997; Różańska 2015).

The course of extensional viscosity curve for entangled polymer fluids remains a more complex one. A standard model anticipates four ranges of changes in extensional viscosity versus the stretching rate in steady uniaxial extensional flow of a monodisperse entangled polymer fluids (Fig. 1b) (Marrucci and Ianniruberto 2004). Occurrence of individual ranges may be explained on the basis of a well-known tube model by Doi and Edwards (1986). With the values for $\dot{\varepsilon}$ below reciprocal reptation time τ_d extensional viscosity has a constant value (Fig. 1b, region I). Within the said range, Trouton ratio equals 3, which indicates that extensional viscosity will be influenced by the same parameters as shear viscosity (chemical structure, molecular weight and molecular weight distribution and for solutions, polymer-solvent interactions (Ferguson et al. 1997)). When the value $\dot{\varepsilon} > \tau_d^{-1}$ reptation is slowed down and tubes oriented parallel to the elongation direction. Within this range the polymer chains are not elongated and extensional viscosity drops with the increase in strain rate (Fig. 1b, region II, strain thinning region). Extending the polymer chains occurs when values $\dot{\varepsilon}$ equal or exceed reciprocal Rouse time τ_R^{-1}, which is demonstrated by the increase in extensional viscosity (Fig. 1b, region III). When full stretching of polymer chains occurs, extensional viscosity has a constant value (Fig. 1b, region IV).

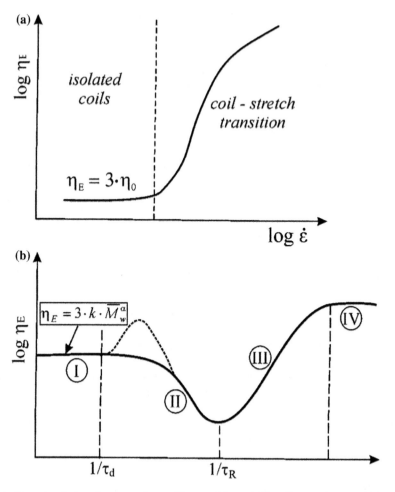

Fig. 1 Extensional viscosity curves for: **a** dilute, and **b** semi-dilute linear polymer solutions

In case of the measurements with regards to transient extensional viscosity of the solutions for some polymers (Kennedy et al. 1995; Różańska 2015) together with the end of the range for constant extensional viscosity (region I, Fig. 1b) a characteristic maximum (dotted line in Fig. 1b) has been observed. The reasons for such an occurrence are not yet clear. This may result from the lack of possibility to maintain equilibrium conditions during the measurements of extensional viscosity.

2.2 The Basic Theory of the Flow of Fluids Through Porous Media

Porosity ϕ (also called voidage) is a basic parameter characterising the porous bed, defined as the fraction of the total volume which is free space available for the flow of fluids. Figure 2 shows the illustrations to characterise the flow of fluid through porous medium, where $\dot{Q}$ is the flow rate through the bed, A is the bed cross sectional area, u_0 is the superficial velocity the total flow rate divided by the cross sectional area and u is the interstitial velocity.

In the case of isotropic porosity (i.e. the same in all directions), the interstitial velocity u is simple related to the superficial velocity u_0 (Holdich 2002):

$$u = \frac{u_0}{\phi} \tag{7}$$

During the fluid flow through porous medium pressure drops occur, which are characterized by the differences between static pressures ΔP. Results of the measurements for static pressures are usually presented in the form of dimensionless resistance coefficient:

$$\Lambda = \frac{d^2 \cdot \phi^3 \cdot (\Delta P / L)}{\eta \cdot u_0 \cdot (1 - d)^2} \tag{8}$$

where d—particle diameter, η—viscosity of the fluid, L—porous medium length.

In a general case, resistance coefficient remains the function of Reynolds number, which has the following form for the flow of non-Newtonian fluids through porous medium:

$$Re = \frac{d \cdot u_0 \cdot \rho}{\eta \cdot (1 - \phi)} \tag{9}$$

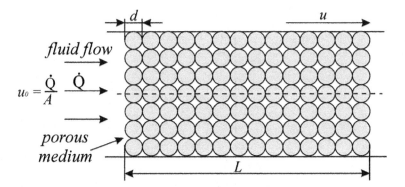

Fig. 2 Illustration of fluid flow through porous medium

where ρ—density of the fluid. In case of Newtonian fluids, the dependence of resistance coefficient from Re number remains a linear functions can be described by Ergun equation (Ergun and Orning 1952):

$$\Lambda = A + B\,Re \qquad (10)$$

where A and B are constants. According to Macdonald et al. (1979) the value of the constant $A = 180$ and the constant $B = 1.8$ (for monodisperse spherical particles). The values of the said constants may differ slightly, depending on the type of the porous medium, its packing and porosity.

With the low values of Re ($Re < 1$) number, the value of resistance coefficient remains constant and the pressure drop is directly proportional to the superficial velocity. This is the so called Darcian regime, where the flow is dominated by viscous forces and the velocity distribution is determined by local geometry (Chhabra et al. 2001). The one-dimensional Darcy equation can be written as:

$$-\frac{\Delta P}{L} = \frac{\eta \cdot u_0}{K} \qquad (11)$$

where $\Delta P/L$ is the pressure drop per unit length along the flow path and K is the permeability of the porous medium.

At high Re numbers inertial effects in the pores dominate over viscous effects and the pressure drop becomes proportional to the square of the superficial velocity (Saez et al. 1994). An exemplary resistance coefficient for water is shown in Fig. 3.

According to Ergun and Orning (1952) the critical value of Re_{kr} number, over which the influence of inertia can be seen, keeps within the range between 3 and 10. Other values of Re_{kr} can be encountered in the literature, for example Scheidegger (1974) Re_{kr} keeps within the range from 0.1 to 75. Hassanizadeh and Gray (1987) suggest accepting $Re_{kr} = 10$ for technical calculations.

Fig. 3 Resistance coefficient in the function of Reynolds number for the flow of water through monodisperse porous medium (Sargenti et al. 1999)

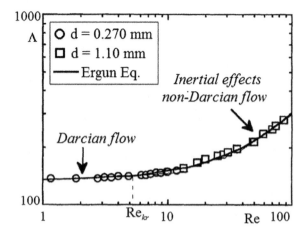

Measurement outcomes for pressure drops during the flow of non-Newtonian fluids through porous media are presented in the literature in two manners (de Castro and Radilla 2017). The first method is based on the presentation of the ratio of resistance coefficient in the function of a properly defined Reynolds number. The second approach takes advantage of the notion of the apparent shear viscosity of the solutions η_a and apparent shear rate $\dot{\gamma}_a$ in the porous medium flow.

As it has been mentioned above, the kinematics in the porous media do not follow a pure shear flow, having components due to both shear and extension. Hence, the resistance coefficient can be expressed as the superposition of a simple shear contribution (Λ_{ss}) and an extensional contribution (Λ_{ext}):

$$\Lambda = \Lambda_{ss} + \Lambda_{ext} \tag{12}$$

Depending on the type of fluids, the flow velocity and the local geometry of the porous medium shear and extension coexist in various proportions that cannot be quantified. Basically, during the flow of polymer solutions with low Reynolds numbers, shear flow dominates, hence it may assumed that $\Lambda_{ss} = \Lambda$, and Λ are proportional to the shear viscosity of polymer solutions, η_s (González et al. 2005).

The apparent shear viscosity of the solutions in porous media flows is most frequently calculated from Darcy equation:

$$\eta_a = -\frac{K}{u_0} \cdot \frac{\Delta P}{L} \tag{13}$$

For packed beds of spheres permeability may be calculated from Kozeny–Carman equation:

$$K = -\frac{\phi^3 \cdot d^2}{36\kappa \cdot (1 - \phi)^2} \tag{14}$$

where κ is the Kozeny–Carman constant, which for beds packed with spherical particles equals 5 (Xu and Yu 2008).

For non-Newtonian fluids shear viscosity remains the function of shear rate $\dot{\gamma}$. The apparent shear rate in the porous medium may be calculated by the use of equation:

$$\dot{\gamma}_{al} = \frac{u_0}{\phi \cdot l} \tag{15}$$

where l is a characteristic length representative of the pore-scale velocity gradients. In line with the suggestion from Patruyo et al. (2002), for the flow of non-Newtonian fluids through porous media of spheres, characteristic length may be described by means of the following equation:

$$l = c \cdot d \tag{16}$$

Constant c is determined experimentally by comparing the shear viscosity curves obtained by means of a rheometer and the apparent shear viscosity curves for the porous media flow with low Reynolds number. Its value for polymer solutions ranges from 0.04 to 0.07. For the purposes of the calculations, the value 0.05 is most frequently recommended (Patruyo et al. 2002; Müller et al. 2004; González et al. 2005; Rojas et al. 2008; Amundarain et al. 2009).

The apparent shear rate is also calculated based on the equation (Perrin et al. 2006; de Castro and Radilla 2017):

$$\dot{\gamma}_{a2} = \alpha \frac{u_0}{\sqrt{\phi \cdot K}} \tag{17}$$

where $\sqrt{\phi \cdot K}$ is the microscopic characteristic length and α is a empirical shift factor. The shift factor α is determined analogically to the constant c.

2.3 Sample Measurement Results

2.3.1 Semi-rigid Polymer Solutions

The examples polymers with semi-rigid (or semi-flexible) chain in aqueous solutions are guar gum, carboxymethylcellulose sodium salt (Na-CMC) (Duda et al. 1983), hydroxyethyl cellulose (Sadowski 1963), xanthan gum (Chauvetau 1982; Sorbie and Huang 1991), scleroglucan (Huang and Sorbie 1993), hydroxypropyl guar (HPG) and hydroxyethylcellulose (HEC). Dependencies of resistance coefficient Λ from the Reynolds number for the said polymers have qualitatively similar course. Figure 4 shows an exemplary dependence for $\Lambda = f(Re)$ obtained for

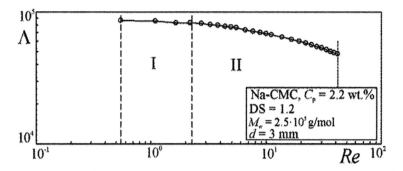

Fig. 4 Shear thinning behavior for Na-CMC solution in water (balls of diameter 3 mm, pipe of diameter 32 mm and length 320 mm) (own data not yet published)

aqueous solution of Na-CMC (M_w = 250,000 g/mol; degree of substitution DS = 1.2).

Two basic regions may be distinguished for it. With low Re numbers (region I) the resistance coefficient is constant, the region is referred to as pseudo Newtonian. With higher values for Re number (region II), the values for resistance coefficient decrease together with the increase in the Reynolds number (shear-thinning Darcy regime). The range of Reynolds numbers within which resistance coefficient has a constant value depends on the type of polymer, its concentration, molecular weight, ionic strength, etc. In case of diluted polymer solutions, similarly to Newtonian fluids, resistance coefficient has a constant value (Tatham et al. 1995; Amundarain et al. 2009; Gonzales et al. 2005).

Figure 5 shows the comparison of the dependences between apparent shear viscosity curves η_a (for calculations of $\dot{\gamma}_{a1}$, the c constant value of c = 0.05 has been assumed), shear viscosity curves η and apparent extensional viscosity curves η_E (obtained with using of opposed nozzles device) for the solutions of Na-CMC (M_w = 700,000 g/mol, DS = 0.9). In all cases, one may observe the shear thinning behaviour, additionally apparent shear viscosity curve and shear viscosity curves overlap. Shear thinning behaviour as well as close values of apparent shear viscosity and shear viscosity demonstrate that the flow resistance is dominated by the shear component.

Concurrence between apparent shear viscosity curves for the flow of semi-rigid polymer solutions in the porous medium and shear viscosity curves occurs only for low shear rates. With high shear rate, the values of η_a remain noticeably higher than η (Fig. 6a, example for aqueous solutions of xanthan gum).

According to González et al. (2005) and Amundarain et al. (2009) the differences in the in the slops of curves $\eta_a = f(\dot{\gamma}_a)$ and $\eta = f(\dot{\gamma})$ may result from the influence of extensional component over the flow resistance. de Castro and Radilla (2017) acknowledged that the deviation from shear-thinning Darcy flow during the flow of xanthan gum solutions is brought about by inertial effects, not by extensional flow.

Fig. 5 Comparison of apparent shear viscosity curves η_a (characteristic length was calculated as $l = 0.05d$), shear viscosity curves η and apparent extensional viscosity curves η_E (obtained using opposed nozzles device) for Na-CMC solutions (M_w = 700,000 g/mol, DS = 0.9) (Różańska and Różański 2017)

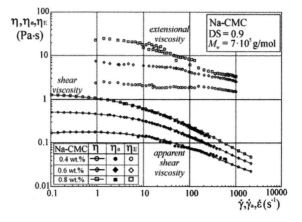

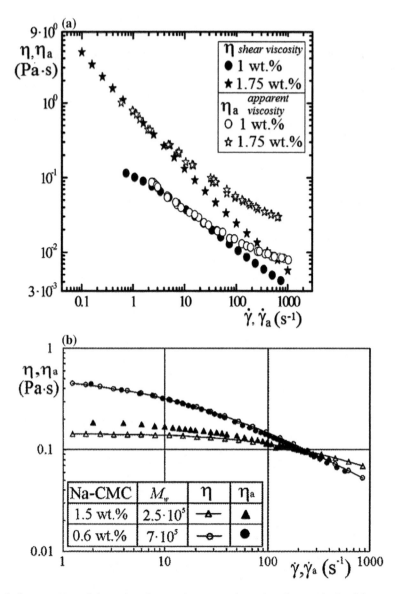

Fig. 6 Superposition of shear viscosity η and apparent shear viscosity η_a calculated from porous media results: **a** for xanthan gum solutions (González et al. 2005), **b** aqueous solution of Na-CMC$_{700}$ and Na-CMC$_{250}$ (Różańska and Różański 2017, own data not yet published)

For semi-rigid polymer solutions the literature suggests a number of correlating equations that allow us to calculate the pressure drop under the flow in the Darcy regime. The relevant listings may be found in the works by Kaur et al. (2011). The most frequently applied modelling fluids include aqueous solutions of Na-CMC

(Fig. 6b) (Chhabra et al. 2001). For non-Darcy regime, the method to calculate the pressure drop was suggested by de Castro and Radilla (2017).

2.3.2 Flexible Polymer Solutions

A different behaviour than for semi-rigid polymer solutions occurs during the extensional flow of polymer solutions with flexible chain structure. Solutions of flexible polymers typically exhibit reduction of shear viscosity with flow rate in shear flows (shear thinning), while in extensional flows substantial increases in extensional viscosity with flow rate (extension thickening). Typical polymers with flexible chains are polyacrylamid (PAM) and poly(ethylene oxide) (PEO). The literature on the subject includes a number of works referring to the properties demonstrated by the solutions of such polymers in shear flow and in extensional flow (Kulicke and Haas 1984; Ghoniem 1985; Różańska et al. 2012; Różańska 2015; Zhang et al. 2016). Figure 7 shows exemplary apparent extensional curves for PEO solutions obtained by using of opposed nozzles device.

During the stretching flow, flexible chains of PEO alter the conformation from a form of a curled ball to a rod form, owing to which the energy dissipation occurs. This is revealed by increased extensional viscosity together with increased deformation rate (Hinch 1977; Fuller and Leal 1980; Chan et al. 1988; Durst and Haas 1988; Prokop et al. 2008). Figure 8 shows the exemplary curves for dependencies of resistance coefficient from Reynolds number for the PEO and PAA solutions. With low values of Re number, the resistance coefficient has a relatively constant

Fig. 7 Shear and extensional viscosity of poly(ethylene oxide) (studies were performed using nozzles opposed device) (Różańska et al. 2012)

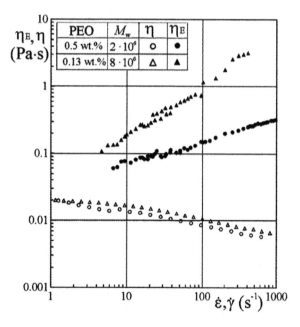

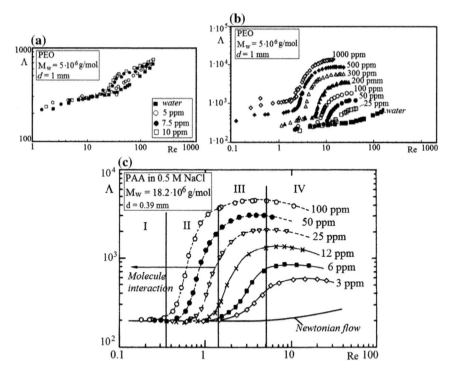

Fig. 8 Resistance coefficients for: **a** aqueous solutions of PEO low concentration, **b** PEO high concentration (Rodriguez et al. 1993), **c** PAA in 0.5 mol NaCl (Haas and Kulicke 1984)

value and the polymer solution behaves exactly as water (region I) (Fig. 8c). Subsequently, one may observe a rapid increase of resistance coefficient (region II) (apparent flow thickening), after which it stabilises at a constant level (Fig. 8c, region III). Deviations of the resistance coefficient from the values measured for pure water may be observed for the solutions with concentrations of 5 ppm PEO (Fig. 8a) and 3 ppm PAA (Fig. 8c). With high Re numbers, the coefficient Λ starts to fall, which is probably related to the mechanical degrading of the polymer (Fig. 8c, region IV).

The research by Haas and Kulicke (1984) also reveals that with very low concentrations of HPAA the value of onset Re_0 number, at which the region II starts, does not depend on the polymer concentration (Fig. 8c). With higher concentration of HPAA (≥ 25 ppm) Re_0 drops together with the increase in polymer concentration, which indicates that molecular interactions occur. A noticeable increase in the polymer concentration also leads to the increased resistance coefficient in region I (Fig. 8c). As within this range flow resistance is dominated by the shear component, the increase in Λ coefficient is brought about by the increase in shear viscosity of PEO solutions (Rodriguez et al. 1993).

As it has already been mentioned before, diminished values of Λ coefficient observed in region IV is probably brought about by mechanical degrading of polymers. High molecular poly(ethylene oxide) and PAA are especially susceptible to mechanical degrading, even with low shear rate. Figure 9 shows the results of resistance coefficient Λ measurements made by Haas and Kulicke (1984) for fresh PAA solutions PAA (Fig. 9, curve A) as well as solutions subject to shear with Re 15 (Fig. 9, curve B) and 20 (Fig. 9, curve C). Initial shear brought about polymer mechanical degrading, which is demonstrated by lower values of Λ coefficient in regions II and III and higher values of onset Re_0 number.

The reasons behind the increased resistance factor within region II have not been fully clarified so far. Critical analysis of the mechanisms for the flow of flexible polymers solutions in porous media suggested by the literature on the subject was performed by Müller and Sáez (1999). Sochi (2010) presented a review in which he described in detail the flow of viscoelastic fluids through porous media.

The first group of hypotheses suggested by the literature is related to the increase in the pressure drop within region II with the elongation nature of the flow in the pores. Elata et al. (1977) explains the increase in flow resistance over Re_0 by the coil-stretch transition of high molecular weight flexible molecules in extensional flow fields. Based on the coil-stretch transition theory, the reasons for the drop in Re_0 together with the increase in concentration cannot be explained. Additionally, Batchelor (1971) noticed that high increase in pressure losses observed cannot be caused by the stretching of isolated macromolecules only. The aforementioned effects may be explained based on the transient network hypothesis suggested by Odell et al. (1988). According to this hypothesis, in extensional flow field occurs the formation of transient networks of polymer molecules. Its formulation paves the way to the noticeable increase in resistance flow. Rodríguez et al. (1993), proved that onset strain rate at which increase in flow resistance occurs, is consistent with that observed in opposed-jets flow when transient entanglement networks are formed.

Another group of literature-suggested hypotheses is connected with the occurrence of apparent flow thickening with unstable flow. Rodríguez et al. (1993) noted

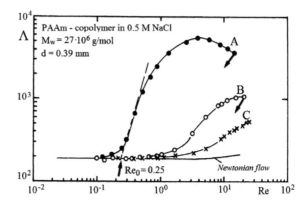

Fig. 9 Mechanical degradation in porous media flow of PAA solutions (Haas and Kulicke 1984)

that with high flow velocities of solutions with low polymer concentration, microscopic velocity profiles in the porous medium may be drastically different than those observed in the viscous flow, with presence of vortices and steep velocity gradients. Therefore, inertia may influence the value of resistance coefficient.

Clarke et al. (2015) and Howe et al. (2015) made the measurements of the pressure drop during the flow of concentrated HPAA solutions (high molecular weight partially hydrolysed polyacrylamides; HPAA concentration $\geq 10c*$, where $c*$ is the overlap concentration) through microfluidic networks and rock cores. The onset of flow thickening in the rock occurred with similar values of apparent shear rate (about 1 s^{-1}), independent of concentration. Images of streak lines made by particle tracking velocimetry (PTV) method in a microfluidic network showed that increase in pressure losses during the HPAA flow is related to the flow instability, whereby shear rate at which the beginning of unstable flow was observed is not dependent on the polymer concentration. Based on the above, Clarke et al. (2015), stated that apparent flow thickening during the flow through porous medium results from elastic turbulence. The term "elastic turbulence" was introduced by Groisman and Steinberg (2000) to describe a widely observed chaotic instability in the flow of viscoelastic fluid at low Reynolds number arising in through a long characteristic relaxation time of the polymer molecules in solution. It must be stated that the occurrence of elastic turbulence clarifies the reasons for the sudden increase in pressure drop during the flow of concentrated flexible polymer solutions in porous media. There is no evidence, that it may be the reason for apparent flow thickening to occur during the flow of dilute solutions.

3 Summary

The following paper discusses the pressure drops occurring during the flow of semi-rigid and flexible polymer solutions in porous media. As the research made by de Castro and Radilla (2017) reveals, the occurrence of a sudden increase in resistance coefficient in non-Darcy regime during the flow of semi-rigid polymer solution ought to be related to inertia effects. Flexible polymer solutions constitute a more complex example. As Clarke et al. (2015) demonstrated for the concentrated solutions of that type of polymers, the occurrence of apparent flow thickening is brought about mainly by elastic turbulence, although viscous dissipation due to extensional flows remains probable as well (Howe et al. 2015). So far there has been no undoubted clarification for the reasons behind a sudden increase in pressure drops during the flow of dilute and semi-dilute flexible polymer solutions. Most frequently it is related to extensional components generated by expansions and contractions within the porous structure. However, there is no sufficient evidence to confirm the relation between high extensional viscosity of such solutions and apparent flow thickening during their flow in porous media.

Acknowledgements This work was supported by PUT research grant No. 03/32/DSPB/0702.

References

Arigo MT, McKinley GH (1998) An experimental investigation of negative wakes behind spheres settling in a shear-thinning viscoelastic fluid. Rheol Acta 37:307–327

Arigo MT, Rajagopalan D, Shapley N et al (1995) The sedimentation of a sphere through an elastic fluid. Part 1. Steady motion. J Non-Newton Fluid 60:225–257

Amundarain JL, Castro LJ, Rojas M et al (2009) Solutions of xanthan gum/guar gum mixtures: shear rheology, porous media flow, and solids transport in annular flow. Rheol Acta 48: 491–498

Batchelor GK (1971) The stress generated in a non-dilute suspension of elongated particles by pure straining motion. J Fluid Mech 46:813–829

Bot ETG, Hulsen MA, Van den Brule BHAA (1998) The motion of two spheres falling along their line of centres in a Boger fluid. J Non-Newton Fluid 79:191–212

Chauvetau G (1982) Rodlike polymer solution flow through fine pores: influence of pore size on rheological behavior. J Rheol 26:111–142

Chen S, Rothstein JP (2004) Flow of a wormlike micelle solution past a falling sphere. J Non-Newton Fluid 116:205–234

Chan RC, Gupta RK, Sridhar T (1988) Fiber spinning of very dilute solutions of polyacrylamide in water. J Non-Newton Fluid 30:267–283

Chhabra RP, Comiti J, Machač I (2001) Flow of non-Newtonian fluids in fixed and fluidised beds. Chem Eng Sci 56:1–27

Clarke A, Howe AM, Mitchell J et al (2015) Mechanism of anomalously increased oil displacement with aqueous viscoelastic polymer solutions. Soft Matter 11:3536–3541

de Castro AR, Radilla G (2017) Non-Darcian flow of shear-thinning fluids through packed beads: experiments and predictions using Forchheimer's law and Ergun's equation. Adv Water Resour 100:35–47

Dehghanpour H, Kuru E (2011) Effect of viscoelasticity on the filtration loss characteristics of aqueous polymer solutions. J Petrol Sci Eng 76:12–20

Doi M, Edwards SF (1986) The theory of polymer dynamics. Oxford University Press, Oxford

Dontula P, Pasquali M, Scriven LE et al (1997) Can extensional viscosity be measured with opposed-nozzle devices? Rheol Acta 36:429–448

Duda JL, Hong SA, Klaus EE (1983) Flow of polymer solutions in porous media: inadequacy of the capillary model. Ind Eng Chem Fund 22:299–305

Durst F, Haas R (1988) Die Charakterisierung viskoelastischer fluide mit hiife ihrer Strömungseigenschaften in Kugeischüttungen. Rheol Acta 21:150–166

Elata C, Burger J, Michlin J et al (1977) Dilute polymer solutions in elongational flow. Phys Fluids 20:49–54

Ergun S, Orning AA (1952) Fluid flow through packed columns. Chem Eng Prog 48:89–94

Ferguson J, Hudson NE, Odriozola MA (1997) The interpretation of transient extensional viscosity data. J Non-Newton Fluid 68:241–257

Ferguson J, Kembłowski Z (1991) Applied fluid rheology. Elsevier Applied Science, New York, London

Fuller GG, Leal LG (1980) Flow birefringence of dilute polymer solutions in two-dimensional flows. Rheol Acta 19:580–600

Gauri V, Koelling KW (1997) Extensional rheology of concentrated poly(ethylene oxide) solutions. Rheol Acta 36:555–567

Ghoniem SAA (1985) Extensional flow of polymer solutions through porous media. Rheol Acta 24:588–595

González JM, Müller AJ, Torres MF et al (2005) The role of shear and elongation in the flow of solutions of semi-flexible polymers through porous media. Rheol Acta 44:396–405

Groisman A, Steinberg V (2000) Elastic turbulence in a polymer solution flow. Nature 405:53–55

Haas R, Kulicke WM (1984) In: Gampert B (ed) The influence of polymer additives on velocity and temperature fields. Springer, Berlin, New York, pp 119–129

Hassanizadeh SM, Gray WG (1987) High velocity flow in porous media. Transp Porous Med 2:521–531

Hinch EJ (1977) Mechanical models of dilute polymer solutions in strong flows. Phys Fluids 20: S22–S30

Holdich RG (2002) Fluid flow in porous media. Fundamentals of particle technology. Loughborough, UK, pp 21–28

Howe AM, Clarke A, Giernalczyk D (2015) Flow of concentrated viscoelastic polymer solutions in porous media: effect of MW and concentration on elastic turbulence onset in various geometries. Soft Matter 11:6419–6431

Huang Y, Sorbie KS (1993) Scleroglucan behavior in flow through porous media: comparison of adsorption and in-situ rheology with xanthan. In: SPE international symposium on oilfield chemistry, New Orleans, Louisiana, Mar 1993, Society of Petroleum Engineer (SPE), p 223

Hudson NE, Ferguson J, Warren BCH (1988) Polymer complexation effects in extensional flows. J Non-Newton Fluid 30:251–266

Jones DM, Walters K, Williams PR (1987) On the extensional viscosity of mobile polymer systems. Rheol Acta 26:20–30

Kaur N, Singh R, Wanchoo RK (2011) Flow of Newtonian and non-Newtonian fluids through packed beds: an experimental study. Transp Porous Med 90:655–671

Kennedy JC, Meadows J, Williams PA (1995) Shear and extensional viscosity characteristics of a series of hydrophobically associating polyelectrolytes. J Chem Soc Faraday Trans 91(5): 911–916

Kulicke WM, Haas R (1984) Flow behavior of dilute polyacrylamide solutions through porous media. 1. Influence of chain length, concentration, and thermodynamic quality of the solvent. Ind Eng Chem Fundam 23(3):308–315

Macdonald IF, El-Sayed MS, Mow K et al (1979) Flow through porous media the Ergun equation revisited. Ind Eng Chem Fund 18:199–215

Marrucci G, Ianniruberto G (2004) Interchain pressure effect in extensional flows of entangled polymer melts. Macromolecules 37:3934–3942

Müller AJ, Sáez AE (1999) The rheology of polymer solutions in porous media. In: Flexible polymer chains in elongational flow. Theory and experiment. Springer, Berlin, pp 335–393

Müller AJ, Torres MF, Sáez AE (2004) Effect of the flow field on the rheological behavior of aqueous cetyltrimethylammonium p-toluenesulfonate solutions. Langmuir 20:3838–3841

Ochowiak M, Broniarz-Press L, Rozanska S, Rozanski J (2012) The effect of extensional viscosity on the effervescent atomization of polyacrylamide solutions. J Ind Eng Chem 18:2028–2035

Odell JA, Miller AJ, Keller A (1988) Non-Newtonian behaviour of hydrolysed polyacrylamide in strong elongational flows: a transient network approach. Polymer 29:179–1190

Patruyo JG, Müller AJ, Sáez AE (2002) Shear and extensional rheology of solutions of modified hydroxyethyl celluloses and sodium dodecyl sulfate. Polymer 43:6481–6493

Perrin CL, Tardy PMJ, Sorbie KS et al (2006) Experimental and modeling study of Newtonian and non-Newtonian fluid flow in pore network micro models. J Colloid Interface Sci 295(2):542–550

Petrie ChJS (2006) Extensional viscosity: a critical discussion. J Non-Newton Fluid 137:15–23

Prokop A, Hunkeler D, DiMari S et al (2008) Water soluble polymers for immunoisolation II. Complex coacervation and cytotoxicity. In: Advances in polymer science, vol 136. Springer, Berlin, pp 1–51

Rojas MR, Müller AJ, Sáez AE (2008) Shear rheology and porous media flow of wormlike micelle solutions formed by mixtures of surfactants of opposite charge. J Colloid Interface Sci 326:221–226

Rodriguez S, Romero C, Sargenti ML et al (1993) Flow of polymer solutions through porous media. J Non-Newton Fluid 49:63–85

Różańska (2015) Rheological properties of aqueous solutions of polyacrylamide and hydroxyethylcellulose in extensional and shear flow. Polymers 60(10):57–63

Różańska S, Broniarz-Press L, Różański J et al (2012) Extensional viscosity and stability of oil-in-water emulsions with addition poly(ethylene oxide). Proc Eng 42:799–807

Różańska S, Różański J (2017) Extensional flow of carboxymethylcellulose sodium salt measured on the opposed-nozzle device. Soft Mater 4:302–314

Sáez AE, Müller AJ, Odell JA (1994) Flow of monodisperse polystyrene solutions through porous media. Colloid Polym Sci 272:1224–1233

Sadowski TJ (1963) Non-Newtonian flow through porous media. PhD thesis, University of Wisconsin, Madison

Sargenti ML, Müller AJ, Sáez AE (1999) Adopted from the rheology of polymer solutions in porous media. In: Nguyen, TQ, Kausch HH (eds), Flexible chain dynamics in elongational flows: theory and experiments. Springer, Berlin, p 340

Scheidegger AE (1974) The physics of flow through porous media, 3rd edn. University of Toronto Press, pp 152–170

Sochi T (2010) Non-Newtonian flow in porous media. Polymer 51:5007–5023

Sorbie KS, Huang Y (1991) Rheological and transport effects in the flow of low-concentration xanthan solution through porous media. J Colloid Interface Sci 145:74–89

Tatham JP, Carrington S, Odell JA et al (1995) Extensional behavior of hydroxypropyl guar solutions: optical rheometry in opposed jets and flow through porous media. J Rheol 39(5):961–986

Trouton FT (1906) On the coefficient of viscous traction and its relation to that of viscosity. Proc R Soc London A Math 77:426–440

Xu P, Yu B (2008) Developing a new form of permeability and Kozeny-Carman constant for homogeneous porous media by means of fractal geometry. Adv Water Resour 31:74–81

Zhang L, Wang J, Zhang Y et al (2016) Rheological behavior of hydrolyzed polyacrylamide solution flowing through a molecular weight adjusting device with porous medium. Chin J Chem Eng 24:581–587

Measuring Techniques and Potential Applications of Interface Rheology

Jacek Różański and Joanna Kmiecik-Palczewska

1 Introduction

To ensure stability of emulsion and foam, it is necessary to use additions of surfactants, proteins or polymers. Many of these substances are adsorbed on the interface, and since their surface concentration is much higher than their bulk concentration, rheological properties of the surface layer are diametrically different from the bulk solution.

The first scientific works on rheological properties of the surface layer were published in the 19th century. However, theoretical foundations of surface rheology (surface dilatational rheology) were developed by Boussinesq (1913) at the beginning of the 20th century. Although since then there have been more than 100 violent years, a rapid increase in the number of scientific papers, especially on surface shear rheology, has been observed for about 10 years. This is mainly related to the development of measurement techniques. Devices that allow making rheological measurements, both during shear deformation as well as dilatational deformation, are currently produced by commercial companies. This chapter discusses measurement methods, examples of test results, and the potential use of surface rheology.

2 Definition

Surface layer can be subject to shear and dilatational deformation (Fig. 1). In general, it can be characterized by both viscous and elastic properties.

J. Różański (✉) · J. Kmiecik-Palczewska
Institute of Chemical Technology and Engineering,
Poznan University of Technology, Poznań, Poland
e-mail: jacek.rozanski@put.poznan.pl

© Springer International Publishing AG, part of Springer Nature 2018
M. Ochowiak et al. (eds.), *Practical Aspects of Chemical Engineering*,
Lecture Notes on Multidisciplinary Industrial Engineering,
https://doi.org/10.1007/978-3-319-73978-6_27

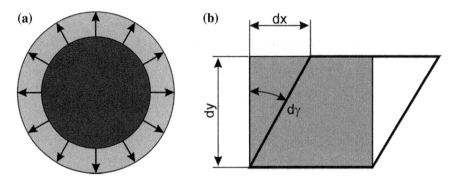

Fig. 1 Scheme of: **a** dilatational, and **b** shear deformations of surface layer

Deformation of the surface layer with elastic properties that undergoes compression and expansion is characterized by the surface dilatational modulus defined by Gibbs (1878):

$$E_d = \frac{d\sigma}{dA/A} = \frac{d\sigma}{d\ln A} \tag{1}$$

whereas during shear the surface shear modulus:

$$G_s = \frac{\tau_s}{\gamma_s} \tag{2}$$

where: σ is the surface tension, A is the area of the interfacial layer, τ_s is the surface shear stress (unit: N/m) and γ_s is the shear strain of the surface layer.

The two surface viscosities have also been defined: surface dilatational viscosity (introduced first by Boussinesq 1913):

$$\eta_d = \frac{\Delta\sigma}{d\ln A/dt} \tag{3}$$

and surface shear viscosity

$$\eta_s = \frac{\tau_s}{\dot{\gamma}_s} \tag{4}$$

where $\dot{\gamma}_s$ is the shear rate. Units of η_s and η_d is (Pa·s·m).

In practice, the surface layer is most often subjected to oscillatory deformation. Measurements of this type allow determining elastic and viscous properties of the surface layer and obtaining information on its microstructure (understanding the interaction between molecules, change of molecular conformations or molecular aggregations). For dilatational deformation, oscillatory changes of the surface layer are induced in accordance with the following equation:

$$A(t) = A_{s,0} + A_0 \sin(\omega t) \tag{5}$$

and surface tension changes are measured, which, in general, can be described by the relation:

$$\sigma(t) = \sigma_{s,0} + \sigma_0 \sin(\omega t + \theta) \tag{6}$$

In Eqs. (5) and (6) $A_{s,0}$ is the initial or reference interfacial area, A_0 is the amplitude of the area oscillations, ω is the frequency imposed, $\sigma_{s,0}$ is the equilibrium reference of interfacial tension, σ_0 is the amplitude of the interfacial tension oscillations, and t is the time. The phase shift θ between the generated change of surface layer A and the interfacial tension allows determining viscous and elastic properties of the surface layer (Fig. 2a). For an ideal elastic and viscus material $\theta = 0°$ and $\theta = 90°$ respectively.

By using a complex notation it can be introduced a complex dilatational modulus:

$$E^* = E_d + iE_\eta = E_d + i\omega\eta_d \tag{7}$$

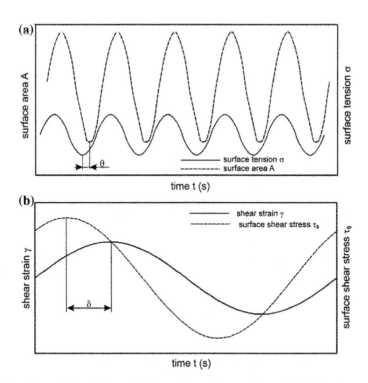

Fig. 2 Time dependences during oscillations of: **a** surface tension and bubble surface area, and **b** surface shear stress and surface strain

Real component E_d, also called dilatational storage modulus and imaginary component E_η, also called dilatational loss modulus, are described by the following equations:

$$E_d = \sigma_0 \cdot \frac{A_{s,0}}{A_0} \cdot \cos \theta = |E^*| \cdot \cos \theta \tag{8}$$

$$E_\eta = \sigma_0 \cdot \frac{A_{s,0}}{A_0} \cdot \sin \theta = |E^*| \cdot \sin \theta \tag{9}$$

where the dilatational viscoelastic modulus of $|E^*|$ is calculated from

$$|E^*| = \sqrt{(E_d)^2 + (E_\eta)^2} \tag{10}$$

Modulus E_d characterizes the elastic properties of interfacial layer, while modulus E_η describes the viscous properties of the interfacial layer during dilatational deformation.

The loss to storage modulus ratio is defined by loss tangent:

$$\tan \theta = \frac{E_d}{E_\eta} \tag{11}$$

Generally, $\tan \theta < 1$ indicates a dominance of elastic fluid properties, and $\tan \theta > 1$ shows a dominance of viscous fluid characteristics.

In the shear flow, during oscillation measurements, phase shift δ is determined between surface shear stress and deformation (Fig. 2b).

By analogy to bulk rheology, the complex surface shear modulus has been defined

$$G_s^* = G_s' + iG_s'' \tag{12}$$

where: G_s' is the surface storage modulus and G_s'' is the surface loss modulus.

The surface storage modulus G_s' and the surface loss modulus G_s'', which are characterized by elastic and viscous properties of the surface layer, are described by equations:

$$G_s' = \frac{\tau_{s,0}}{\gamma_{s,0}} \cdot \cos \delta \tag{13}$$

$$G_s'' = \frac{\tau_{s,0}}{\gamma_{s,0}} \cdot \sin \delta \tag{14}$$

where $\tau_{s,0}$ is the amplitude surface stress and $\gamma_{s,0}$ is the amplitude surface strain.

3 Measurement Methods

3.1 Dilation Rheology

Experiments used to measure dynamic surface tension can simultaneously be used for rheological measurements during compression and expansion of the surface layer. The Langmuir trough with an oscillating barrier is a classic device used for this type of research. Changes in surface tension over time are recorded using a Wilhelmy plate. This method can be applied to a frequency of approximately 0.2 Hz (Derkach et al. 2009).

At present, the most commonly used techniques to make surface dilatational rheology measurements are the oscillating bubble method and the oscillating drop method (Fig. 3). In these methods harmonic changes of the surface of a drop or a gas bubble are induced. Surface tension and surface area can be determined based on the image of a drop (bubble) or capillary pressure measurements and volume of a drop (bubble). The surface tension is calculated based on Young-Laplace equation. A certain disadvantage of the oscillating bubble method and the oscillating drop method is the limited frequency range in which they can be carried out. By measuring surface tension changes based on a drop image, the range of attainable frequencies falls from $1 \cdot 10^{-3}$ to $2 \cdot 10^{-1}$ Hz, while using a capillary pressure measurement from approximately $1 \cdot 10^{-1}$ Hz up to approximately 1 Hz (Noskov 2010). At higher frequencies, the shape of the drop begins to significantly deviate from the spherical one.

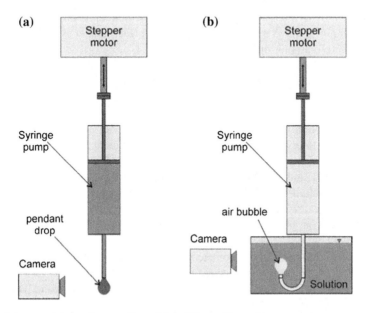

Fig. 3 Schemes of the: **a** drop profile and **b** bubble profile tensiometry set ups

Thus, the assumption that deformation takes place only in the radial direction is no longer valid, and the results obtained can have a significant error. Classic methods of measuring rheological properties during dilatational deformation also include methods based on the measurements of the characteristics of capillary waves. Their advantage is the possibility to make measurements at high frequencies (up to 1000 Hz), which would not be possible for other methods. Currently, these techniques are rarely used, mainly due to difficulties in interpreting measurement results. Often, negative values of surface dilatational viscosity are obtained which is at variance with the second law of thermodynamics (Noskov 2010).

3.2   Shear Rheology

There are three basic groups of experiments in which rheological measurements are made during shear deformation of the surface layer: indirect methods; direct methods and microrheology. Indirect measurements consist in placing tracer particles on the surface of liquid and observing the trajectory of its movement. In direct methods, torque and angular velocity are measured during the rotation of specially designed elements placed on the liquid surface. Among microrheological methods, the passive ones are the most commonly used in surface layer research as they consist in recording thermal motions of inoculum particles located at the interface. The most well-known indirect devices are a canal surface viscometer and a rotating wall knife-edge surface viscometer, as well as a deep-channel surface viscometer. At present, commercial surface rheometers are produced, which allow for a direct measurement of surface shear stress, which is why indirect methods are rarely used. A description of indirect methods can be found in (Edwards et al. 1991; Krägel and Derkatch 2009), among others.

If direct methods are concerned, different forms of measurement systems are used which are in contact with the surface layer (Fig. 4). The most commonly used are bicone, double wall-ring and Du Noüy ring (Vandebril et al. 2010).

In the past, such systems were suspended on thin torsion wire. Now, they are connected to traditional rotary rheometers equipped with an air-magnetic bearing. When using a rotary rheometer, it is possible to calculate surface tension and deformation rate based on the measured torque and angular frequency. Bicone, double wall-ring and Du Noüy ring can be used to make measurements at the gas-liquid and liquid-liquid interfaces. There are also specialist structures of surface rheometers, an example of which is a device in which the movement of a magnetic needle located on liquid surface between Helmholtz coils (Fig. 5) is recorded (Shahin 1986; Brooks et al. 1999). The movement of the needle is recorded using a digital camera. Magnetic rod rheometers are characterized by very high sensitivity. It is possible to measure surface shear stress in the range from 10^{-8} to 10^{-4} N/m. This type of device can be used for measurements at the gas-liquid and liquid-liquid interfaces. With the use of movable barriers during measurement, it is possible to change the surface concentration of surface active agents.

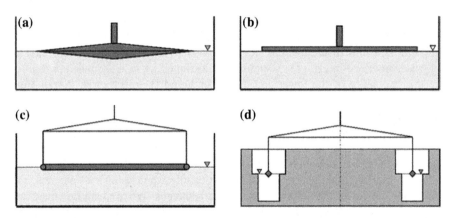

Fig. 4 Scheme of geometries for direct measurements of surface stress: **a** biconical disk, **b** disk, **c** Du Noüy ring, **d** double wall-ring

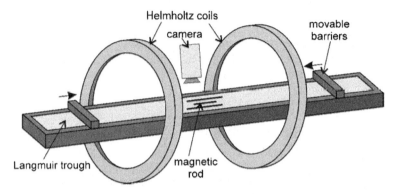

Fig. 5 Scheme of magnetic rod rheometer

The mentioned measurement systems cause not only the movement of the surface layer but also of deeper liquid layers. Thus, the total measured drag is the result of the forces induced by the movement of both the surface layer, as well as the subphase which directly adheres thereto. The relation of surface drag to subphase drag is called the Boussinesq number, which is expressed by the equation (Vandebril et al. 2010):

$$Bo = \frac{surface\ drag}{subphase\ drag} = -\frac{\eta_s \cdot \frac{v}{L_1} \cdot P_1}{\eta \cdot \frac{v}{L_2} \cdot A_2} = \frac{\eta_s}{\eta \cdot L} \tag{15}$$

where: η is the bulk viscosity (Pa·s), v is a characteristic velocity (m/s), L_1 and L_2 are the characteristic length scales over which the characteristic velocity decays at the interface and in the subphases, respectively (m), P_1 is the contact perimeter between the surface probe and the interface (m), and A_2 is the contact area between the geometry and the surrounding subphases (m^2). The ratio $(A_2 \cdot L_1)/(P_1 \cdot L_2)$ has the units of length and defines a characteristic length scale L, which will be dependent on the dimensions of the measurement geometry (Vandebril 2010). In the case of oil–water interface in Eq. (15) η should be substituted by $\eta_1 + \eta_2$, where η_1 and η_2 are the viscosities of the lower and the upper liquid (Derkach et al. 2009). High values of Bo number ($Bo \gg 1$) mean that the measured drag is mainly caused by the surface layer, so surface viscosity is easy to calculate. If Bo values $\ll 1$, the measurement system is not suitable for determining surface viscosity of a given liquid. Currently, for the most commonly used methods of measuring the surface shear viscosity (bicone, double wall-ring and magnetic rod rheometer), the lowest values of characteristic length L will be achievable when using the magnetic rod rheometer, however, the solution has its disadvantages. When using one magnetic rod, measurements in a limited range of surface viscosity changes can be made. For this reason, it is necessary to use magnetic rods of different dimensions. With bicone it is possible to make measurements over a wide range of surface viscosities, simultaneously for this value system Bo numbers will be much lower than when using a magnetic rod rheometer at the same viscosity ratio value η_S/η. The double wall-ring geometry is an attempt to combine the advantages of both solutions. For this system, the characteristic values of the L dimension are comparable to values obtained for the magnetic rod rheometer, while at the same time it has the advantages of bicone.

4   Examples of Research Results

Most of the research results that have been presented so far cover the rheological properties of surface layers of surfactants, proteins, polymers, and mixtures of proteins and polymers with surfactants (Langevin and Monroy 2010; Lucassen-Reynders et al. 2010). The rheological properties of the surface layer of these systems depend on many factors. The most important are the diffusion rate of surfactant molecules towards the direction of the interfacial surface, and interactions between molecules adsorbed at the interface. The current state of knowledge on this subject matter can be found in (Bos and Vliet 2001; Miller and Liggieri 2009; Krägel and Derkatch 2010; Narsimhan 2016) reviews. This chapter discusses examples of research results.

Surface viscosity measurements of low molecular weight surfactants are most often made under conditions of dilatational deformation. In shear flow of surface viscosity of solutions of these compounds it is usually very low, often unmeasurable. Figure 6 shows the complex dilatational modulus correlation in the function of frequency for aqueous solution of oxyethylated surfactant $C_{14}EO_8$ (noninionic surcitrate) obtained by Fainerman and co-workers (Fainerman et al. 2008) using bubble profile analysis tensiometer. The characteristic property of surfactants is the ability to form micelles in solutions. Their presence in a solution influences the dynamics of formation of the surface layer. The data in Fig. 6 shows that the highest

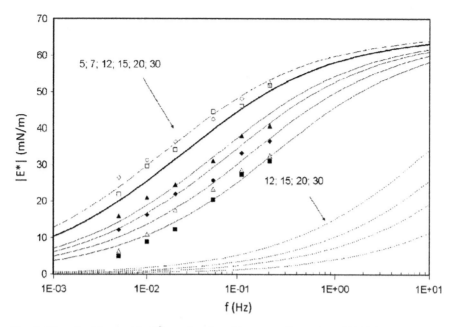

Fig. 6 Viscoelasticity module $|E^*|$ as a function of frequency f for various C14EO8 concentrations below the cmc ($\Diamond$5 µmol/dm3), at cmc ($\Box$7 µmol/dm^3), and above the cmc at 12 ($\blacktriangle$), 15 ($\blacklozenge$), 20 ($\triangle$), and 30 µmol/dm^3 ($\blacksquare$) (Fainerman et al. 2008)

values of viscoelatic modulus $|E^*|$ are achievable at a concentration of the surfactant close to or slightly below the critical micelle concentration (cmc). Increasing the surface area concentration above the cmc results in a decrease in viscoelatic modulus $|E^*|$. Its value will depend not only on the diffusion rate of molecules towards the surface, but also on the dynamics of micelle formation.

When the interfacial surface is expanded, the adsorption of surfactant molecules from the bulk solution takes place thereon. Thus, their concentration in the sub-phase adhering to the interfacial surface is lower than in the state of equilibrium, which disintegrates part of the micelles and simultaneously forces them to diffuse from the bulk solution. During compression of the surface layer, the situation is reversed. In this case, monomer desorption, increased concentration of the sur-factant in the bulk solution, and formation of new micelles will occur. Thus, dilatational rheology provides information not only on the dynamics of surface layer formation but also on micelles themselves.

Shear flow measurements have been used in research on the surface properties of protein solutions and their mixtures with surfactants and polymers, as well as nanoparticles. High values of surface viscosity characterize aqueous solutions β-lactoglobulin (BLG), hence the solutions of this protein are often used as model liquids. Oscillatory shear measurements allow following the process of forming the structure of the surface layer of protein solutions. Figure 7 shows the dependence of the surface storage modulus G'_s and the surface loss modulus G''_s on time for

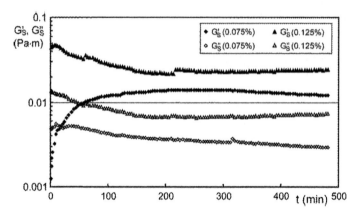

Fig. 7 Dependence of G'_s and G''_s modules on time function t for BLG solutions (Broniarz-Press et al. 2015)

aqueous BLG solutions obtained in shear flow (pH = 6.3, tests were made using a bicon system connected to the Physica MCR 501 rheometer). For protein concentration of 750 ppm, values of the surface storage modulus G'_s clearly increase as a function of time, reaching a maximum, and then they slightly decrease. The initial growth of G'_s modulus growth demonstrates that the surface layer formation process is controlled by BLG diffusion from the bulk solution. Depending on the type of protein, it can last from a few to several dozen hours (Bos and Vliet 2001). Increased BLG concentration to 1250 ppm results in module values G'_s and G''_s being very high at the beginning, and then decreasing and reaching a constant value. Occurrence of the maximum value of modulus G'_s for a solution of 750 ppm and a drop in its value for concentration at 1250 ppm is the result of unfolding BLG molecules adsorbed on the interfacial surface.

Oscillatory measurements in the shear flow of surface layer, as in the case of bulk fluids, can be made in linear and nonlinear viscoelastic regimens. The linear viscoelasticity range is determined on the basis of the dependence of the surface storage modulus G'_s and the surface loss modulus G''_s in the function of strain (stress) amplitude (stress or strain sweep experiments) (Fig. 8a). Based on this graph, it is also possible to determine surface yield stress $\tau_{s,y}$. As in the case of bulk rheology, the analysis of nonlinear measurements is made based on Lissajous curves. In the range of linear viscoelasticity, measurements of modules G'_s and G''_s as a function of frequency sweep experiments are made as well (Fig. 8b). The results of this experiment allow obtaining some information on the microstructure of the surface layer. For example, an approximate parallel course of dependence and as a function of angular frequency for BLG solutions provides funding for the formation of gel-like films.

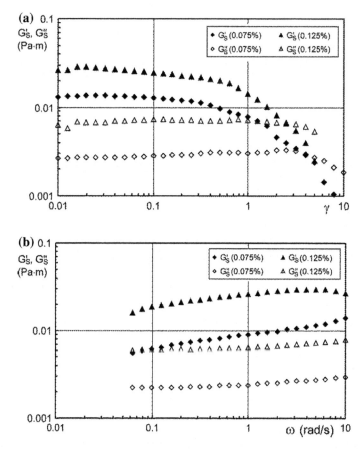

Fig. 8 Examples of: **a** strain sweep experiments, **b** frequency sweep experiments for BLG solution (Broniarz-Press et al. 2015)

5 Possible Practical Use

Rheological research on surface layer are made for diluted solutions of such substances as proteins, surfactants, polymers and mixtures thereof. Solutions of this type are found in food and cosmetic products in which surface-active substances are used as emulsifiers or foaming agents because of their amphiphilic character.

Examples of food products with complex rheological properties of the surface layer are milk, instant coffee and beer (Piazza et al. 2008; Maldonado-Valderrama and Patino 2010; Drusch et al. 2012; Dan et al. 2013; Broniarz-Press et al. 2014). Multiphase systems are also common in the chemical industry. In literature on the subject, attention has been paid to the possibility of a relationship between surface rheology and flotation, microflotation and microfiltration (Dukhin et al. 2009; He 2016). In many food products and industrial processes, foams play an important role. Some foams are impermanent materials, which is why their rheological properties are difficult to be determined by methods used in classic rheology. Hence the idea to use surface rheology to analyse factors affecting the stability and mechanical properties of foam. Unfortunately, according to analysis made by Erni and co-workers (Erni et al. 2009), at present it cannot be clearly stated that there is a relationship between surface rheology of the solutions and foam stability. Of course, along with the increase in protein concentration (for example, ovalbumin) the surface shear viscosity and foam lifetime increase as well. Simultaneously, low molecular weight surfactants also stabilize foams and emulsions, although the surface shear viscosity and surface dilatational viscosity of their solutions are very low. Lexis and Willenbacher (Lexis and Willenbacher 2014) compared the results of research on the surface rheology of solutions (surface shear and dilatational elastic moduli) and bulk rheology of foam (apparent yield stress and storage modulus) obtained for casein solutions, whey protein isolate and two surfactants: nonionic Triton X and anionic sodium dodecyl sulphate.

In general, more flexible foam was formed from protein solutions then surfactants. At the same time, the surface layer of protein solutions was characterized by greater elasticity than the surface layer of low molecular weight surfactants. It was found that in the case of protein solutions, the increase in surface storage modulus was not always accompanied by an increase in bulk storage modulus of foam. As in the case of foams, the relationship between surface rheology and stability and bulk rheology of emulsions is not fully explained (Georgieva et al. 2009; Maldonado-Valderrama and Patino 2010).

Another area of potential practical use of surface rheology is medical diagnostics. The advantage of this type research is the possibility to record changes in properties of surface layers with very little change in the concentration of general active substance. Kazakov and co-workers (Kazakov et al. 2009) discussed the published results of measurements of dynamic surface tensions and surface rheological parameters of various human liquids, such as blood serum, urine, breathing air condensates, cerebrospinal liquor and others. Research on samples of human liquids taken from healthy and ill patients showed a relationship between surface rheology and kidney stones, rheumatism and pulmonary diseases. Only dilational rheology is used in this research. This is mainly due to the need to use a relatively large amount of liquid in the case of surface shear measurements, which is most often impossible in the case of human liquids. Scheuble and co-workers (Scheuble et al. 2014) used surface shear rheology to monitor the properties of medium-chain triglyceride oil/water layer characteristics under various gastric digestion steps. The authors showed that increased emulsion stability in the gastric environment might lead to homogeneously distributed lipid droplets in the stomach, which could be faster transported to the duodenum and sensed by the body leading to a faster satiety signal.

In the work by Çelebioğlu and co-workers (Çelebioğlu et al. 2017) it was also suggested that the rheological properties of the surface layer of emulsion can influence their perception. However, there has been no evidence of such a relationship so far.

6 Conclusion

Surface rheology allows collecting information on the kinetics of formation and structure of the surface layer being formed at the liquid-liquid and gas-liquid inter-face. The results of this research are very interesting for cognitive reasons, but they have relatively little practical significance. It seems that the biggest challenge is to establish a relation between dynamic interfacial properties and multiphase system properties. This will allow using surface rheology in the description of industrial processes.

Acknowledgements This work was supported by PUT research grant No. 03/32/DSPB/0702.

References

Bos MA, Vliet T (2001) Interfacial rheological properties of adsorbed protein layers and surfactants: a review. Adv Colloid Interfac 91:437–471

Boussinesq MJ (1913) Sur l'existence d'une viscosité seperficielle, dans la mince couche de transition séparant un liquide d'un autre fluide contigu. Ann Chim Phys 29(8):349–357

Broniarz-Press L, Różański J, Różańska S et al (2014) Rheological properties of liquid surface layer in selected commercial beers. Food Sci Tech Qual 95:43–52

Broniarz-Press L, Różański J, Kmiecik J (2015) Reologia warstw powierzchniowych – zastosowanie, metody pomiaru, przykładowe wyniki. Inż Ap Chem 54(6):306–307

Brooks CF, Fuller GG, Frank CW et al (1999) An interfacial stress rheometer to study rheological transitions in monolayers at the water–air interface. Langmuir 15(7):2450–2458

Çelebioğlu HY, Kmiecik-Palczewska J, Lee S et al (2017) Interfacial shear rheology of β-lactoglobulin—Bovine submaxillarymucin layers adsorbed at air/water interface. Int J Biol Macromol 102:857–867

Dan A, Gochev G, Krägel J et al (2013) Interfacial rheology of mixed layers of food proteins and surfactants. Curr Opin Colloid In 18:302–310

Derkach SR, Krägel J, Miller R (2009) Methods of measuring rheological properties of interfacial layers (experimental methods of 2D rheology). Colloid J 71(1):5–22

Drusch S, Hamann S, Berger A et al (2012) Surface accumulation of milk proteins and milk protein hydrolysates at the air–water interface on a time-scale relevant for spray-drying. Food Res Int 47:140–145

Dukhin SS, Kovalchuk VI, Aksenenko EV et al (2009) Influence of surface rheology on particle bubble interaction in flotation. In: Miller R, Liggieri L (ed) Interfacial Rheology. Leiden, Boston, pp 567–613

Edwards DA, Brenner H, Wasan DT (1991) Interfacial transport processes and rheology. Butterworth-Heinemann, Oxford

Erni P, Windhab EJ, Fischer P (2009) Interfacial rheology in food science and technology. In: Miller R, Liggieri L (eds) Interfacial rheology. Leiden, Boston, pp 614–653

Fainerman VB, Petkov JT, Miller R (2008) Surface dilational viscoelasticity of C14EO8 micellar solution studied by bubble profile analysis tensiometry. Langmuir 24(13):6447–6452

Georgieva D, Schmitt V, Leal-Calderon F et al (2009) On the possible role of surface elasticity in emulsion stability. Langmuir 25(10):5565–5573

Gibbs JW (1878) On the equilibrium of heterogeneous substances. Am J Sci 16(96):441–458

He Z, Miller DJ, Kasemset S, Wang L et al (2016) Fouling propensity of a poly(vinylidene fluoride) microfiltration membrane to several model oil/water emulsions. J Membrane Sci 514:659–670

Kazakov VN, Knyazevich VM, Sinyachenko OV et al (2009) Interfacial rheology of biological liquids: application in medical diagnostics and treatment monitoring. In: Miller R, Liggieri L (eds) Interfacial rheology. Leiden, Boston, pp 519–566

Krägel J, Derkatch SR (2009) Interfacial shear rheology—an overview of measuring techniques and their applications. In: Miller R, Liggieri L (eds) Interfacial rheology. Leiden, Boston, pp 373–420

Krägel J, Derkatch SR (2010) Interfacial shear rheology. Curr Opin Colloid Interface Sci 15: 246–255

Langevin D, Monroy F (2010) Interfacial rheology of polyelectrolytes and polymer monolayers at the air–water interface. Curr Opin Colloid Interface Sci 15(4):283–293

Lexis M, Willenbacher N (2014) Yield stress and elasticity of aqueous foams from protein and surfactant solutions—The role of continuous phase viscosity and interfacial properties. Colloids Surf A 459:177–185

Lucassen-Reynders EH, Benjamins J, Fainerman VB (2010) Dilational rheology of protein films adsorbed at fluid interfaces. Curr Opin Colloid Interface Sci 15(4):264–270

Maldonado-Valderrama J, Patino JMR (2010) Interfacial rheology of protein–surfactant mixtures. Curr Opin Colloid Interface Sci 15(4):271–282

Miller R, Liggieri L (eds) (2009) Interfacial rheology. Leiden, Boston

Narsimhan G (2016) Characterization of interfacial rheology of protein-stabilized air-liquid interfaces. Food Eng Re 8:367–392

Noskov BA (2010) Dilational surface rheology of polymer and polymer/surfactant solutions. Curr Opin Colloid Interface Sci 15(4):229–236

Piazza L, Gigli J, Bulbarello A (2008) Interfacial rheology study of espresso coffee foam structure and properties. J Food Eng 84:420–429

Scheuble N, Geue T, Windhab EJ et al (2014) Tailored interfacial rheology for gastric stable adsorption layers. Biomacromol 15:3139–3145

Shahin GT (1986) The stress deformation interfacial rheometer, Ph.D. Thesis, University of Pennsylvania, Philadelphia

Vandebril S, Franck A, Fuller GG et al (2010) A double wall-ring geometry for interfacial shear rheometry. Rheol Acta 49(2):131–144

An Effective Production of Bacterial Biosurfactant in the Bioreactor

Wojciech Smułek, Agata Zdarta and Ewa Kaczorek

1 Introduction

The last decades have been the period of intensive development of many branches of biotechnology. The growing demand for environmentally friendly materials, which are alternatives to synthetic compounds, contributes to the improvement of microbial culture-based production techniques. They allow to obtain chemical compounds with high-yield in relatively mild process conditions.

Biotechnological production is most often carried out in bioreactors, whose construction is a development of solutions used in classical chemical reactors. At the same time, the specificity of working with microorganisms means that a number of specific factors must be taken into account in carrying out the biotechnological process. The purpose of this chapter is to bring closer the issues related to the selection of optimal bioprocessing conditions, with particular focus on the production of biosurfactants, natural surfactants produced by certain strains of bacteria and fungi.

2 Effective Production in a Bioreactor

Different factors affect the effectiveness of microbial production. They can be divided into three main groups (Fig. 1). The first group includes factors related to the growth conditions of the microbial strain and the carbon source used for this purpose. The second group consists of parameters of the performance of the microbial culture and the bioreactor settings. The third group creates the parameters

W. Smułek · A. Zdarta · E. Kaczorek (✉)
Institute of Chemical Technology and Engineering,
Poznan University of Technology, Poznań, Poland
e-mail: ewa.kaczorek@put.poznan.pl

© Springer International Publishing AG, part of Springer Nature 2018
M. Ochowiak et al. (eds.), *Practical Aspects of Chemical Engineering*,
Lecture Notes on Multidisciplinary Industrial Engineering,
https://doi.org/10.1007/978-3-319-73978-6_28

of the isolation of the product from the culture. Each of these groups of factors has a significant impact on the cost of production and the quality of the product obtained (Sharma et al. 2011).

2.1 Microorganism

Only some microorganisms, inhabiting every soil and water ecosystems, are capable of producing useful compounds, like e.g. biosurfactants. Their synthesis by microbial cells is dependent on many environmental and physiological factors. Every microorganism has optimal growth conditions. However, the most convenient conditions for the cells, manifesting itself in the growth of biomass, are not very often optimal for the production of desired chemical compounds (Reis et al. 2013).

The high ratio of carbon to nitrogen and the type of sources of these elements, the concentration of macro and microelements, the degree of oxygenation and the pH of the culture are the main factors influencing the production of biological surfactants (Krzyczkowska and Białecka-Florjańczyk 2012). Different sources of nitrogen can act as inhibitors or activators. Additionally, the low solubility of the carbon sources can intensify the secretion of surface active compounds (Jamal et al. 2012). Among the carbon sources used in the production of the biosurfactants are vegetable oils (like olive, soybean or corn oil) and carbohydrates. Because of the costs, waste products are of great interest, e.g. from the food industry and glycerin, which is a by-product of biodiesel production (Krzyczkowska and Białecka-Florjańczyk 2012; Kaloorazi and Choobari 2013).

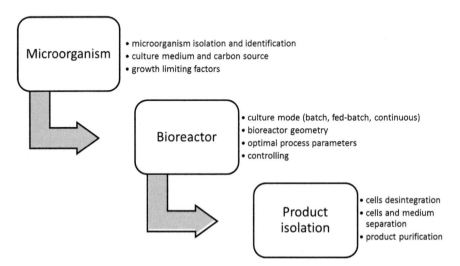

Fig. 1 Main stages of bioproduct production in bioreactor

Depending on the conditions of the process, the biosurfactant synthesis may take place according to different mechanisms involving their production by microorganisms under limited growth conditions or in the presence of lipophilic precursors (Gumienna and Czarnecki 2010). Finally, the scale of biosurfactant secretion is strictly connected with the physiological state of the cell. The metabolic processes change also with the cell age and the culture growth phase. The optimal period of the biosurfactant production is specified for given microorganism strain (Mouafi et al. 2016).

The kinetics of microorganism growth (Fig. 2) is described by the Monod equation (Eq. 1), which is one of the most important in biotechnology and environmental engineering:

$$\mu = \mu_{max} \frac{[S]}{k_s + [S]} \tag{1}$$

μ is the specific growth rate of the microorganisms (g dm^{-3} s^{-1}), μ_{max} is the maximum specific growth rate of the microorganisms (g dm^{-3} s^{-1}), S is the concentration of the limiting substrate for growth (g dm^{-3}), k_s is the "half-velocity constant" (g dm^{-3})—the value of S when $\mu/\mu_{max} = 0.5$. The μ_{max} and k_s are empirical coefficients and differ between microorganism species and culture conditions (Metcalf & Eddy, Inc. 2003).

Simultaneously, the rate of substrate utilization r_{su} is directly related to the specific growth rate μ:

$$r_{su} = -\mu X/Y \tag{2}$$

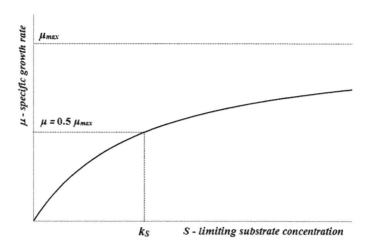

Fig. 2 The mathematical model of microorganism growth kinetics (Metcalf & Eddy, Inc. 2003)

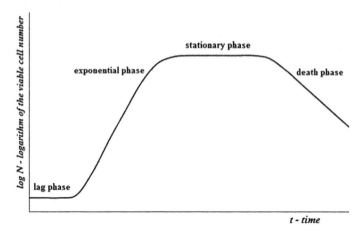

Fig. 3 Microorganism growth curve

where: X—the total biomass (when μ is normalized to the total biomass, g dm^{-3}),
Y—the yield coefficient (g dm^{-3}) (Maier 2009).

In the microorganisms growth four main stages/phases (Fig. 3) can be distinguished: lag phase, exponential phase, stationary phase and death phase. The lag phase is the time, when microorganisms adapt to new environmental conditions and cells multiplication does not occur. During the second, exponential phase cells double intensively, until any limiting factor (e.g. concentration of the toxic product) is not present. Then, the stationary phase starts, when the rate of cell doubling is equal to the rate of cell death. Finally, when dying of the cells dominates (e.g. caused by a lack of the carbon source), the death phase starts (Zwietering et al. 1990).

2.2 Cell Culture Conditions

Production of biosurfactants in a bioreactor is a complex process requiring careful control of process conditions. Firstly, the culture mode (batch, fed-batch, continuous) should be chosen. Each mode is characterized by different kinetics of biomass growth, concentration of substrates and products or by-products (Fig. 4).

Continuous cultures are carried out in bioreactors called chemostats. The constant volume is provided by continuous addition fresh medium and continuous removal culture liquid containing nutrients, metabolic and products (Fig. 4a). Adjusting the media delivery speed allows to maintain a steady growth rate of microorganisms at a constant level and a physiological steady state of the cell, which is mostly desired. There are several variations of the continuous bioreactors, like the turbidostat (when feed stream rate is regulated to maintain constant culture turbidity), the auxostat (when pH, oxygen or selected substrate/product

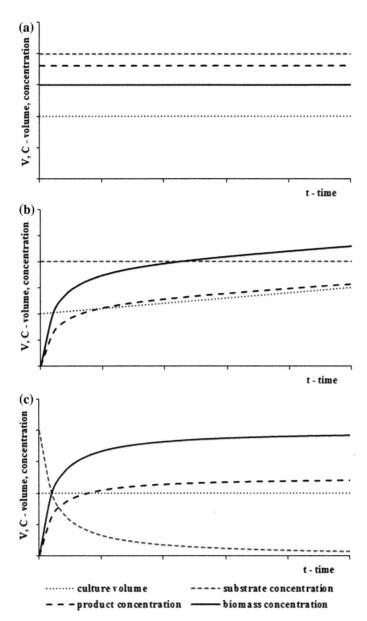

Fig. 4 Changes of main process parameters in bioreactor during: **a** continuous, **b** fed-batch and **c** batch cultures

concentration are constant value) and the retentostat (when culture liquid without biomass is removed from the bioreactor, the biomass concentration is maintained by the amount of limiting nutrient in feed stream). An advantage of the continuous process is a reduction of time devoted to non-productive actions, like sterilization. However, in that case, the risk of culture contamination and microorganisms mutations is increasing (Ziv et al. 2013).

In the fed-batch processes the stable feed stream of substrates allows to reach very high concentrations of the desired secondary metabolites (Mulligan 2005). The advantage of that culture mode is a possibility to control the concentration of fed-substrate at desired (very often relatively low) levels (Fig. 4b). This makes it possible to avoid undesirable phenomena such as substrate inhibition or metabolism changes in cells occurring in high substrate concentration. Moreover, the fed-batch processes is used, when high biomass concentration is expected (Chisti 2013).

The batch process in a bioreactor is an easy way to conduct the process and to provide sterile conditions. It should be noted, however, that dynamic changes in the amount of biomass, concentration of sorbates and products have occurred. The classic pattern of changes in periodic culture is shown in Fig. 4c. In addition, the batch process involves (with respect to continuous and semi-continuous culture) high cost and long time spent on preparatory and post-run activities such as emptying and filling bioreactors or its sterilization and cleaning. These activities may in some cases take more time than culturing (Chisti 2013).

2.3 Bioreactor Geometry

Among the most popular types of bioreactors used in the laboratory and industrial biotechnological production the following should be indicated:

- stirred tank bioreactors
- airlift bioreactors
- photobioreactors
- bioreactors with packed bed
- membrane bioreactors
- fluidized bed bioreactors.

The popularity of stirred tank bioreactors is due to their numerous advantages, like simplicity of construction, easy control or easy adaptation to multiphase runs. The airlift bioreactors are distinguished by relatively easy cleaning and sterilizing operations. The bed bioreactors are characterized by the ease of separation of microorganisms from medium culture. In turn, the membrane bioreactors allow to separate the extracellular metabolites in situ and facilitate the conduction of the continuous process (Chisti 2013).

The number of phases (at least three: culture liquid medium, air and microorganisms) should be considered, because the carbon source (e.g. vegetable oil) and can be an additional phase in bioreactor. Hence, providing the homogeneity in the bioreactor is crucial and especially the kind of impeller and mixing speed determine mass and heat transfer. However, the values of the shear stress in the vicinity of the impeller cannot damage bacterial cells. Discussing the bioreactors impellers, the most typical are Rushton turbine, pitched-blade impellers for shear-sensitive cells, gentle marine-blade impellers and special impellers for packed-bed bioreactors (Mirro and Voll 2009) (Fig. 5). Another aspect of the bioreactor geometry is the presence and shape of baffles and heaters, which should take into account biofilm formation as well as effective sterilization of the bioreactor before culture initiation (Hines et al. 2010). An interesting alternative, allowing sterilization procedure to be avoided, is the use of single-use (disposable) bioreactors (Fakruddin 2012).

Moreover, the dosing of carbon source and other supplements should be set carefully. The optimal oxygenation can be regulated by a way of providing air (or other gas mixture) and its flow rate. What is important, exchange of all substrates, metabolism products or gases have to be conducted in a sterile manner and prevent the contamination of the culture, especially with foreign microorganisms (Fakruddin 2012). A frequent problem to arise during biosurfactant production is foam formation, which should be controlled. Not only can the foam disturb the gases exchange but can also lead to clogged supply lines and sensors. The addition of chemical antifoam agents also leads to decreased microbial metabolism. Hence, special attention should be given to the compatibility level of the intensification of the gas–liquid operation in bioreactors (Delvigne and Lecomte 2010).

Finally, controlling all reactor operating parameters requires a series of sensors with fast response times and software for clear presentation and recording of current temperature, oxygenation, biomass concentration, redox potential etc. (Harms et al. 2002). The all internal equipment, like sensors, has to be resistant to sterilization conditions as well as microorganisms activity. Hence, as it could be noticed, the requirements for effective breeding of microorganisms in bioreactors are very complex and must take into account phenomena inexistent in classical chemical reactors (Kronemberger et al. 2008).

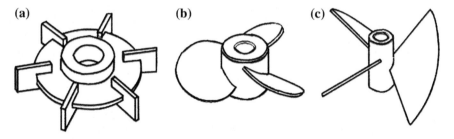

Fig. 5 The most common impellers used in the bioreactors: **a** Rushton turbine, **b** marine-blade impeller, **c** pitched-blade impeller (Mirro and Voll 2009)

2.4 Product Isolation

The last group of aspects to be considered includes these referring to the isolation of the product from the culture. Very often this is a multi-stage process involving (Ottens et al. 2013):

- disintegration of the cells (if the desired product is an intercellular compound or one of the cell building materials, like phosphorolipids)—using sonicators, ball mills, homogenisators
- cell and culture medium separation—mainly by filtration or centrifuging of the culture
- purification (preceded by concentration, if necessary)—by water evaporation or freeze drying, membrane filtration or ultrafiltration, precipitation or extraction (liquid-liquid or liquid-solid).

The most popular methods of isolating the biosurfactants include different typed of liquid-liquid extractions (with organic solvents like ethyl acetate) (Vecino et al. 2015). The purification of the product, if it is necessary, is conducted by preparative chromatography and membrane techniques. However, it should be considered that the purification stage can significantly increase the cost of the product. Hence, for purposes that do not require a high purity biosurfactant (e.g. crude oil bioremediation), it can be used in an unmanaged form (Gogoi et al. 2016; Varjani and Upasani 2017).

2.5 Increasing Production Scale

Underestimating the differences between the scales of laboratory and industrial production is a common mistake made in the planning of biotechnological process. Apart from the problems encountered in the transfer of mass and energy, consideration should be given to the aspects related to the cultivation of living micro-organisms. Sensitivity of the cells significantly limits the range of process parameters, like flow rates and heating medium temperature. Among the most typical problems which may result in the cell destruction, the following aspects should be mentioned (Kossen 1994):

- presence of local overheating regions (relatively high temperature on the heating-element)
- high shear stress in the neighborhood of rotors and impellers
- high pressure in pipelines.

Moreover, when the scale of production is increased, it is important to take into account whether the adopted solutions protect the culture from contamination. Sterilization of the culture and provision of sterility for feed and discharge streams should be effective and low cost. An laboratory equipment can be relatively easy sterilized by autoclaving, but it is complicated in larger scale (Garcia-Ochoa and Gomez 2009).

The maximum size of the bioreactor is limited by a number of unfavorable phenomena and structural difficulties (Noorman 2013). These include:

- the largest possible volume should ensure homogeneity of the culture
- the large volume of the bioreactor increases the response time of the camera to changing process conditions and it also increases the warm up and cooling time of the bioreactor
- the pressure above the liquid, increased to provide a higher oxygen solubility, is limited by the mechanical strength of the apparatus
- the air flow rate and air pressure are limited by the cost of the compressor
- the increase in investment and operating costs cannot exceed the profitability of production.

Given the aforementioned limitations, a common solution used to increase the scale of production is the use of batteries of lower volume bioreactors. This also allows for a reduction in production risk, because a failure of one apparatus or contamination of one culture does not exclude the use of cultures from other bioreactors. The use of bioreactor batteries can, however, significantly increase the investment costs of production (Noorman 2013).

3   A Case Study—Production of Biosurfactant for Hydrocarbons Biodegradation

Surface active compounds are widely used in practically all industries ranging from the extraction of crude oil and metal ores, through the machine industry to the food and pharmaceutical industries. This contributes to the wide and large-scale production of surfactants (Rosen and Kunjappu 2012). At the same time, however, the environmental pollution of this group of compounds is still growing. The negative impact of surfactants on macro- and microorganisms is mainly related to the amphiphilic structure of their molecules and the adsorption at interfacial boundaries. They can disturb the mass transport, including the exchange of gases and nutrients, between cells and their environment. Moreover, by penetrating into the phospholipid layers, which are an essential building block for any cell, surfactants may also contribute to cell damage and in consequence to cells' death. This entails high surfactant toxicity. Moreover, surfactants produced by chemical synthesis are usually characterized by low biodegradability and tendency to accumulate in sediments and soil (Britton 1998; Cserháti et al. 2002). In nature there are a number of chemicals that can be an alternative to synthetic surfactants. These include saponins, present in the tissues of certain plant species, and biosurfactants, which are extracellular products of certain bacterial and fungi strains (Mulligan et al. 2014).

In this study, the production of biosurfactant in a lab-scale bioreactor is presented. After selecting a bacterial strain producing surfactants and selecting a carbon source, the most advantageous conditions for conducting the culture in the bioreactor were found. Then, the surface active properties of isolated biosurfactant were evaluated.

3.1 Experimental

Individual bacterial strains were isolated from the soil samples and were identified based on the 16S rDNA assay (Guzik et al. 2009) and for rapid identification of biosurfactant-producing strains the haemolysis test was used (Thavasi et al. 2011). As a result, the *Ochrobactrum* sp. M2B strain was selected for further experiments.

Production of biosurfactant was carried out in a bioreactor BIOSTAT B Plus (Sartorius Stedim, Germany). The bioreactor has double-jacketed glass vessel (1.5 L of maximum working volume) with four baffles, a thermostat system and stirrer shaft with single mechanical seal. It is equipped with a top drive with direct coupling agitation system and 6-blade Rushton turbine and ring sparger. The schematic drawing of the apparatus is shown in Fig. 6.

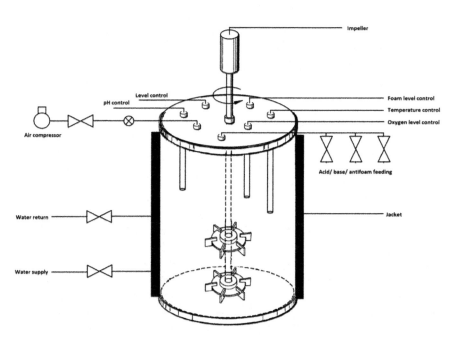

Fig. 6 The scheme of the lab-scale bioreactor BIOSTAT B Plus

One of the main parameters determining the production of biosurfactants is the selection of a suitable carbon source. The main attention is paid to its low cost and low toxicity (Moya Ramírez et al. 2016). Hence, in the research conducted, it was decided to use sucrose (white beet sugar), rapeseed oil and sunflower oil (at concentrations 0.5, 1.0 or 2.0%). In all cases, these sources of carbon were purchased as food products. In addition, the effect of the composition of the culture medium and the concentration of additional nitrogen source (yeast extract) on the production of biosurfactant was compared. Based on the review of the literature, three culture media described by Dobslav and Engesser (2012), Wei et al. (2008) and Siskin and Trocenko (Śliwka et al. 2009), respectively. Additionally, the different parameters of the bioreactor settings were tested: temperature (15–30 °C, step 5 °C), impeller rotation speed (0–150 rpm, step 50 rpm), air flow rate (0.1, 0.2. or 0.5 L min^{-1}). Every 4 days, 30 ml of the culture were collected, centrifuged and the surface tension of the supernatant was measured. Surface tension measurements were carried out using the du Nouy method using a platinum ring.

After finding the most favorable process conditions, whole liquid culture was centrifuged (5000 × g, 15 min, 20 °C). The biosurfactant was isolated by liquid-liquid extraction, where the organic phase was a mixture of chloroform and methanol (2:1 v/v). After evaporation of the organic solvents the resulting precipitate was lyophilized. The biosurfactant thus obtained was used for further investigation.

In order to evaluate the composition of the obtained biosurfactant its infrared spectrum (FT-IR Vertex 70 spectrometer, Bruker, Germany) were performed. The composition of the biosurfactant was supplemented with the assessment of carbohydrate content (Albalasmeh et al. 2013) and proteins (Okutucu et al. 2007) by spectrophotometric methods (UV-Vis spectrometer, V-650, Jasco Inc., USA). The fatty acid content was analyzed (Pegasus 4D GCxGC TOF-MS, Leco Corp., USA) according to the procedure described by Luna et al. (2012). Finally, the surface activity of the biosurfactant have been characterized by determining the adsorption isotherm based on the surface tension measurements of the various concentrations of the biosurfactant water solution.

The final step of the study was devoted to the application the produced biosurfactant in enhancing hydrocarbon biodegradation. Bacterial cultures, supplemented with mixture of aliphatic hydrocarbons were carried out analogously as described by Kaczorek et al. (2015). The quantitative analysis of the biodegradable compounds was performed on a gas chromatograph coupled to a mass spectrometer (Pegasus 4D GCxGC TOF-MS, Leco Corp., USA) with a BPX-5 column (28 m, 250 μm, 0.25 μm). All analyses carried out were performed in triplicate, calculating the mean of the measurements obtained and the standard deviation. Statistica 12.0 software (StatSoft, Poland) was used to statistically analyse the results.

3.2 Results and Discussion

In order to indicate the most convenient conditions for biosurfactant production by
Ochrobactrum sp. M2B strain the surface tension of bacterial cultures was measured. In the first step the cultures differed by the mineral medium composition as
well as carbon source (Table 1). The following was observed, a kind of carbon
source had the most important influence on surface tension decrease. In the cultures
conducted in the presence of saccharose (white sugar) the surface tension did not
decline below 52 mN m^{-1}. The cultures with vegetable—sunflower and rapeseed
oils showed the surface tension between 30 and 35 mN m^{-1}. The greatest decrease
in surface tension (down to 29.9 mN m^{-1}) was observed at 21 day of bacterial
growth in the culture in the Wei medium with rapeseed oil. The most convenient
bioreactor settings were as followed: 20 °C, 21 days, 100 rpm, culture in the Wei
medium with 1% of rapeseed oil.

The next stage of the study included the analyses to determine the chemical
structure of the biosurfactant. In the infrared spectrum of the biosurfactant, signals
characteristic for the hydroxyl and carbonyl groups and single carbon-oxygen
bonds. The protein content in the biosurfactant below 1% was established. Among
identified fatty acids the most dominated were palmitoleic, stearic and oleic acids.
Moreover, the saturated arachidic acid and some amounts pentadecylic and non-
adecylic acids were observed. The analysis indicated that the analyzed biosurfactant
can belong to glycolipds group of compounds.

Noparat et al. (2014), after analysis of FTIR and H^1 NMR spectra, concluded
that *Ochrobactrum anthropi* 2/3 strain growing on palm oil decanter cake produced
the biosurfactant of glycolipid nature, which correspond with the results in presented study, However, in the case of *Ochrobactrum* intermedium CN3 strain
(grown on 4% glycerol) after FTIR analysis of biosurfactant allowed to ascertain its
character of lipopeptide (Bezza et al. 2015). It suggests that the strains of
Ochrobactrum genus may produce the biosurfactants of different chemical nature.
Then, the measurements of the surface tension of the biosurfactants solution
allowed to determine its critical micellisation concentration, which amounted to
0.2 g L^{-1}. At this concentration, the surface tension was 31.6 mN m^{-1}.

The biodegradation test indicated that the biosurfactant can be successfully used
in hydrocarbon bioremediation, allowing to increase the biodegradation especially
the persistent hydrocarbons with odd number of carbon atoms (like tridecane).

4 Conclusions

The bioreactor-scale production of the valuable products is complicated, complex
and requires a broad perspective on the various factors that affect the efficiency of
the process. It is necessary to combine the knowledge of a biotechnologist and the
experience of a chemical engineer. The research carried out within the framework

of the following study allowed to identify a new bacterial strain capable of biosynthesis of surfactants. Because of the low requirements for growth conditions and no risk to humans, it may be an interesting material for industrial biotechnology. At the same time, the use of rapeseed oil as a substrate for biosurfactant production with significant potential as an emulsifier and a surface tension reducing agent has been demonstrated.

Acknowledgements The authors acknowledge for financial support from the National Centre of Science (Poland) awarded by the decision number DEC-2015/19/N/NZ9/02423.

References

Albalasmeh AA, Berhe AA, Ghezzehei TA (2013) A new method for rapid determination of carbohydrate and total carbon concentrations using UV spectrophotometry. Carbohyd Polym 97(2):253–261

Bezza FA, Beukes M, Chirwa EMN (2015) Application of biosurfactant produced by *Ochrobactrum intermedium* CN3 for enhancing petroleum sludge bioremediation. Process Biochem 50(11):1911–1922

Britton LN (1998) Surfactants and the environment. J Surf Det 1:109–117

Chisti Y (2013) Bioreaktory. In: Ratledge C, Kristiansen B (eds) Podstawy biotechnologii. Wydawnictwo Naukowe PWN, Warszawa

Cserháti T, Forgács E, Oros G (2002) Biological activity and environmental impact of anionic surfactants. Environ Int 28(5):337–348

Delvigne F, Lecomte J-P (2010) Foam formation and control in bioreactors. In: Flickinger MC (ed) Encyclopedia of industrial biotechnology. Wiley, New York

Dobslaw D, Engesser KH (2012) Degradation of 2-chlorotoluene by *Rhodococcus* sp. OCT 10. Appl Microbiol Biotechnol 93(5):2205–2214

Fakruddin M (2012) Biosurfactant: production and application. J Petrol Environ Biotechnol 3:124

Garcia-Ochoa F, Gomez E (2009) Bioreactor scale-up and oxygen transfer rate in microbial processes: an overview. Biotechnol Adv 27(2):153–176

Gogoi D, Bhagowati P, Gogoi P et al (2016) Structural and physico-chemical characterization of a dirhamnolipid biosurfactant purified from *Pseudomonas aeruginosa*: application of crude biosurfactant in enhanced oil recovery. RSC Adv 6(74):70669–70681

Gumienna M, Czarnecki Z (2010) Rola mikroorganizmów w syntezie związków powierzchniowo czynnych. Nauka Przyroda Technologie 4:1–14

Guzik U, Greń I, Wojcieszyńska D et al (2009) Isolation and characterization of a novel strain of *Stenotrophomonas maltophilia* possessing various dioxygenases for monocyclic hydrocarbon degradation. Braz J Microbiol 40(2):91–285

Harms P, Kostov Y, Rao G (2002) Bioprocess monitoring. Curr Opin Biotechnol 13:124–127

Hines M, Holmes C, Schad R (2010) Simple strategies to improve bioprocess pure culture processing. Pharm Eng 10(3):1–6

Jamal P, Nawawi WMFW, Nawawi W et al (2012) Optimum medium components for biosurfactant production by *Klebsiella pneumoniae* WMF02 utilizing sludge palm oil as a substrate. Aust J Basic Appl Sci 6(1):100–108

Kaczorek E, Smulek W, Zgoła-Grześkowiak A et al (2015) Effect of Glucopon 215 on cell surface properties of *Pseudomonas stutzeri* and diesel oil biodegradation. Int Biodeter Biodegrad 104:129–135

Kaloorazi NA, Choobari MFS (2013) Biosurfactants: properties and applications. J Biol Today's World 2:235–241

Kossen NWF (1994) Scale-up. In: Galindo E, Ramirez OT (eds) Advances in bioprocess engineering. Kluwer Academic, Dordecht

Kronemberger FA, Santa Anna LM, Fernandes AC et al (2008) Oxygen-controlled biosurfactant production in a bench scale bioreactor. Appl Biochem Biotechnol 147(1–3):33–45

Krzyczkowska J, Białecka-Florjańczyk E (2012) Biotechnologiczna synteza związków powierzchniowo czynnych i przykłady ich praktycznego zastosowania. Żywność. Technologia. Jakość 4:5–23

Luna D, Posadillo A, Caballero V et al (2012) New biofuel integrating glycerol into its composition through the use of covalent immobilized pig pancreatic lipase. Int J Mol Sci 13 (8):10091–10112

Maier RM (2009) Bacterial growth. In: Pepper IL, Gerba CP, Gentry T, Maier RM (eds) Environmental microbiology, 2nd edn. Elsevier Inc., London

Metcalf & Eddy, Inc. (2003) Wastewater engineering: treatment and reuse, 4th edn. McGraw-Hill, New York

Mirro R, Voll K (2009) Which impeller is right for your cell line? BioProcess Int 7(1):52–57

Mouafi FE, Abo Elsoud MM, Moharam ME (2016) Optimization of biosurfactant production by *Bacillus brevis* using response surface methodology. Biotechnol Rep (Amst) 9:31–37

Moya Ramírez I, Altmajer Vaz D, Banat IM et al (2016) Hydrolysis of olive mill waste to enhance rhamnolipids and surfactin production. Bioresour Technol 205:1–6

Mulligan CN (2005) Environmental applications for biosurfactants. Environ Poll 133:183–198

Mulligan CN, Sharma SK, Mudhoo A (2014) Biosurfactants. Research trends and applications. CRC Press, Taylor and Francis Group, London

Noorman HJ (2013) Wymiana masy. In: Ratledge C, Kristiansen B (eds) Podstawy biotechnologii. Wydawnictwo Naukowe PWN, Warszawa

Noparat P, Maneerat S, Saimmai A (2014) Utilization of palm oil decanter cake as a novel substrate for biosurfactant production from a new and promising strain of *Ochrobactrum anthropi* 2/3. World J Microbiol Biotechnol 30(3):865–877

Okutucu B, Dınçer A, Habib Ö et al (2007) Comparison of five methods for determination of total plasma protein concentration. J Biochem Biophysic Meth 70(5):709–711

Ottens M, Wesselingh JA, van der Wielen LAM (2013) Procesy wydzielania i oczyszczania. In: Ratledge C, Kristiansen B (eds) Podstawy biotechnologii. Wydawnictwo Naukowe PWN, Warszawa

Reis RS, Pacheco GJ, Pereira AG et al (2013) Biosurfactants: production and applications, biodegradation. In: Chamy R (ed) Life of Science. InTech, Rijeka

Rosen MJ, Kunjappu JT (2012) Surfactants and interfacial phenomena, 4th edn. Wiley, New Jersey, Hoboken

Sharma D, Saharan BS, Sahu RK (2011) A review on biosurfactants: Fermentation, current developments and perspectives. Genetic Eng Biotechnol J 2011:1–14

Śliwka E, Kołwzan B, Grabas K et al (2009) Influence of rhamnolipids from *Pseudomonas* PS-17 on coal tar and petroleum residue biodegradation. Environ Prot Eng 35(1):139–150

Thavasi R, Sharma S, Jayalakshmi S (2011) Evaluation of screening methods for the isolation of biosurfactant producing marine bacteria. J Petrol Environ Biotechnol S-1(1):1–6

Varjani SJ, Upasani VN (2017) Critical review on biosurfactant analysis, purification and characterization using rhamnolipid as a model biosurfactant. Bioresour Technol 232:389–397

Vecino X, Bustos G, Devesa-Rey R et al (2015) Salt-free aqueous extraction of a cell-bound biosurfactant: a kinetic study. J Surf Det 18(2):267–274

Wei YH, Cheng CL, Chien CC et al (2008) Enhanced di-rhamnolipid production with an indigenous isolate *Pseudomonas aeruginosa* J16. Process Biochem 43(7):769–774

Ziv N, Brandt NJ, Gresham D (2013) The use of chemostats in microbial systems biology. J Vis Exp 80:50168

Zwietering MH, Jongenburger I, Rombouts FM et al (1990) Modeling of the bacterial growth curve. Appl Environ Microbiol 56(6):1875–1881

Reduction of Energy Consumption in Gas-Liquid Mixture Production Using a Membrane Diffuser and HE-3X Stirrer

Waldemar Szaferski

1 Gas-Liquid Systems in Industrial Use

The effects of gas dispersion in the mixer, expressed by, for example, the degree of gaseous phase retention in the mixture (i.e. gas holdup) or the size of the interfacial surface, depend on a number of physical characteristics of the process. The type of secondary circulation that has a major influence on the gas phase dispersion is important. In mass transfer processes of two-phase liquid-gas dispersions, the main mass transfer resistance is on the liquid side. The rate of transport from the gas phase to the surrounding liquid depends to a large extent on the hydrodynamic conditions in the apparatus. Especially important here are the dimensions of the gas bubbles, the rate of movement and the time they reside in the liquid. The mechanical mixing of this system leads to more intensive processes. The hydrodynamic conditions produced in the mixed biphasic system are determined primarily by the stirrer structural parameters, the physical parameters of the fluids and the kinetics of the processes conducted. Each stirrer produces the original liquid stream in the apparatus, the so-called primary circulation, which generates circulation throughout the apparatus along the circulation loops (called secondary circulation). As a result of analyzing energy dissipation in the fluid circulation streams in typical construction designs with mechanical stirrers it has been shown that the energy of the secondary circulation that determines the mixing effect, is much smaller than the primary one (Brauer 1985; Doran 2013).

With bioreactors we can carry out biotechnological processes in a number of industries, such as pharmaceutical, food, chemical, cosmetic, environmental, energy, fodder and many other related industries where various multiphase gas-liquid mixing systems are employed. The commonly used reactor type is a

W. Szaferski (✉)
Institute of Chemical Technology and Engineering Faculty of Chemical Technology,
Poznan University of Technology, Poznań, Poland
e-mail: waldemar.szaferski@put.poznan.pl

© Springer International Publishing AG, part of Springer Nature 2018
M. Ochowiak et al. (eds.), *Practical Aspects of Chemical Engineering*,
Lecture Notes on Multidisciplinary Industrial Engineering,
https://doi.org/10.1007/978-3-319-73978-6_29

mechanical agitated tank (i.e. mixer) used in alkylation, nitration, chlorination, sulfonation, hydrocarbon oxidation, catalytic hydrogenation of vegetable oil, aerobic fermentation, waste water treatment and other similar processes (Nagata 1975; Ledakowicz et al. 1984; Pohorecki 2002; Liwarska-Bizukojc and Ledakowicz 2003).

Neutralization of industrial and municipal waste water also includes processes using mixing technology. In the aeration processes used for the treatment of municipal and industrial waste water the primary component is oxygen. There are two basic and currently used aeration systems:

- system based on fine-bubble membrane diffusers,
- systems based on jet diffusers (injectors).

From the point of view of mixing technology, the aeration process can be divided into several physicochemical systems:

- gas (air)-liquid (effluent), where air is dispersed in the sewage and as a result, oxygen is transferred to the sewage,
- liquid (wastewater)-solid (sludge) that delivers oxygen from the sewage to the sludge particles that is the sewage component, resulting in generation of an extra biomass,
- sewage mixing system with sludge across the entire tank, which allows for the desired amount of dissolved oxygen to be obtained in the sewage.

Diffusers have been found to be used in aeration of sewage in averaging and activated sludge chambers in household, municipal and industrial sewage treatment plants and in water purification with air and pure oxygen.

2 Constructions in Gas-Liquids Processes

The process of mixing liquids with gas externally supplied under pressure to a system is referred to as pressure aeration (Stręk 1981; Lin 2008; Zhang et al. 2014; Ding et al. 2016). The gas introduced into the liquid filled mixer enters the under pressure region generated around the rotating agitator as a result of the liquid jet ripping off its outer edge and then is decomposed into bubbles. The size of the generated gas bubbles is determined by the tangential stresses occurring in the gas stream surrounding the stirrer. Their diameter is smaller, the thinner the gas stream and the lower the height of the stirrer. Therefore, stirrers producing high tangential stresses and those whose height compared to other types of stirrers is the lowest, i.e. turbine disc, perforated disc, propeller, are best for mixing liquids with gas supplied to the pressurized system.

Due to the wide spectrum of applications for liquids with complex physicochemical properties, the most commonly used stirrer is the standard turbine-disc stirrer with six straight vanes. The first one who had done very thorough research for this type of stirrer was Rushton and it is commonly called Rushton turbine after his name. The versatility of this type of stirrer was observed during mixing of various multiphase systems such as liquid-liquid, gas-liquid and gas-liquid-solid. In the case of liquid aeration, the agitator disc traps the gas and then the vanes gradually break up large bubbles into smaller ones.

In the processes where there is less demand for gas flow and no sterility is required, self-suction stirrers can be used (Heim et al. 1995; Poncin et al. 2002; Kurasiński et al. 2010). These stirrers can be an alternative to conventional ones as they do not require any gas supply. Gas is sucked into the system from the surface of the liquid via a hollow shaft. Aeration occurs only when the rotational rate is high enough ($n > n_p$). These minimum rates are different for various stirrers, the more streamlined is the shape of a stirrer (Kamieński 2004).

The most popular self-suction stirrers with hollow shafts include prismatic (triangular, rectangular, pentagonal), hexagonal with channels, disc, chamber, double prismatic, double prismatic with funnel, tubular and stirrers with projections (Heim et al. 1995; Poncin et al. 2002).

Particular attention should be paid to conical stirrers that meet the requirements of stirrers producing gas-liquid biphasic systems, as they show a high hydraulic efficiency and performance. Their pumping efficiency is one quarter higher than that of conventional turbine stirrers. They enable efficient dispersing of significant amounts of gas in the liquid, which is greater than that observed with other stirrers operating under similar conditions. The beneficial distribution of the peripheral velocity of the blade ends, its lower gradients and hence the shear stresses, pre-dispose them to be used additionally in bioreactors where there is a danger of destroying the structure of microorganisms under high stresses and high rotation rate. The most commonly used conical stirrers are conical, semi-conical and disk stirrers (Stręk 1981; Dyląg and Talaga 1991; Kamieński 2004).

Kuncewicz and Stelmach (2010) have shown that, from the point of view of the amount of gas retained in the liquid, the self-suction disk stirrers give way to turbine designs with an external gas supply through the bubbler but surpass both horizontal and vertical surface aerators.

Pressure aeration of liquid in the mixer can be made in three ways. Most often, a pressurized gas supply is used by means of a single tube positioned centrally under the stirrer. Often, the tube's end has a variety of nozzles or bubbles, which can exist in the form of perforated spiral or ring tubes. A common way is to provide air through a hole in the bottom of the tank or a system of multiple openings that are evenly arranged (Holland and Chapman 1966; Nagata 1975; Błasiński and Młodziński 1976; Loiseau et al. 1977; Frijlink et al. 1990; Hiruta et al. 1997).

The modern designs of gas distributors differing in structure are the membrane diffusers. Depending on the size of the bubbles they emit, they can be divided into thick-bubble, medium-bubble and fine-bubble (Gawroński 1999; Łomotowski and Szplindor 1999). The size of air bubbles for thick bubbles is greater than 10 mm. They have found use in the pipe systems with drilled openings with a diameter of more than 3 mm. Currently they are mainly used in hydraulic jacks (mammoth pumps), but due to their low efficiency in the aeration process they are often replaced by other diffusers. Medium-bubble diffusers with bubble size in the range of $d_p = 3$–10 mm, are constructed similarly to thick-bubble ones, with a diameter of 0.5–2.0 mm. Fine-bubble diffusers with air bubble size of $d_p < 3$ mm are most popular. Fine-bubble diffusers are now a standard component of compressed air systems for aeration of wastewater.

The diffusers discussed here are made in two material versions, as membrane or ceramic diffusers. Membrane diffusers are resistance to clogging and suitable for cyclic operation. The disk membrane is made of an EPDM (ethylene-propylene diene monomer) elastomer with laser openings. EPDM is a homogeneous material throughout the cross-section, with no additives or plasticizers that will get oxidized or rinsed over time. At the same time it is extremely sensitive to UV rays. It has the so called "material memory", that is, it always returns to its original shape and its extensibility is about 300%. It is flexible in low and high temperatures (does not crack or crumble under temperature fluctuations). The membrane assumes a proper shape in a vacuum that completely eliminates the possibility of dead space. Modern laser technology enables to cut openings in the membrane with the highest precision.

3   Experimental Set-up and Scope of Research

Basic measurements of the gas volume in the liquid and the mixing power were made on a experimental set-up constructed for this purpose, the main component of which was a mixer consisting of a flat bottom tank (made of methyl poly-methacrylate) with an active volume of 0.0192 m^3 with four baffles, mechanical stirrer and gas distributor. The aeration system was composed of a fan, a shell-and-tube heat exchanger for air temperature control, the air supply pressure regulator, the gas flow regulator and the bubbler in the form of different membrane diffusers (Fig. 1).

Experimental tests included membrane diffusers with different membrane diameters of 50–270 mm and different diameter of air bubbles, and the system was equipped with Rushton (RT) mechanical disc-turbine stirrer, Chemineer HE-3 (HE-3) and modified HE-3X stirrer (HE-3X) (Fig. 2). The disk present in the Rushton turbine is well performing in mechanical mixers to disperse the gas. In contrast, the HE-3 stirrer, according to the literature and the manufacturer (Chemineer), performs the process at low mixing power. The idea of the HE-3X agitator was to combine these two features, i.e. good dispersion of gas and low mixing power.

The stirrer was leveled at 1/3 the height of the liquid in the apparatus against the bottom of the apparatus with a diameter of 0.100 m. The internal diameter of the apparatus and the height of the liquid were equal and amounted to 0.290 m.

Experimental studies were conducted in two stages. In the first step, the amount of gas retained in the liquid (ε) and the energy input for the individual liquid-air two phase systems and different aeration systems in the form of membrane diffusers at

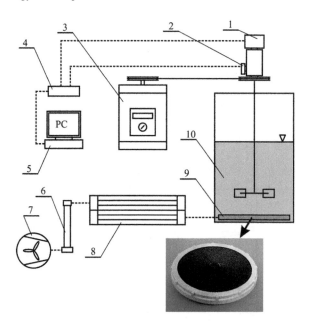

Fig. 1 Diagram of experimental set-up used in experimental tests: 1—system of clamping and bearing of the stirrer shaft, 2—turning sensor, 3—motor and inverter, 4—torque meter, 5—computer, 6—rotameter, 7—ventilator, 8—heat exchanger, 9—gas distributor, 10—tank

constant volume flow rate (Q_g) were determined. In the second step, the amount of gas retained in the liquid at a constant stirring speed (N) for a variable volume gas flow rate in the range of 0–3 m³/h was determined. After starting the stirrer drive and setting the constant of rotation rate, the gas flow control valve was used to adjust the amount of gas entering the apparatus and, when determining the appropriate gas flow rate, the height of the liquid level in the apparatus was measured from its scale.

4 Gas in Liquid Holdup Coefficient

In liquid aeration, the volume of the gas phase in the mixer is not constant as opposed to liquid-liquid or liquid-liquid-solid systems where the volume of the dispersed phase is directly due to the material balance. The gas holdup is, as already mentioned, the basic hydrodynamic parameter determining the efficiency of the process.

Gas holdup (ε) is a basic measure of the efficiency of the gas retained in a liquid. The amount of gas retained at a time in the liquid depends on many parameters, namely: the stirrer's design, its rotational speed, the gas flow rate and the structure of the dispersing system in apparatus. The gas holdup ratio defined as the volume share

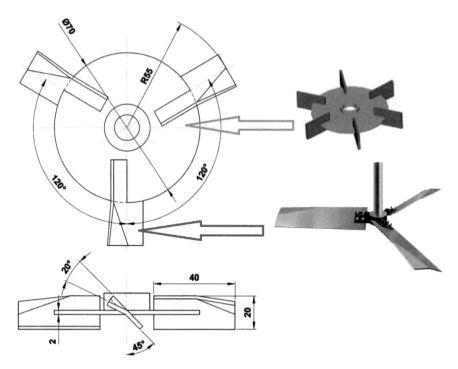

Fig. 2 Scheme and concept of HE-3X stirrer

of gas in the two phase mixture has long been used in the literature (Foust et al. 1944; Vlček et al. 1969; Loiseau et al. 1977) and is described by the equation:

$$\varepsilon = \frac{V_G}{V_M} = \frac{V_M - V_L}{V_M} = \frac{H_M - H_L}{H_M} \tag{1}$$

where V—volume (m^3) of gas (G), mixture (M), liquid (L), H—height (mm).

The first tests were conducted for diffusers with diameters from 50 to 270 mm and continuous medium was distilled water. Among them were fine-bubble diffusers that produced air bubbles below 3 mm and thick-bubble diffusers which produced air bubbles of more than 10 mm. Better results, with respect to the amount of trapped gas and unit mixing power, are obtained with diffusers whose membranes produce smaller air bubbles and the stirrer also breaks them down into even smaller and distributes evenly throughout the apparatus even at small rotational rates. The distribution of gas in almost the entire stirrer in the case of the distributor 50 is achieved only at high stirrer rotation rates for both fine and thick bubbles. The introduction of the HE-3X stirrer did not affect the obtained results and came to the same conclusions as previously obtained for the Rushton turbine (Broniarz-Press et al. 2005, 2006a, b, c, d; Broniarz-Press and Szaferski 2005, 2008). In addition, it was observed that small diffusers supplied the

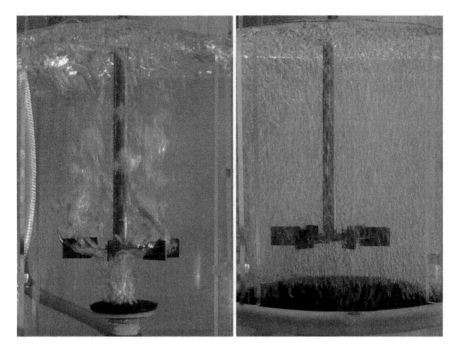

Fig. 3 Comparison of membrane diffusers with a diameter of 50 mm (left) and 270 mm (right)

gas like a single opening in the bottom and therefore were discarded in further tests. Further experimental research focused on the 270 mm diffuser whose design caused a reduction in the so-called dead zones in the lower part of the stirrer (Fig. 3) and the gas could be contacted more quickly with the continuous phase. The diffuser diameter was close to the inside diameter of the apparatus. In the paper by Pawełczyk (1988), a similar solution based on perforated bottom was presented as a gas distributor in a stirrer. In addition, the membrane diffusers are characterized by low air flow resistance, and by using ethylene-propylene diene (EPDM) membranes, silicone and polyurethane they operate like a check valve thus improving the safety of the apparatus.

Distilled water and aqueous solutions of Rokrysol WF1 polyacrylamide at concentrations of 0.1–0.3% were used for the tests. The density of tested solutions ranged from 997 to 999 kg/m^3. Because of the complex rheological properties of the polymer solutions, appropriate tests were performed using the Physica MCR-501 rotary rheometer.

The results of the rheological tests are presented in Table 1. All analyzed systems proved to be exponent fluids because their technical flow curves were described by equations in general form (Kembłowski 1985):

Table 1 Rheological parameters of Eq. (1) for mixtures used in experimental tests

Rokrysol WF1 concentration in aqueous solution [%]	K'	m
0.1	0.0906	0.649
0.2	0.164	0.594
0.3	0.322	0.549

$$\tau_w = K' \cdot \dot{\gamma}_w^m \qquad (2)$$

where τ_w—shear stress (Pa), $\dot{\gamma}_w$—shear rate (1/s), K'—consistency ratio (Pa s), m—flow ratio (–).

The dependence of the gas retention coefficient (ε) in distilled water and the aqueous solution of Rokrysol WF1 polyacrylamide from the Reynolds number for the varying gas flow rate (Q_g) in the range of 0.133–0.833 dm^3 s^{-1} was analyzed. For all tested mixtures, the gas holdup ratio was dependent on the Reynolds number. With a constant gas flow rate along with an increase in the stirrer rotation rate and also the Reynolds numbers, the gas holdup coefficient in the two-phase mixture increases (Fig. 4). Similar is a dependence for the process conducted at the constant frequency of the Rushton turbine and the variable amount of gas fed to the apparatus. The gas holdup coefficient increases with increasing volume of gas flow rate (Fig. 4), as well as with increasing concentration of aqueous polymer solution in the range of 0.1–0.3%.

The influence of the type of stirrer was also analyzed by comparing the standard Rushton turbine-disc stirrer and the modified HE-3X stirrer for the same media and gas volume flow ranges. It has been shown that the Rushton turbine stirrer achieves higher amounts of gas retained in the mixture than HE-3X. As a result of the

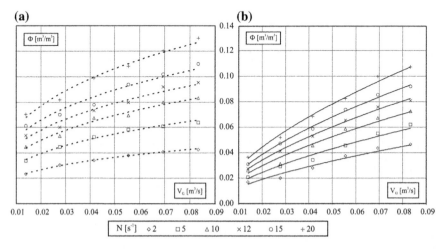

Fig. 4 Exemplary dependencies of gas holdup in the mixture in function of volumetric gas flow rate for 0.1% polyacrylamide aqueous solution for analyzed stirrers: **a** RT and **b** HE-3X

Table 2 Comparison of correlation coefficients of Eq. (3) for tested stirrers

Stirrer type	Rokrysol WF1 [%] concentration in aqueous solution	A	B	C
HE-3X	0.1	0.0158	0.188	0.62
	0.2	0.0149	0.157	0.55
	0.3	0.0141	0.121	0.46
RT	0.1	0.00915	0.111	0.35
	0.2	0.00910	0.102	0.30
	0.3	0.00900	0.095	0.26

approximation of measurement data for aqueous polymer solutions, a correlation equation was obtained (Eq. 3), the coefficients of which are compared in Table 2.

$$\varepsilon = (A \cdot N + B) \cdot Q_g^C \tag{3}$$

In the case of results of experimental tests conducted for continuous phase such as distilled water and the identical volume of gas flow rate, the amount of gas retained in the mixture was correlated as the following equations:

- for Rushton turbine (RT):

$$\varepsilon = 0.232 Q_g^{-0.0142N + 0.430} \tag{4}$$

- for HE-3X:

$$\varepsilon = 0.232 Q_g^{-0.0143N + 0.423} \tag{5}$$

5   Gas-Liquid Hold-up Coefficient in Contour Maps

By conducting the aeration process in two ways, i.e. at a constant volumetric gas flow rate and a variable rotational rate, then at a constant rotational rate and variable volumetric flow rate, we can conclude that increasing the mixing frequency is more effective. The stirrer not only distributes air in the liquid volume of the apparatus but also exposes its suction of air from above the surface of the liquid.

From the point of view of industrial applications, we want to provide apparatus operation under certain preferred conditions. In order to facilitate the selection of process parameters, the distilled water contour diagram (Fig. 5) and the aqueous

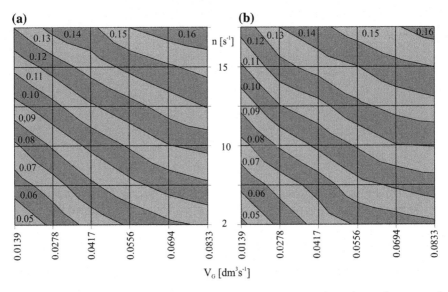

Fig. 5 Gas holdup (ε) as a function of the controlled parameters (volumetric gas flow rate and rotational rate of the stirrer) for distilled water **a** RT, and **b** HE-3X

polyacrylamide solutions (Fig. 6) were generated for the membrane membrane-based aerator system.

As shown by contour maps (Figs. 5 and 6), in which the contour represents the amount of gas being fed, when determining the rotation frequency and the type of stirrer used, the amount of gas holdup can be determined. The obtained information about the gas holdup coefficient in the mixture refers not only to the gas distributor in the apparatus but also to the "suction" of the gas through the stirrer. The residence time of the gas will affect the amount of oxygen that moves locally into the liquid. Similar observations were reported in aeration using a single tube (Vermeulen et al. 1955), which confirms that the measurements carried out are correct. It should be noted, however, that comparing the results with another work (Broniarz-Press et al. 2005) is difficult or even impossible due to the aeration system used. The contour maps can be used in aeration process design for other systems such as gas-liquid-liquid systems (Szaferski and Mitkowski 2016).

6 Mixing Power Characteristics in Aeration with Membrane Diffusers

An important parameter characterizing the mechanical mixing of the liquid-gas system is the mixing power necessary to ensure the preset hydrodynamic conditions. Mixing power depends on the diameter of the stirrer, the diameter of the

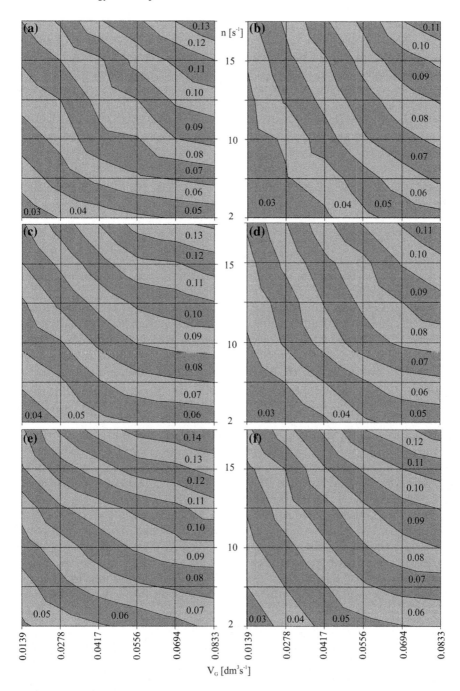

Fig. 6 Gas holdup (ε) as a function of the controlled parameters (volumetric gas flow rate and rotation frequency) for aqueous polyacrylamide solutions: **a** RT 0.1, **b** HE-3X 0.1, **c** RT 0.2, **d** HE-3X 0.2, **e** RT 0.3, **f** HE-3X 0.3

mixer, the height of the liquid in the tank, the width of the baffle, the geometric characteristics of the stirrer, the stirrer distance from the bottom of the tank, the viscosity of the liquid, the density of the mixture, the rotational rate of stirrer and the earth gravity.

Introducing gas to the mixed liquid results in a decrease in mixing power, with the exception of some stirrers in respect of low gas consumption and low rotation where power can increase.

An analysis of the relationship of the Newton number and the Reynolds number was performed to determine the characteristics of the diffuser-stirrer tested design solutions. Due to the non-Newtonian nature of aqueous solutions of polyacrylamide, the Reynolds number for the mixing process (Re_m) is defined by the equation proposed by Goloz and Pawlushenko (1967):

$$Re_m = (4\pi)^{1-m} \frac{N^{2-m} \cdot D^2 \cdot \rho}{K} \tag{6}$$

where m, K—coefficients presented in Table 1, N—rotational speed of impeller (s^{-1}), D—diameter of the impeller (m), ρ—density (kg/m^3).

For a process carried out in distilled water, the average power of 0.54 (HE-3), 0.60 (HE-3X) and 5.5 (Rushton turbine) was obtained within the range of Reynolds number of 20,000 up to 200,000. The modification made did not directly affect the power output, and indirectly impacted the mixing power in relation to the HE-3 stirrer. Please note that the HE-3X stirrer is to be an alternative to the Rushton turbine stirrer for generating gas-liquid mixtures. The resulting power is much lower, which guarantees a reduction in the energy consumption of the process.

The measurement results for the gas-liquid mixture confirmed the reduction of energy consumed by the modified HE-3X stirrer compared to the other stirrers. In the case of Rokrysol WF1 aqueous solution processes, within 1000–3000 ppm, the power for the individual stirrers with the values given in Table 3 was found in the range of Reynolds numbers from 1000 to 16,000 and for permanent gas flow rate of 0.278–0.833 dm^3 s^{-1} and rotational rate variable of 2–20 s^{-1}.

It has been observed that the more amount of gas is fed to the system and the concentration of aqueous solutions of polyacrylamide increases, the power decreases. The reason for this is observed in the gas distributor's design. The active surface of the diffuser is close to the inner diameter of the stirrer. Air bubbles change the physical parameters of the solution and at the same time move almost throughout the apparatus. The stirrer partially uses its suction properties by bringing

Table 3 Power numbers obtained (Ne) for Rokrysol WF1 polyacrylamide aqueous solutions

Rokrysol WF1 concentration [%] in aqueous solution	TR	HE-3	HE-3X
0.1	1.6–1.7	0.25–0.35	0.5–0.75
0.2	1.7–1.9	0.35–0.45	0.6–0.75
0.3	1.9–2.1	0.4–0.55	0.6–0.8

gas from the outside and also spreads air bubbles into areas where the diffuser is unable to deliver gas.

Based on visual observation, it was noted that the modified HE-3X stirrer was more stable when producing a mixture and that no additional vibrations were observed in the mixer as was the case with other mixers, resulting in much higher repeatability of results.

7 Conclusions

The complexity of industrial installations brings about the necessity to look for mechanical and operational solutions that will reduce the energy consumption of the process while maintaining its maximum efficiency. Introduction of membrane diffusers to mechanical mixers producing gas-liquid mixtures is meaningful as showed in the work. Their effectiveness results from the fact that they emit fine bubbles evenly over the entire surface of the membrane and mixer bottom. Stirrers used in these apparatus are based on disc-shaped stirrers with straight vanes. They generate high power, which translates into high power required to produce the mixture. The HE-3X stirrer, even though it does not have such pumping properties as the Rushton turbine stirrer, lowers the energy demands. Just in connection with a membrane diffuser that protects the process in respect of gas demand in the apparatus, the HE-3X stirrer can be an alternative to other mixers of this type.

Acknowledgements This work was supported by PUT research grant number 03/32/DSPB/0702.

References

Błasiński H, Młodziński B (1976) Aparatura przemysłu chemicznego. WNT, Warszawa

Brauer H (1985) Ein- und Mehrphasen Strömungen. Biotechnology, vol 2. Weinheim

Broniarz-Press L, Szaferski W (2005) Effect of truncated cone gas distributor type on gas hold-up and power consumption in a mixer. In: Proceeding of 7th World congress of chemical engineering, Paper No. P42-059:1–9, Glasgow

Broniarz-Press L, Szaferski W (2008) Aeration using membrane diffusers in a tank equipped with two impellers on a common shaft. Chem Proc Eng 29(1):61–73

Broniarz-Press L, Szaferski W, Sadowska J (2005) Effect of the geometry of plate membrane diffusers on hold-up and power consumption during liquid aeration in an agitated tank. Inż Apar Chem 44(36):17–21

Broniarz-Press L, Borowski J, Szaferski W (2006a) Mixing in two-phase systems in the presence of non-Newtonian media. Polish J Chem Techn 8(1):54–60

Broniarz-Press L, Szaferski W, Ochowiak M (2006b) Analiza przydatności dystrybutorów membranowych gazu do napowietrzania cieczy w mieszalnikach. Inż Ap Chem 45(37):31–33

Broniarz-Press L, Szaferski W, Ochowiak M (2006c) Analiza stopnia zatrzymywania gazu w mieszanych układach nienewtonowskich. Inż Ap Chem 45(37):18–20

Broniarz-Press L, Szaferski W, Ochowiak M (2006d) The effect of the construction of membrane diffusers on the gas hold-up in Newtonian systems. Polish J Chem Technol 8(4):42–44

Chemineer. http://www.chemineer.com/literature

Ding A, Liang H, Li G et al (2016) Impact of aeration shear stress on permeate flux and fouling layer properties in a low pressure membrane bioreactor for the treatment of grey water. J Memb Sci 510:382–390

Doran PM (2013) Bioprocess engineering principles. Academic Press, London

Dyląg M, Talaga J (1991) Hydrodynamika mechanicznego mieszania układu trójfazowego ciecz–gaz–ciało stałe. Inz Chem i Proc 12(1):3–25

Foust C, Mack DE, Rushton JH (1944) Gas-liquid contacting by mixers. Ind Eng Chem 36:517–522

Frijlink JJ, Bakker A, Smith JM (1990) Suspension of solid particles with gassed impellers. Chem Eng Sci 45(7):1703–1718

Gawroński R (1999) Procesy oczyszczania cieczy. Oficyna Wydawnicza Politechniki Warszawskiej, Warsaw

Gluz MD, Pavlushenko IS (1967) Power consumption in agitation of non-Newtonian liquids. J Appl Chem USSR 40:2475

Heim A, Krasawski A, Rzyski E et al (1995) Aeration of bioreactors by self-aspirating impellers. Chem Eng J Biochem Eng J 58:59–63

Hiruta O, Yamamura K, Takebe H et al (1997) Application of Maxblend Fermentor® for microbial processes. J Ferment Bioeng 83:79–86

Holland FA, Chapman FS (1966) Liquid mixing and processing in stirred tanks. Reinhold Publishing Co., New York

Kamieński J (2004) Mieszanie układów wielofazowych. WNT, Warszawa

Kembłowski Z (1985) Podstawy teoretyczne inżynierii chemicznej i procesowej. WNT, Warsaw

Kuncewicz C, Stelmach J (2010) Zastosowanie autokorelacji do określania prędkości pęcherzyków gazu na podstawie zdjęć z podwójną ekspozycją. Inż Ap Chem 49(1):63–64

Kurasiński T, Kuncewicz C, Stelmach J (2010) Distribution of values of the volumetric mass transfer coefficient during aeration of liquid using a self-aspirating impeller. Chem Proc Eng 31(3):505–514

Ledakowicz S, Nettelhoff H, Deckwer WD (1984) Gas-liquid mass transfer data in a stirred autoclave reactor. Ind Eng Chem Fundam 13(4):510–512

Lin C (2008) A negative-pressure aeration system for composting food wastes. Bioresour Technol 99:7651–7656

Liwarska-Bizukojc E, Ledakowicz S (2003) Estimation of viable biomass in aerobic biodegradation processes of organic fraction of municipal solid waste (MSW). J Biotechnol 101:165–172

Loiseau B, Midoux N, Charpentier JC (1977) Some hydrodynamics and power input data in mechanically agitated gas-liquid contactors. AIChE J 23(6):931–935

Łomotowski J, Szplindor A (1999) Nowoczesne systemy oczyszczania ścieków. Arkady, Warsaw

Nagata S (1975) Mixing. Principles and applications. Kodansha Ltd., Tokyo

Pawełczyk R (1988) Barbotaż z równoczesnym mieszaniem mechanicznym. IV. Wpływ mocy doprowadzanej doukładu. Inz Chem i Proc 3:599–619

Pohorecki R (2002) Fluid mechanical problems in biotechnology. Biotechnologia 2:60–77

Poncin S, Nguyen C, Midoux N et al (2002) Hydrodynamics and volumetric gas–liquid mass transfer coefficient of a stirred vessel equipped with a gas-inducing impeller. Chem Eng Sci 57:3299–3306

Stręk F (1981) Mieszanie i mieszalniki. WNT, Warszawa

Szaferski W, Mitkowski PT (2016) Aeration of liquid-liquid systems with use of various agitators in mixer equipped with membrane diffuser. Chem Eng Technol 39(12):2370–2379

Vermeulen T, Williams GM, Langlosi GE (1955) Interfacial area in liquid-liquid and gas-liquid agitation. Chem Eng Prog 51:85–94

Vlček J, Steidl H, Kudrna V (1969) Holdup of dispersed gas fed continuously into a mechanically agitated liquid. Coll Czechoslov Chem Commun 34:373–386

Zhang Q, Liu S, Yang C et al (2014) Bioreactor consisting of pressurized aeration and dissolved air flotation for domestic wastewater treatment. Sep Purif Technol 138:186–190

Dyląg M, Talaga J (1991) Hydrodynamika mechanicznego mieszania układu trójfazowego ciecz–gaz–ciało stałe. Inz Chem i Proc 12(1):3-25

Atomizers with the Swirl Motion Phenomenon

Sylwia Włodarczak, Marek Ochowiak and Magdalena Matuszak

1 Introduction

Liquid atomization consists in liquid breakup which can be induced by powers created naturally or artificially. The aim of liquid atomization is to achieve small or satellite droplets. Atomizers are precise subassemblies designed to achieve the defined efficiency under the given conditions. In order to achieve long-lasting, effective and optimum performance spray system must be considered as a whole and a plan of its estimation, monitoring and exploitation have to be developed. When the spray system doesn't work at an optimum level, then the problems increasing costs and working time may appear. These problems may include: quality controls, unplanned production downtimes, increased exploitation costs, increased consumption costs of chemicals, water or electricity and a negative influence on the environment. By optimizing of atomization system you can achieve: the maximum efficiency, permanent and low exploitation and maintenance costs, excellent quality control, lack of unplanned work downtimes, minimum influence on environment (Nasr et al. 2002; Laryea and No 2003; Wimmer and Brenn 2013; Lan et al. 2014; Ochowiak 2016).

In practice a few kinds of atomizers can be distinguished which differ mainly with a source of the delivered energy causing the disintegration of the streams. Figure 1 shows their basic division. Atomizers can also be divided regarding their construction. Atomizer selection depends on the kind of process it is implemented in and the process requirements (Lefebvre 1989).

The next part of the chapter discusses in detail pressure-swirl atomizers, hybrid effervescent-swirl atomizers and also effervescent atomizers, which modified, provide atomizers with the swirl motion phenomenon.

S. Włodarczak (✉) · M. Ochowiak · M. Matuszak
Institute of Chemical Technology and Engineering,
Poznan University of Technology, Poznań, Poland
e-mail: sylwia.wlodarczak@put.poznan.pl

© Springer International Publishing AG, part of Springer Nature 2018   437
M. Ochowiak et al. (eds.), *Practical Aspects of Chemical Engineering*,
Lecture Notes on Multidisciplinary Industrial Engineering,
https://doi.org/10.1007/978-3-319-73978-6_30

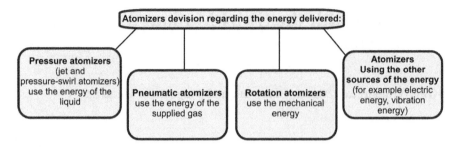

Fig. 1 Atomizers division regarding the energy delivered (Lefebvre 1989)

2 Pressure-Swirl Atomizers

The pressure-swirl atomizers use the energy of the liquid itself to the atomization process, which is used to produce a swirl inside the atomizer. Additionally, the atomizer can be equipped with a swirling insert placed in the atomizer axis which forces the arising of powers introducing the liquid into rotational movement. The produced swirl maintains the air core inside a chamber. The stability of a gas core (steady shape or height) is one of the most important features distinguishing pressure-swirl atomizers. Liquid swirls inside the swirl chamber owing to the centrifugal force. Circumferential velocity of the liquid flow increases while moving to the outlet. Then the centrifugal force increases too. A thin liquid film closes the outlet and breaks apart into droplets. Swirls at the atomizer outlet influence favorably droplets dispersion and the propagation of aerosol itself (Lee et al. 2010). The proper use of swirl movement in the liquid stream is one of the ways to improve atomization process and its optimization (Som and Mukhejee 1980; Yule and Widger 1996; Nonnenmacher and Piesche 2000; Amini 2016).

3 Hybrid Effervescent-Swirl Atomizers

Effervescent atomizers belong to twin-fluid atomizers with internal mixing. Liquid flows into the atomizer through the upper inlet and flows inside the perforated pipes to the central outlet. The gas is then dispensed into the annular chamber around the central perforated pipe. The gas pressure is slightly higher than liquid pressure. The gas, in the form of bubbles, flows through holes in the central pipe to the liquid stream. The amount of the dosed gas is lower in comparison to other twin-fluid atomizers. The internal flow in this type of atomizers is more complicated than in other one- or twin-fluid atomizers. The effervescent twin-fluid mixture flows down and it's "thrown away" through the outlet. A quick increase of bubbles at the atomizer's outlet improves atomization by quick liquid stream breakup.

Due to the way of gas injection two construction types are distinguished: inside-out and outside-in. The inside-out effervescent atomizer uses a technique where gas, in form of bubbles, is introduced into the liquid volume from aerator placed in the middle part of the atomizer's mixing chamber. Gas flows inside the perforated pipe, and bubbles get out to the surrounding liquid. In the outside-in atomizer, air is introduced into the mixing chamber from the annular space surrounding this chamber through small holes in the pipe.

Effervescent atomization ensures bigger advantages in comparison to atomization in traditional twin-fluid atomizers. Effervescent atomizers are able to produce aerosol droplets with dimensions highly independent of the atomizer outlet diameter, due to the low pressure required to produce a twin-liquid flow. The applied flow rates are considerably lower than required in other atomizers types. The outlet diameters are bigger than in other atomizers types and this fact considerably solves clogging problems and facilitates aerosol production. The aerosols produced by the effervescent atomizers help in lower pollution due to the air presence in atomizer's core. The atomizer can be used to spray different liquids without any impact on the process efficiency. While leaving the outlet, the flow rates are considerably lower than in traditional atomizers, thereby the outlet erosion decreases in case of atomization of liquids containing solids. Effervescent atomizers are simple, durable and reliable and also easy to maintain as well as cheap (Chen and Lefebre 1994; Wade et al. 1999; Ochowiak and Broniarz-Press 2008; Ochowiak 2012b).

Effervescent-swirl atomizers combine features of pressure-swirl and effervescent atomizers. Swirl motion introduction to effervescent atomizers aims at additional weakness of flow ring structure and production of bubble-film structure. It affects aerosol quality, improves atomization process and ensures its optimization. Moreover gas phase flow improves the atomization process. In the result the achieved aerosol may be characterized by smaller droplet diameters. The application of the effervescent-swirl atomizer allows to lets increase effectiveness of gas-liquid mixing. The flow character in the atomizer is similar to the process that can be seen in a typical effervescent atomizer. However, the mixing force of twin-flow system is bigger, what results from rotation of liquid, gas or liquid-gas mixture in the mixing chamber. Figure 2 shows an exemplary of effervescent-swirl atomizer construction. Liquid inlet is placed tangentially to the inner wall of the atomizer. Stream swirl increases as the liquid stream flow rate increases. In the atomizer axis of symmetry an aerator is placed which helps to introduce gas in form of bubbles into the mixing chamber (Ochowiak 2012a, 2016; Ochowiak et al. 2015).

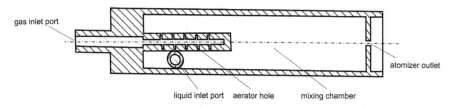

Fig. 2 Exemplary construction of a hybrid effervescent-swirl atomizer (Ochowiak 2012a)

4 Practical Implementation of Atomizers

The predominant feature of pressure-swirl atomizers is their high energy-efficiency. In many industrial processes, pressure-swirl atomizers are used because of their reliability and possibility to achieve droplets of small dimensions combined with high performance (Lefebre 2000; Wimmer and Brenn 2013). The most noticeable importance of pressure-swirl atomizers results from their wide industrial use in processes of combustion (cars and planes), drying (Tratnig et al. 2009; Huntington 2004), painting, vaporizing, hydrating, cooling (Li et al. 2006; Nasr et al. 2006), air conditioning and plant protection (Yule and Widger 1996; Halder et al. 2004; Endalewa et al. 2010; Belhadef et al. 2012; Chaudhari and Kulshreshtha 2013).

Agricultural applications are one of the most important ranges where pressure-swirl atomizers are used. Improper sprayers selection or low precision impact the treatment quality and they can even destroy plants. Thus, it is important to use such an equipment which will decrease the use of crop protection products, which means their decreased emission to the natural environment. Sprayer's working parameters must be selected in such a way to let them take into consideration weather conditions, accurate spraying parameters (inlet pressure), spray liquid properties, equipment type and its technical conditions. The accurate adjusting of working parameters results in an even distribution of the spraying liquid and a suitable coverable of the sprayed surface. Weather conditions, especially wind which drifts satellite droplets, considerably influence the protective treatment effectiveness. Wind can even make the spraying impossible. This adverse phenomenon can be eliminated by using special anti-drift atomizers or an auxiliary air stream. In case of pressure atomizers, which include pressure-swirl atomizers, pressure increase causes the droplets size to decrease. Then the risk of drift and vaporization is higher. Excessive operating pressure also may make the droplets bounce off the leaves surface. In spite of that, smaller droplets are more effective in plants protection, because they guarantee better surface coverage of plants. Larger droplets are able to cover a bigger surface being under agricultural application treatment and are more difficult to drift, but they flow down from leaves easier (Miller and Butler Ellis 2000; Szewczyk 2010).

Power engineering is a field where the majority of processes connected with liquid atomization are used. This group includes such processes as combustion, cooling and cleaning. Combustion can take place in furnaces with different kinds of burners. A fuel atomizer is the main burner's element. Combustion processes also takes place in energetic gas turbines. The gas turbine consists of three basic parts: a compressor, a combustion chamber and a turbine. The atomizer is placed in the combustion chamber (in front of a flue tube). Its task is to deliver fuel to a stabilizing swirl zone. The atomization influences the efficient work of combustion chambers. Small diameter droplets are necessary to produce a sufficient quantity of vapor for combustion process support. In gas turbines spray angle size impacts considerably the ignition efficiency, limits of flame and emission of pollution.

For petrol engines with a direct injection, an optimum spray angle lets steady combustion process, giving the engine maximum torque with minimum exhaust gas emission. In gas turbines, pressure-swirl and pneumatic atomizers are used. The excessive exhaust gas emission, mainly NO_X, is the most important problem regarding combustion in gas turbines. In order to solve this problem, an injection of properly prepared water is carried out, or vapor is delivered to a combustion zone to decrease flame temperature. It also supports a process of mixing fuel with air (Moon et al. 2009; Park et al. 2009; Moon et al. 2010; Li et al. 2011; Agarwal et al. 2014; Prakash et al. 2014; Sivakumar et al. 2015).

Effervescent atomizers are also primarily used in the combustion processes, among all in gas turbines, furnaces and boilers. They are also used when fuel injectors with the capability to handle a variety of fuels are required (Sovani et al. 2000, 2001). Effervescent atomizers produce the aerosol of smaller diameters (compared to other twin-fluid atomizers), which helps to reduce pollution. They also meet the requirements regarding work in a wide range of flow rates. Moreover, they are able to atomize fuels of different physical properties (they enable atomization of liquid with high viscosity). They have large outlets that eliminate clogging problems. A low liquid flow rate decreases erosion problems which can appear in case of solids suspended in waste. A low liquid flow rate also decreases droplets velocity and this fact enables to design smaller furnaces. The possibility to work at lower injection pressure affects limitations of losses and engine efficiency increases. Effervescent atomizers application contributes to limiting water and air contamination (by hydrocarbon fuels or volatile organic compounds) (Sovani et al. 2001).

5 Effect of Atomizer Geometry Over the Characteristics of Aerosol Produced

Following the technology development, better construction solutions to improve atomization process for the specific application are searched for. Depending on the industry branch and a scale on which the atomizer is to be used, there are different requirements. For example, the specific size of the droplets obtained. Such solutions are sought that allow to meet users' requirements and, at the same time, reduce energetic losses and ensure as high environment safety as possible. To achieve this, all possible geometric parameters of the atomizer are modified (e.g. shape of outlet or swirl chamber). It is also necessary to conduct further research (studies) to state the importance of atomizer's geometry impact on the parameters of the atomized stream (Som and Mukherjee 1980; Datta and Som 2000).

5.1 The Outlet Diameter, Length and Shape

The outlet diameter d_0, its length l_0 and shape have the most importance for aerosol producing (Lee 2008; Lee et al. 2010). The best atomization is for small outlets but, in practice, difficulty to keep fluids without solids usually limits a minimum outlet size to about 0.3 mm. The majority of research has been conducted for atomizers with cylindrical outlets (Sharma and Fang 2015). The outlet diameter impacts the size of the spray angle. Studies by Rizk and Lefebvre (1985) present that the greater outlet diameter, the wider spray angle. This correlation was confirmed by results obtained of Datta and Som (2000), Broniarz-Press et al. (2010) and Ochowiak et al. (2011).

Jedelsky et al. (2008) analyzed impact of the outside-in effervescent atomizer construction on the spray angle value. As the ratio of the outlet length to diameter decreased, the spray angle increase was observed. The spraying quality worsens when, as the result of use, the atomizer outlet changes its shape because of erosion (Gajkowski 2000). Research for rectangle, square and triangle outlet shapes are conducted. Their application results in quicker liquid stream disintegration and lets achieve a wider spray angle compared with cylindrical outlets (Sharma and Fang 2015). Ochowiak et al. (2015) analyzed the impact of the effervescent-swirl atomizer outlet shape on the spray angle value. Studies were conducted with regards to cylindrical-shaped (at different ratios of outlet length to diameter l_0/d_0), conical, reversed cone and profiled outlets. The widest spray angles were achieved for the profiled outlet while the smallest ones for the cylindrical (Ochowiak et al. 2015).

The outlet diameter also affects droplet diameters. The larger the outlet diameter is (at the same pressure value), the larger the droplets size becomes. Broniarz-Press et al. (2016) analyzed the impact of the outlet shape diameter and l_0/d_0 ratio on droplet size distribution and Sauter mean diameter. Sauter mean diameter (SMD) is presented by the following correlation:

$$SMD = \frac{\sum_{i=1}^{i=j} n_i d_i^3}{\sum_{i=1}^{i=j} n_i d_i^2} \tag{1}$$

The droplet diameters achieved in atomization in the atomizers with cylindrical, conical and profiled outlets were compared. The least droplet diameters were achieved for conical outlets, and slightly larger for profiled outlets. In case of cylindrical ones, the least SMD values were for $l_0/d_0 = 1.01$, and the largest SMD values were for $l_0/d_0 = 4.13$. Figure 3 presents the dependence of Sauter mean diameter on the l_0/d_0 ratio and on injection pressure (P). SMD value decreases with l_0/d_0 ratio decrease and injection pressure increase (Broniarz-Press et al. 2016). SMD value decrease together with l_0/d_0 decrease is assigned to a quicker spray breakup which is accompanied by higher turbulence intensity. In connection to that, the outlet length is one of the most important parameters taken into account in atomizer's designing (Lee 2008). In the atomizer's outlet a phenomenon of jet contraction occurs, which disappears with the outlet length.

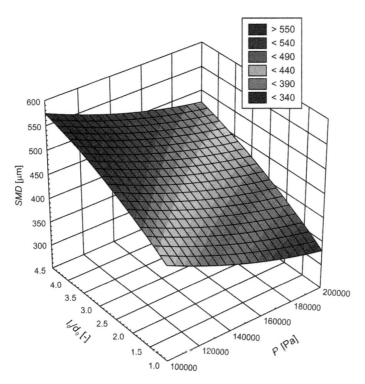

Fig. 3 Dependence of Sauter mean diameter on l_0/d_0 value and injection pressure (Broniarz-Press et al. 2016)

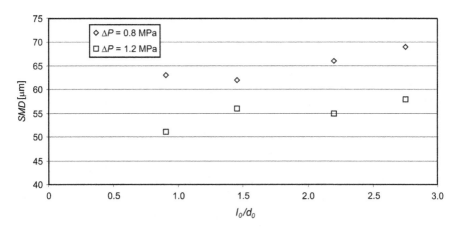

Fig. 4 Dependence of Sauter mean diameter on l_0/d_0 value for different injection pressures (Rashad et al. 2016)

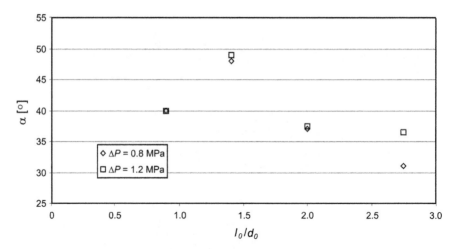

Fig. 5 Dependence of the spray angle on l_0/d_0 ratio for different injection pressures (Rashad et al. 2016)

Rashad et al. (2016) observed Sauter mean diameter increase with outlet length to diameter ratio increase from 0.81 to 2.69. The dependence of *SMD* value on l_0/d_0 is presented in Fig. 4.

The value of l_0/d_0 ratio also influences the spray angle size. Rashed et al. (2016) presented that the spray angle increases along with the increase of l_0/d_0 ratio from 0.81 to 1.44 and subsequently it decreases with l_0/d_0 ratio increase from 1.44 to 2.69. The optimum ratio value of the outlet length to diameter is stated for $l_0/d_0 = 1.44$. The dependence of the spray angle on l_0/d_0 value is presented in Fig. 5.

5.2   The Diameter, Length and Shape of the Mixing Chamber

The increase of the mixing chamber diameter causes the increase of the spray angle. This phenomenon can be directly connected to changes of liquid flow rate and tangential and axial flow rates. At the same time, small changes of the spray angle with the change of the swirl chamber length are observed. Rashed et al. (2016) didn't present a significant impact of the ratio of swirl chamber length to diameter (Fig. 6). The most favorable results were achieved for $L_s/D_s = 3.75$. After exceeding this value, the spray angle rapidly was decreasing.

They had similar conclusions for Sauter mean diameter value. After exceeding the optimum value $L_s/D_s = 3.75$, *SMD* value was rapidly increasing it is presented in Fig. 7 (Rashad et al. 2016).

Som and Mukherjee (1980) presented that the spray angle increases along with the increase in the ratio of outlet diameter to swirl chamber diameter, with the

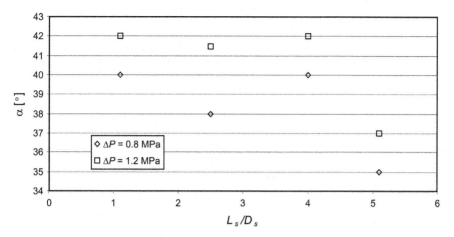

Fig. 6 Dependence of the spray angle on L_s/D_s ratio for different injection pressures (Rashad et al. 2016)

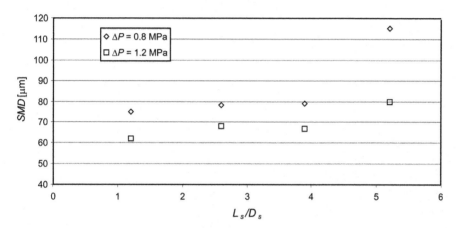

Fig. 7 Dependence of Sauter mean diameter on L_s/D_s ratio for different injection pressures (Rashad et al. 2016)

increase of swirl chamber angle or with decrease of the ratio of swirl chamber length to diameter. This phenomenon is attributed to a mechanism of hydrodynamics change of the flow in the atomizer with the defined geometry. Resistance increase on the atomizer caused by a change of any atomizer geometric parameter decreases swirling power inside the atomizer and finally decreases values of other parameters, such as the spray angle, which is directly connected with the swirl flow force. Figure 8 shows the dependence of the spray angle on Reynolds number for different ratio values of outlet diameter to swirl chamber diameter (d_o/D_s). Reynolds number value is determined in the equation:

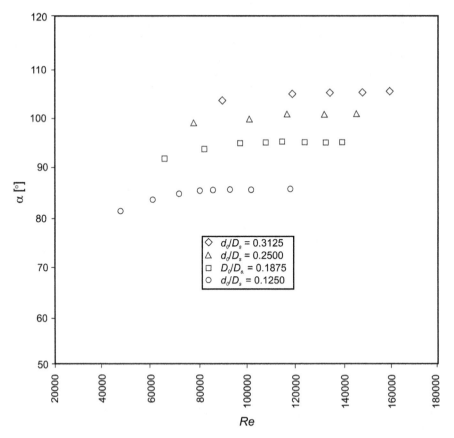

Fig. 8 Dependence of the spray angle on Reynolds number for different values of d_o/D_s ratio (Som and Mukherjee 1980)

$$Re = \frac{w_c d_o \rho_c}{\eta_c} \qquad (2)$$

where w_c—liquid velocity, ρ_c—liquid density, η_c—liquid viscosity.

Jedelsky et al. (2008) analyzed impact of the outside-in effervescent atomizer construction on the spray angle size. They showed that along with decreasing diameter and length of the swirl chamber, the spray angle increased.

Datta and Som (2000), on the basis of numerical expectations of flow liquid in simplex atomizer, stated the impact of liquid flow rate, atomizer's outlet diameter, tilting of swirl chamber interior walls, and tangential inlets surfaces on the spray angle value. Simplex atomizer is a single-stage atomizer of a simple construction which finds application in various fields. They showed that the spray angle increases along with increase of liquid flow rate, outlet diameter, swirl chamber angle and also with decrease of tangential inlets surfaces.

5.3 Other Geometric Parameters

The characteristic parameter describing the swirl atomizer operation is the swirl atomizer geometric index K, which includes the most important atomizer geometric dimensions:

$$K = \frac{A_p}{D_s d_0} \tag{3}$$

where A_p—surface of inlets.

As the atomizer geometric index increases, the spray angle and droplets size decrease (Yule and Widger 1996; Datta and Som 2000). The internal atomizer geometry controls gas-liquid flow inside of the atomizer and may strongly influence the atomizer and process effectiveness. The aerator geometry plays an important role for effervescent atomizers and effervescent-swirl atomizers. Wang et al. (1989) analyzed the impact of aerator holes dimensions and number on the mean drop size in inside-out atomizer, and Roesler and Lefebvre (1989) conducted studies for outside-in atomizer. It was proved that aerators with many holes produce drops with slightly narrower distributions than aerators with one hole and the same effective hole surface. Wade et al. (1999) also presented the atomizer geometry impact on Sauter mean diameter value. The aerator surface increase and aerator diameter decrease cause rise of the spray angle value (Jedelsky et al. 2008).

6 Impact of Liquid Properties and Atomizer Operating Parameters on the Resulting Aerosol Characteristics

Liquid properties considerably influence the jet breakup while leaving the atomizer. A special meaning is attributed mainly to viscosity and surface tension. Liquid density impact on the spray process is marginal.

Viscosity and surface tension raise energy volume necessary to liquid spray. The increase of each of these parameters usually causes the droplet size increase. Dorfner et al. (1995) and Wang et al. (2010) observed Sauter mean diameter increase along with viscosity and surface tension increase. Sovani et al. (2001), during researches conducted at the increased pressure (11–13 MPa), observed mean drop diameter increase along with the viscosity increase. Broniarz-Press et al. (2010) and Ochowiak and Broniarz-Press (2009) analyzed impact of polymer addition, that means viscosity as well, on Sauter mean diameter in case of spraying with effervescent atomizer. It was observed that the liquid viscosity increase caused the drop diameter increase, but the jet breakup was becoming more difficult. Sallevelt et al. (2015) checked the spray process of biodiesel, plant oil and glycerin in similar operational conditions using an atomizer. It was shown that the original breakup process quickly worsened with the liquid viscosity increase. Higher viscosity makes the jet breakup longer, and ligaments are visible in the measurement place. Moreover, these data indicate that *SMD* increases as the surface tension increases.

The surface tension should not influence the spray angle value and it is usually confirmed in experimental research. The higher liquid viscosity, the smaller spray angle (Ochowiak et al. 2015). Wang et al. (2010) conducted research of diesel spraying and two kinds of biodiesel at high pressure (up to 300 MPa). The achieved spray angles for biodiesel were smaller than for diesel. Much higher biodiesel viscosity leads to lower spray angle value and weak aeration of the liquid jet. The viscosity impact on spray angle value for water and water glycerin solutions with concentration from 10 to 90% was analyzed by Yao et al. (2012). They observed smaller mean spray angles for fluids with higher viscosity. In case of fluids with lower viscosity, the spray angle was increasing very quickly in the initial phase of spray and for fluids with the highest viscosity the liquid jet breakup was slight (Yao et al. 2012). Basak et al. (2013) analyzed the impact of liquid viscosity on the spray angle dimension. In the research plant oil of higher viscosity and biodiesel of lower viscosity were used. Bigger values of the angle were observed for the liquid of lower viscosity.

Stelter et al. (2002) and Park and Harrison (2008) checked the impact of viscoelastic fluids properties on aerosol generating in different types of atomizers. The checked fluids included solutions of polymers showing non-Newtonian liquids properties. It was agreed that flexibility of polymer chains had a significant importance for drop creating. The increase of extensional viscosity causes the breakup delay of the liquid spreading out of the atomizer outlet. The extensional viscosity makes forming of fibres among drops. It contributes to Sauter mean diameter increase and decrease of droplets quantity participation.

The increase of injection pressure raises the spray angle. Higher injection pressures can cause slight decrease of the angle value (Hamid and Atan 2009). The drop mean diameter is strongly connected with the injection pressure. The majority of research (Sovani et al. 2001; Broniarz-Press et al. 2016) shows that drop mean diameter of aerosol decreases as the injection pressure increases. The impact of the injection pressure is usually more visible at low injection pressures than at high injection pressures.

The flow rate increase causes the drop size rise (Nonnenmacher and Piesche 2000). In case of twin-fluid atomizers also ratio value of gas mass flow rate to liquid mass flow rate (*GLR*) matters a lot:

$$GLR = \frac{\dot{m}_g}{\dot{m}_c} \tag{4}$$

Ochowiak et al. (2015) analyzed the spray process in typical inside-out effervescent atomizers. They showed that as *GLR* value increased the spray angle initially raised, and then reached the maximum with *GLR* = 0.07 and subsequently it decreased with further *GLR* increase. Ochowiak and Broniarz-Press (2008) observed also the impact of gas and liquid flow rate on Sauter mean diameter.

The research was conducted for the process of water spray in effervescent atomizers with one-phase- and twin-fluid rates.

7   Factors Disturbing the Spray Process

During the spray process various problems may appear and they may include:

- High temperature—some fluids have to be sprayed at high temperature or it is necessary to apply high temperature of the environment, which can result in material softening or its break if the atomizer is not built from materials resistant to high temperatures.
- Caking (bearding)—the material accumulates on the interior and outer edge of the outlet because of liquid vaporization. A solid lawyer remains and obscures the outlet or the interior flow canals. Accumulation of the material near the atomizer outlet has a detrimental effect on the atomizer work.
- Accidental damage—the outlet or the atomizer damage may be caused by the accidental scratch because of inaccurate tools for cleaning or accidental dropping.
- Clogging—the unwanted solid particles may block the outlet, which results in the limited flow and the aerosol homogeneity is disturbed.
- Improper assembly—some atomizers demand careful re-assembly after cleaning, thanks to which interior elements, such as seals, sealing rings and interior blades, are accurately leveled. Inaccurate settings may cause leakage and ineffective spray. The excessive tightening of the overlays on the atomizer may cause a thread to destroy.

8   Summary

The atomization process is influences by many factors: geometric dimensions, liquid properties and operational conditions. Generally, in case of the atomizers and hybrid effervescent-swirl atomizers the decrease of atomizer outlet diameter, decrease of ratio of outlet diameter to length and outlet profiling cause the spray angle to increase and Sauter mean diameter to decrease. Except for the outlet, the mixing chamber has the significant impact on the atomization process. The mixing chamber diameter's increase worsens the spray process. The impact of the chamber length is not unequivocal but in majority of cases the length increase affects negatively the atomization process. Application of the conical mixing chamber allows to gain more favorable results in comparison to the cylindrical chamber. Moreover, the spray angle and drop diameter values depend, among others, on the aerator geometry. In case of the atomizers the most favorable parameters of the sprayed stream are gained for the ratio $l_o/d_o = 1.44$ and $L_s/D_s = 3.75$. The typical effervescent atomizer is about 0.1 m long and has a diameter of 0.05 m. The mixing chamber diameter keeps within the range of 0.005–0.025 m and the outlet diameters are between 0.0001 and 0.006 m. Moreover, the atomizer outlet profiling influences positively the atomization process. Except for geometric dimensions, liquid properties, mainly viscosity and surface tension, have the important meaning for the atomization process. Viscosity increase causes the droplet diameters to increase and the spray angle to decrease. Raising the surface tension leads to Sauter mean diameter increasing. The significant influence of the surface tension on the spray angle has not been shown. The process operating parameters have an impact on the parameters of the sprayed stream too. They mainly include injection pressure and also gas and fluid flow rates. In case of effervescent atomizers, the best results are achieved for $GLR \approx 0.07$. Designing new types of atomizers, their application and operating conditions should mainly be taken into consideration.

Acknowledgements This work was supported by PUT research grant no. 03/32/DSPB/0702.

References

Agarwal AK, Dhar A, Gupta JG et al (2014) Effect of fuel injection pressure and injection timing on spray characteristics and particulate size–number distribution in a biodiesel fuelled common rail direct injection diesel engine. Appl Energ 130:212–221

Amini G (2016) Liquid flow in a simplex swirl nozzle. Int J Multiphase Flow 79:225–235

Basak A, Patra J, Ganguly R et al (2013) Effect of transesterification of vegetable oil on liquid flow number and spray cone angle for pressure and twin fluid atomizers. Fuel 112:347–354

Belhadef A, Vallet A, Amielh M et al (2012) Pressure-swirl atomization: modeling and approaches. Int J Multiphase Flow 39:13–20

Broniarz-Press L, Ochowiak M, Woziwodzki S (2010) Atomization of PEO aqueous solutions in effervescent atomizers. Int J Heat Fluid Flow 31:651–658

Broniarz-Press L, Włodarczak S, Matuszak M et al (2016) The effect of orifice shape and the injection pressure on enhancement of the atomization process for pressure-swirl atomizers. Crop Prot 82:65–74

Chaudhari K, Kulshreshtha D (2013) Design and experimental investigation of 60° pressure swirl nozzle for penetration length and cone angle at different pressure. Int J Adv Eng Technol 1:76–84

Chen SK, Lefebvre AH (1994) Spray cone angles of effervescent atomizers. Atom Sprays 4:291–301

Datta A, Som SK (2000) Numerical prediction of air core diameter, coefficient of discharge and spray cone angle of a swirl spray pressure nozzle. Int J Heat Fluid Flow 21:412–419

Dorfner V, Domnick J, Durst F et al (1995) Viscosity and surface tension effects in pressure-swirl atomization. Atom Sprays 5(3):261–285

Endalewa AM, Debaer C, Ruttenb N et al (2010) Modelling pesticide flow and deposition from air-assisted orchard spraying in orchards. A new integrated CFD approach. Agric For Meteorol 15:1383–1392

Gajkowski A (2000) Technika ochrony roślin. Wydawnictwo AR, Poznań

Halder MR, Dash SH, Som SK (2004) A numerical and experimental investigation on the coefficients of discharge and the spray cone angle of a solid cone swirl nozzle. Int J Therm Sci 28:297–305

Hamid AHA, Atan R (2009) Spray characteristics of jet–swirl nozzles for thrust chamber injector. Aerosp Sci Tech 13:192–196

Huntington DH (2004) The influence of the spray drying process on product properties. Dry Technol 22:1261–1287

Jedelsky J, Landsmann M, Jicha M et al (2008) Effervescent atomizer: influence of the operation conditions and internal geometry on spray structure; study using PIV-PLIF. In: Proc of the 22th ILASS, pp 1–8

Lan Z, Zhu D, Tian W et al (2014) Experimental study on spray characteristics of pressure-swirl nozzles in pressurizer. Ann Nucl Energy 63:215–227

Laryea GN, No SY (2003) Development of electrostatic pressure-swirl nozzle for agricultural applications. J Electrostat 57:129–142

Lee SG (2008) Geometrical effects on spray characteristics of air-pressurized swirl flows. J Mech Sci Technol 22:1633–1639

Lee EJ, Oha SY, Kim HY et al (2010) Measuring air core characteristics of a pressure-swirl atomizer via a transparent acrylic nozzle at various Reynolds numbers. Exp Therm Fluid Sci 34:1475–1483

Lefebvre AH (1989) Atomization sprays. Hemisphere Publishing Corporation, New York

Lefebvre AH (2000) Fifty years of gas turbine fuel injection. Atom Sprays 102:251–276

Li BQ, Cader T, Schwarzkopf J et al (2006) Spray angle effect during spray cooling of microelectronics: experimental measurements and comparison with inverse calculations. Appl Therm Eng 26:1788–1795

Li T, Nishida K, Hiroyasu H (2011) Droplet size distribution and evaporation characteristics of fuel spray by a swirl type atomizer. Fuel 90:2367–2376

Li Z, Wu Y, Yang H (2013) Effect of liquid viscosity on atomization in an internal-mixing twin-fluid atomizer. Fuel 103:486–494

Miller PCH, Butler Ellis MC (2000) Effects of formulation on spray nozzle performance for applications from ground-based boom sprayers. Crop Prot 19:609–615

Moon S, Abo-Serie E, Bae C (2009) Air flow and pressure inside a pressure-swirl spray and their effects on spray development. Exp Therm Fluid Sci 33:222–231

Moon S, Abo-Serie E, Bae C (2010) Liquid film thickness inside the high pressure swirl injectors: real scale measurement and evaluation of analytical equations. Int J Therm Sci 34:113–121

Nasr GG, Sharief RA, Yule AJ (2006) High pressure spray cooling of a moving surface. J Heat Transfer 128:752–760

Nasr GG, Yule AJ, Bendig (2002) Industrial sprays and atomization. Springer, London

Nonnenmacher S, Piesche M (2000) Design of hollow cone pressure swirl nozzles to atomize newtonian fluids. Chem Eng Sci 55:4339–4348

Ochowiak M (2012a) Koncepcja atomizera pęcherzykowo-wirowego z analiza oporów przepływu. Inż Ap Chem 51(6):360–361

Ochowiak M (2012b) The effervescent atomization of oil-in-water emulsions. Chem Eng Process 52:92–101

Ochowiak M (2016) The experimental studies on atomization for conical twin-fluid atomizers with the swirl motion phenomenon. Chem Eng Proc 109:32–38

Ochowiak M, Broniarz-Press L (2008) Atomization performance of effervescent atomizers with gas-liquid internal mixing. Pol J Chem Tech 10(3):38–41

Ochowiak M, Broniarz-Press L (2009) Wpływ średnicy dyszy na średnią średnicę kropli w rozpylaczach typu pęcherzykowego. Inż Ap Chem 48(5):81–82

Ochowiak M, Broniarz-Press L, Woziwodzki S et al (2011) The analysis of silica suspensions atomization. Int J Heat Fluid Flow 32:1208–1215

Ochowiak M, Broniarz-Press L, Różańska S et al (2015) Characteristics of spray angle for effervescent-swirl atomizers. Chem Eng Proc 98:52–59

Park GY, Harrison GM (2008) Effects of elasticity on the spraying of a non-Newtonian fluids. Atom Sprays 18(3):243–271

Park SH, Kim HJ, Suh HK et al (2009) A study on the fuel injection and atomization characteristics of soybean oil methyl ester (SME). Int J Heat Fluid Flow 30(1):108–116

Prakash RS, Gadgil H, Raghunandan BN (2014) Breakup processes of pressure swirl spray in gaseous cross-flow. Int J Multiphase Flow 66:79–91

Rashad M, Yong H, Zekun Z (2016) Effect of geometric parameters on spray characteristics of pressure swirl atomizers. Int J Hydr Energy 41:15790–15799

Rizk NK, Lefebvre AH (1985) Internal flow characteristics of simplex swirl atomizers. J Propul Power 1(3):193–199

Roesler TC, Lefebvre AH (1989) Studies on aerated-liquid atomization. Int J Turbo Jet Eng 6:221–230

Sallevelt JLHP, Pozarlik AK, Brem G (2015) Characterization of viscous biofuel sprays using digital imaging in the near field region. Appl Energ 147:161–175

Sharma P, Fang T (2015) Spray and atomization of a common rail fuel injector with non-circular orifices. Fuel 153:416–430

Sivakumar D, Vankeswaram SK, Sakthikumar R et al (2015) Analysis on the atomization characteristics of aviation biofuel discharging from simplex swirl atomizer. Int J Multiph Flow 75:88–96

Som SK, Mukherjee SG (1980) Theoretical and experimental on the coefficient of discharge and spray cone angle of swirl spray atomizing nozzle. Acta Mech 32:79–102

Sovani SD, Chou E, Sojka PE (2000) High pressure effervescent atomization: effect of ambient pressure on spray cone angles. Fuel 80:427–435

Sovani SD, Sojka PE, Lefebvre AH (2001) Effervescent atomization. Prog Energy Combust Sci 27:483–521

Stelter M, Brenn G, Durst F (2002) The influence of viscoelastic fluid properties on spray formation from flat-fan and pressure-swirl atomizers. Atom Sprays 12:299–327

Szewczyk A (2010) Analiza ustawienia, parametrów i warunków pracy rozpylacza w aspekcie jakości opryskiwania upraw polowych. Monografie XCVII, Wrocław

Tratnig A, Brenn G, Strixner T et al (2009) Characterization of spray formation from emulsions by pressure-swirl atomizers for spray drying. J Food Eng 95:126–134

Wade RA, Weerts JM, Sojka PE et al (1999) Effervescent atomization at injection pressures in MPa range. Atom Sprays 9:651–667

Wang XF, Chin JS, Lefebvre AH (1989) Influence of gas injector geometry on atomization performance of aerated-liquid nozzles. Int J Turbo Jet Eng 6:271–280

Wang X, Huang Z, Kuti OA (2010) Experimental and analytical study of biodiesel and diesel spray characteristics under ultra-high injection pressure. J Heat Fluid Flow 4:659–666

Wimmer E, Brenn G (2013) Viscous flow through the swirl chamber of a pressure-swirl atomizer. Int J Multiph Flow 53:100–113

Yao S, Zhang J, Fang T (2012) Effect of viscosities on structure and instability of sprays from a swirl atomizer. Exp Therm Fluid Sci 39:158–166

Yule AJ, Widger IR (1996) Swirl atomizers operating at high water pressure. Int J Mech Sci 38:981–999

Process Data Modeling—New Challenges for Education of Chemical Engineers

Szymon Woziwodzki and Igor Ośkiewicz

1 Fundamentals of Process Design

Process design is a multi-stage process, usually with a high level of difficulty and a big challenge for a group of chemical and process engineers. These challenges are related to the requirements of the customer regarding the cost, time spent, and even the tools that the project is designed with. This places great demands on design offices that have to adapt to the requirements of qualified design engineers who are familiar with design methodology and advanced design tools such as PDS software or process simulation software (Mitkowski et al. 2016).

Generally, the design process can be divided into four stages (Fig. 1), however it can include more stages (Fig. 2) (Pahl et al. 2007; Moran 2015; Mitkowski et al. 2016).

In the first stage of Identifying, there is a consideration of the task requested by the customer. Preliminary analysis is carried out, specifications and standards are determined, location, geotechnical and meteorological conditions are analyzed as well as location, its influence and health and environmental requirements (Moran 2015). This allows determining customer's expectations and limiting conditions for the project. Designers also review the available documentation provided by the customer, geodetic conditions, or carry out inventory directly at the construction site. The latter is usually carried out using a laser scan, which generates a cloud of points for the object of inventory. All these actions allow us to choose the right method for further design.

In the second stage, the design itself is carried out. At this stage, process engineers perform calculations that take into account the possible scenarios of the operating conditions of the installation (1D design). Typically, they use tools for

S. Woziwodzki (✉) · I. Ośkiewicz
Institute of Chemical Technology and Engineering,
Poznan University of Technology, Poznań, Poland
e-mail: szymon.woziwodzki@put.poznan.pl

© Springer International Publishing AG, part of Springer Nature 2018
M. Ochowiak et al. (eds.), *Practical Aspects of Chemical Engineering*,
Lecture Notes on Multidisciplinary Industrial Engineering,
https://doi.org/10.1007/978-3-319-73978-6_31

Fig. 1 Basic stages of
process design

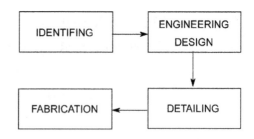

process simulation, such as ChemCad, Aspen HYSYS, gPROMS, ICAS. In the
second phase of this step, other disciplines (other designers) are also involved,
complementing each other such as automation engineers, piping specialists, electrical engineers and civil engineers. The result of their work is to create documentation (2D design—process flowcharts (PFD or P & ID) and 3D model of the
entire installation. This requires designers to use professional design tools such as
AVEVA Plant, Bentley Microstation, Intergraph SmartPlant, Autodesk Autocad
Plant 3D, Bentley AutoPipe, Inter-graph Caesar II.

The next step is to refine the details. At this stage, the support structures that are
necessary for the implementation of the process are usually designed and the design
documentation is created (Mitkowski et al. 2016).

Changes are also possible in the project, which sometimes requires the execution
of multiple computational loops, resulting from, for example, HAZOP risk analysis
(Moran 2015; Mitkowski and Bal 2015). This approach allows for a final design
review prior to the fabrication phase, as it is easier to change a line on the flowsheet
than to move the physical pipeline on construction site. This action reduces project
costs and shortens the time of its delivery.

In the last stage, the installation is built, followed by the start-up phase. It is
carried out under the supervision of the relevant inspection authorities, e.g. in
Poland it is the Technical Inspection Authority. In the course of further operation,
security procedures may also be updated.

Design requires the cooperation of many groups of designers. Therefore, it is
necessary to ensure an adequate exchange of data between them. Sample information flow is shown in Fig. 3.

This also applies to documentation. Engineers spend some time creating documentation or creating the right file format for the project or for other designer
groups and disciplines. All this wastes time and extends project delivery time. Data
management software such as AVEVA Engineering, Bentley ProjectWise,
Intergraph SmartPlant Foundations are used to streamline workflow and improve
data management. As a result, work on the project can run in parallel, which in turn
leads to shorter project delivery times, reduced costs and improved quality (Pahl
et al. 2007). This becomes particularly important in view of the new BIM design
philosophy.

Fig. 2 Major steps for creating installation (Mitkowski et al. 2016)

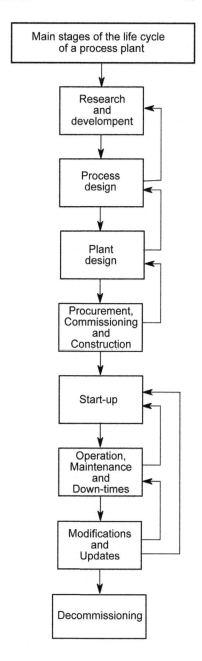

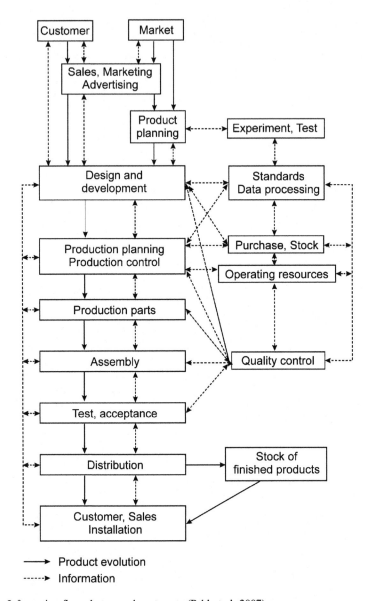

Fig. 3 Information flows between departments (Pahl et al. 2007)

2 BIM—New Concept of Design

BIM (Building Information Modeling) means managing information about the object. This term first appeared at the beginning of the 21st century and was related to the flow of information on buildings in the construction industry. BIM is a process that generates object models at a specific level of detail.

However, BIM is not only a 3D model, but also information and data about the elements of this object. This is how a common base is obtained. It is expected that construction costs will have been reduced by 33% and project delivery time by 50% by 2025, according to HM Government Strategy (2013). This poses new challenges in implementing the BIM methodology and concerns not only the construction industry but also all other industries including chemical and process engineering. These challenges are also set for the education system of engineering staff in universities around the world.

The current leader of BIM implementation is the United Kingdom. According to the UK BIM Maturity Model (GCCG 2011), four levels of BIM modeling can be distinguished:

Level 0: CAD format, unmanaged, paper or electronic documentation,
Level 1: managed 2D or 3D CAD compatible with BS 1192,
Level 2: managed 3D environment with non-geometric data. Data are contained in closed file formats managed by the software for data exchange,
level 3: managed 3D environment with IFC (Open BIM) file data, managed via BIM server.

It is also possible to use additional levels:

BIM 4D: BIM documentation along with element installation time data
BIM 5D: BIM 4D and element cost data
BIM 6D: BIM 5D with information needed to manage the facility

The UK Government has implemented the standards necessary to implement the BIM standard: BS 1192: 2007+A2: 2016, PAS 1192-2: 2013, PAS 1192-3: 2014, BS 1192-4: 2014, PAS 1192-5: 2015, BS ISO 16739: 2016, CIC BIM Protocol 2013, Uniclass 2015 (KPMG 2016).

2.1 BIM in Poland

BIM in Poland is currently at an initial level and the Ministry of Infrastructure and Construction considers if it will be implemented. Autodesk BIM research among Polish companies (Autodesk 2015) shows that the current awareness is 46% and among the most knowledgeable groups are architects and construction and installation designers (Fig. 4).

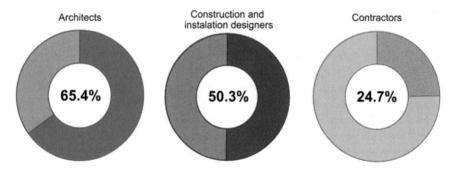

Fig. 4 BIM awareness studies across different occupational groups (based on the Autodesk report "BIM—Polish perspective, study report", 2015)

The expert opinion on the possibility of implementing the BIM methodology in Poland, the Building Information Modeling developed by KPMG in Poland at the request of the Ministry of Infrastructure and Construction shows that 63% of Polish companies are not able to deliver a public procurement contract under BIM mode (KPMG 2016). Among the biggest barriers to implementing BIM are lack of specialized software and lack of competent staff (KPMG 2016) as well as too low design costs and the lack of commonly applicable standards (Autodesk 2015). The study also indicates that the dominant form of design documentation is the paper form and 2D documentation in PDF, DWG or other CAD format (Fig. 5).

This documentation is usually provided by e-mail or by traditional fax or letter (Fig. 6). Only a small portion of respondents creates the BIM documentation.

As the activities needed to promote BIM in Poland, the most frequently mentioned ones are those that have the potential to overcome competence barriers (BIM education, creation of standards and BIM libraries), and increasing BIM awareness and its benefits.

Fig. 5 Forms of submitted design documentation. *Source* Expertise on the implementation of BIM methodology in Poland, building information modeling developed by KPMG in Poland commissioned by the Ministry of Infrastructure and Construction; with permission of KPMG Poland

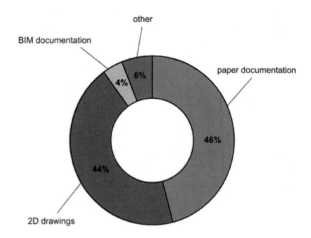

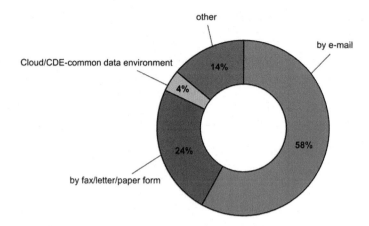

Fig. 6 Method of transmission of design information. *Source* Expert opinion on Implementation of BIM methodology in Poland, building information modeling developed by KPMG in Poland commissioned by the Ministry of Infrastructure and Construction; with permission of KPMG Poland

The universities play a special role in educating engineers, who are going to design industrial objects based on BIM in the future.

In the case of training of chemical and process engineers it is necessary to familiarize students with the ways and methods of designing different installations in the view of BIM.

For process engineers and designers, the concept of BIM is not entirely unfamiliar, as the methodology of plant and industrial design is based on PIM (Plant Information Modeling) techniques that are not fully implemented at universities.

The advancement of the BIM design methodology at universities training civil engineers and chemical engineers is similar to that of Autodesk and KPMG (Autodesk 2015; KPMG 2016). For chemical and process installations, the most common form of documentation is paper or electronic as PDF/DWG. Hence this level can be defined as Level 0. There is therefore a need to change the way projects are implemented by introducing 3D modeling of plant and process equipment and design management information.

Facing this, the Faculty of Chemical Technology at the Poznan University of Technology has introduced computer-aided design education classes where chemical engineers and process engineers learn the BIM/PIM design methodology and find out how to use the design tools that enable them to carry out chemical projects. These classes were established in cooperation with the AVEVA Gmbh Branch in Poland, which provided the AVEVA Plant platform (Mitkowski et al. 2014, 2016).

3   Data Model Concept

The role of a process engineer is to create a P&ID diagram describing the technical implementation of the industrial process. During its work a considerable amount of design data is produced: process, mechanical and strength. These data are also used by other designers, departments and disciplines. It is therefore extremely important to allow easy access to them for all design groups and to create an appropriate management system for them.

Project information management is divided into three main stages (Fig. 7): data model setup, status definition and datasheets.

3.1   Data Model Setup

In the first step, a database for engineering data has been created and the user defined element types (UDET) and user defined attributes (UDA) are grouped. In order to preserve the order, the process, mechanical, electrical and control data were distinguished (Fig. 8) as separate engineering databases.

During design, process engineering students generate not only process data, but also carry out apparatus strength calculations, such as wall thickness calculations (mechanical data), choose apparatus equipment, such as engines for stirred vessels (electrical data) or choose the basic control-measurement apparatus (control data).

Therefore, the number of engineering databases can be increased accordingly depending on the advancement and complexity of the process design.

Within each database, design data typically has to be assigned to individual UDAs and UDETs. UDETs are elements that reflect specific parts of equipment, instruments, piping, valves or fittings. UDAs are data, information, and values that

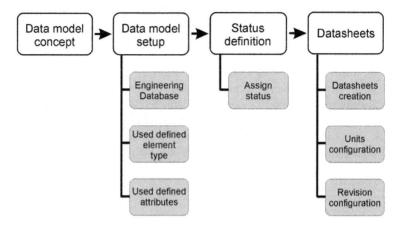

Fig. 7 Data model concept

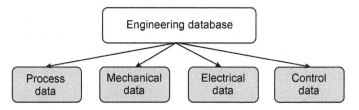

Fig. 8 Engineering databases

describe all the elements of the P&ID project. These are both process data such as volumetric flow rates for the pump or tank capacity, electrical information i.e. the voltage at which the compressor works, mechanical and construction data on, for example, the type of dishes mounted in the tank or reactor or the material class of the piping being designed.

With this solution, a user can add additional attributes through the administrator and not rely on only those that are preliminary loaded into the project, such as name, diameter or description. This is important because it does not restrict users and administrators in any way and allows them to manage the program in their own way.

3.2 Status Definition

Project status creation is dictated by the progress of the work on the element, the equipment data sheet, or the entire installation. This allows controlling and informing both designers and the customer about the progress of a single element and the whole project. This is especially important when the customer for whom the installation is being carried out requires information about the progress of the design work. Statuses are defined both by name and by percentage from 0 to 100%. Statuses are linked together and there is the possibility of promotion and demotion by one level only (Fig. 9).

Status called *Identifying* defines the time from the beginning of work on the project to the installation design and corresponds to the first stage of the design process.

In the case of promotion from *Identifying* status, an element or whole project will advance to the Designing status, which is the first of two statuses in the second step of engineering design.

Designing status begins when the analysis is completed and ends when a designer or a group of designers finishes designing of element or a project and send it to acceptation.

Another status, *Waiting for acceptance*, defines the time at which an element or a project waits for being accepted by more experienced designers or project managers. The percentages for the above described statuses are 25 and 60% respectively. The next two statuses: *Accepted* and *Approved* are related to the third stage

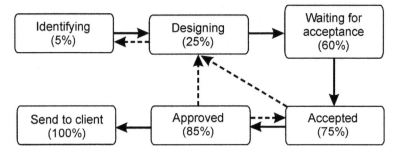

Fig. 9 Exemplary status definitions

of the design process, that is refining of details. They were described as percentages of: 75 and 85% respectively. The last defined status of the project is issued at 100% status, corresponding to the last stage of the design-creation and construction process. At this stage, the entire project is issued to the client and the construction company at the installation site.

3.3 Datasheets

The work of a process engineer is not limited to creating a P&ID diagrams, but also a 2D documentation containing the required process data. Hence, it is important to indicate during training of process engineers that the P&ID diagram is not just a drawing, but also the data that is assigned to a particular element such as equipment, nozzle, pipeline or valve. These data are usually presented as datasheets and contain process, mechanical and electrical data, etc. The datasheets form is not standardized, and each discipline can create its own datasheet (Fig. 10) as needed.

Datasheets contain large amounts of data, so it becomes important to manage them accordingly. These data should be once entered and immediately made available to all interested departments, what allows saving time on the whole project. As a result, designers use the advanced software i.e. AVEVA Diagrams, AVEVA P&ID software, Bentley OpenPlant PID, Intergraph SmartPlant P&ID, Autodesk AutoCad P&ID and Autodesk Plant 3D.

3.4 Revisions

Revisions allow managing changes in P&ID documentation as well as CAD installation drawings.

The method of configuring a revision is essentially customary, however, the simplicity of markings is recommended. The method used to review the

POLITECHNIKA POZNAŃSKA

WYDZIAŁ TECHNOLOGII CHEMICZNEJ
Instytut Technologii i Inżynierii Chemicznej
Zakład Inżynierii i Aparatury Chemicznej
ul. Berdychowo 4, 60-965 Poznań, tel. +48 61 665 2154, fax +48 61 665 2789

KARTA DANYCH REAKTORA ZBIORNIKOWEGO

NAZWA	OBSZAR	KLASA MATERIAŁOWA

WYMIARY PODSTAWOWE

Średnica wewnętrzna [mm]	Wysokość części cylindrycznej [mm]	
Średnica zewnętrzna [mm]	Pojemność dennicy dolnej [m³]	
Pojemność rzeczywista [m³]	Pojemność dennicy górnej [m³]	**PRZEGRODY**
Pojemność nominalna [m³]	Wysokość dennicy dolnej [mm]	Liczba przegród
Wysokość [m]	Wysokość dennicy górnej [mm]	Szerokość [mm]

DANE PROCESOWE

Typ mieszanego układu	☐ jednofazowy	☑ wielofazowy	☐ trójfazowy	☐ inny

Składnik	Nazwa	Objętość [m3]	Lepkość [Pas]	Gęstość [kg/m3]	K [Pas]	m	x_s
Ciecz 1 [D]							
Ciecz 2 [C]							
Gaz							
Ciało stałe							

Objętość mieszana [m3]	Temperatura pracy [°C]	Max. temperatura pracy [°C]
Wysokość słupa cieczy [mm]	Ciśnienie pracy [MPa]	Max. ciśnienie pracy [MPa]
Średnica cząstki/kropli/pęcherza [μm]	Powierzchnia międzyfazowa [m²/m³]	

DANE KONSTRUKCYJNE

Typ dennicy górnej	elipsoidalna ▼	Orientacja mieszalnika	pozioma ▼	
Typ dennicy dolnej	stożkowa ▼	Właz ☐ Zastosowano ☐ Brak	Typ włazu	
Rodzaj kołnierzy górnych (GK)	z szyjką ▼	Rodzaj kołnierzy dolnych (DK)	brak ▼	Ilość 1 ▼
Masa zbiornika pustego [kg]	Łapy ☑ Tak ☐ Nie	Rozmiar W180	Ilość 4	
Masa zbiornika zalanego [kg]	Średnica zewnętrzna GK	Średnica zewnętrzna DK		

OGRZEWANIE/CHŁODZENIE MIESZALNIKA (G/CH)

Typ wymiennika	brak ▼	Czynnik G/CH	
Temperatura		Ciśnienie czynnika	MPa

MIESZADŁO

Typ mieszadła	Średnica [m]	
Obroty [1/s]	wys.zawieszenia	[m]

SILNIK ELEKTRYCZNY

Typ silnika	Moc [kW]	Klasa
Obroty [rpm]	Zasilanie [V]	

UKŁAD PRZENIESIENIA NAPĘDU

Typ	☐ Falownik ☐ Motoreduktor	Symbol
Sprzęgło	TKN [Nm]	

WERSJA DOKUMENTU | **LISTA KRÓĆCÓW**

Nr	Data	Nazwisko	Nazwa	Typ	DN	PN
			K1			
			K2			
			K3			
			K4			
UWAGI			K5			
			K6			
			K7			
			K8			

www.fct.put.poznan.pl

Fig. 10 Exemplary datasheet (in polish)

documentation during the course of the design classes was as follows: A1, A2, A3, …, B1, B2, B3, …, C1, C2, C3 …. The first letter meant changes after being non-accepted following the 'Waiting for acceptance' status and return to the

Designing status. The digits, on the other hand, stand for the next version in the task assigned.

The above system (including statuses) can also be used for process data. In situations where a process project is complex and requires, for example, time-consuming CFD simulation, it is possible to make references to next versions of the parameters that are important for the installation and update them smoothly.

Managing the process data in the view of BIM becomes important in the designing and training of engineering staff. This process is complex, time-consuming (system configuration) and often requires large financial resources to purchase specialized software. However, the software development towards BIM is irreversible. In Poland this process is in the initial phase but will most likely be based on the experiences of Great Britain. For this reason, it is now important to train and familiarize the engineering staff with BIM design methodology.

Acknowledgements This work was supported by PUT research grant no. 03/32/DSPB/0702.

References

Autodesk (2015) BIM - polska perspektywa. Raport z badania. Autodesk, Warszawa

GCCG (2011) Building information modelling (BIM) working party strategy paper. Government Construction Client Group

HM Goverment (2013) Industrial strategy: government and industry in partnership. Construction 2025. HM Goverment

KPMG (2016) Building information modeling. Ekspertyza dotycząca możliwości wdrożenia metodyki BIM w Polsce wykonana na zlecenie Ministerstwa Infrastruktury i Budownictwa. KPMG, w Polsce

Mitkowski PT, Bal SK (2015) Integration of fire and explosion index in 3D process plant design software. Chem Eng Technol 38:1212–1222

Mitkowski PT, Woziwodzki S, Jabłoński G (2014) Jesteśmy pierwsi. Unikatowa w polsce formuła zajęć - projektowanie obiektu przemysłowego. Głos Politech 23:16–18

Mitkowski PT, Woziwodzki S, Jabłoński G (2016) Komputerowe wspomaganie projektowania obiektów przemysłowych. Nowy obszar w kształceniu inżynierów w Polsce. Przem Chem 95:8–11

Moran S (ed) (2015) An applied guide to process and plant design. Butterworth–Heinemann, Oxford

Pahl G, Beitz W, Feldhusen J, Grote KH (2007) Engineering design—A systematic approach, 1st edn. Springer, Berlin

British Standards

BS 1192-4:2014 Collaborative production of information. Fulfilling employer's information exchange requirements using COBie. Code of practice; free copy: http://shop.bsigroup.com/forms/PASs/BS-1192-4-2014/

BS 1192:2007+A2:2016 Collaborative production of architectural, engineering and construction information. Code of practice; free copy: http://shop.bsigroup.com/forms/PASs/BS-1192-2007/

BS EN ISO 16739:2016 Industry Foundation Classes (IFC) for data sharing in the construction and facility management industries. http://www.bsigroup.com

PAS 1192-2:2013 Specification for information management for the capital/delivery phase of construction projects using building information modelling; free copy: http://shop.bsigroup.com/forms/PASs/PAS-1192-2/

PAS 1192-3:2014 Specification for information management for the operational phase of assets using building information modelling; free copy: http://shop.bsigroup.com/forms/PASs/PAS-1192-3/

PAS 1192-5:2015 Specification for security-minded building information modelling, digital built environments and smart asset management; free copy: http://shop.bsigroup.com/forms/PASs/PAS-1192-5/

Building information model (BIM) protocol: Standard protocol for use in projects using building information models. Agreement and contract conditions CIC/BIM Pro, Construction Industry Council, 2013. http://www.bimtaskgroup.org/bim-protocol/

Universal classification system for the construction industry (Uniclass 2015), the NBS.com, https://toolkit.thenbs.com/articles/classification#classificationtables. Retrieved 26 December 2017

Polish Standards

PN-EN ISO 29481-2:2016-12 Building information models—Information delivery manual—Part 2: Interaction framework

PN-EN ISO 12006-3:2016-12 Building construction—Organization of information about construction works—Part 3: Framework for object-oriented information

PN-EN ISO 16739:2016-12 Industry Foundation Classes (IFC) for data sharing in the construction and facility management industries

Printed by Printforce, the Netherlands